AF271426

Drive to Survive

Drive to Survive

The Art of Wheeling the Rig

Chris Daly

Copyright © 2020 by
Fire Engineering Books & Videos
110 S. Hartford Ave., Suite 200
Tulsa, Oklahoma 74120 USA

800.752.9764
+1.918.831.9421
www.FireEngineeringBooks.com

Senior Vice President: Eric Schlett
Operations Manager: Holly Fournier
Sales Manager: Josh Neal
Managing Editor: Mark Haugh
Production Manager: Tony Quinn
Developmental Editor: Chris Barton
Cover Designer: Brandon Ash
Book Designer and Compositor: Robert Kern, TIPS Technical Publishing, Inc.
Indexers: Lauren Westbrook and Madison Haynes

Library of Congress Cataloging-in-Publication Data Available on Request

Daly, Chris
Drive to Survive: The Art of Wheeling the Rig
ISBN 978-1-59370-483-4
eISBN 978-1-59370-616-6

Printed in the United States of America

3 4 5 6   26 25 24 23

This book is dedicated to several people, starting with my beautiful wife and daughters. They patiently support my ongoing mission to reduce the number of emergency vehicle crashes through research, training, and the years it took to author this book. Regardless of how much time or how long the travel, they always understand. This book is also dedicated to my Dad, who gave me my first crash reconstruction textbook at 11 years of age...and my Mom, who told me to pick a different major.

Finally, this book is dedicated to all the first responders who stand watch every day to protect and serve throughout the world. Most importantly, this book is dedicated to those first responders who gave their lives as part of this mission. Through their supreme sacrifice, the rest of us must come to understand how to prevent a similar tragedy from happening in the future.

Contents

Preface   xvii

Acknowledgments   xxi

1   Fire Apparatus Stopping Distance 1
   Introduction   1
   Perception and Reaction Time   2
   Drive Defensively   4
   Perception and Reaction Distance   4
   Braking Distance:
      Understanding Energy   6
   Braking Distance:
      Understanding Drag Factor   7
   Braking Distance:
      The Skid to Stop Formula   10
   Total Stopping Distance   11
   How Changing Conditions Affect Skid Distance   12

2   Sight Distance 17
   Overview of Sight Distance   17
   Stopping Sight Distance   17
   Stopping Sight Distance and Speed Limits   21
   Sight Distance and Emergency Response   23

3   G-Force and Curve Dynamics 27
   The Dangers of Curves   27
   G-Force   27
   G- Force and Vehicle Handling   28
   Rollover or Spinout   29
   Understanding Rollover Thresholds   30

G-Force While Driving   33
Steering-Induced Rollovers   37
Weight Shift   37
Air Force Testing   40
Slosh   40
Baffles   41
Case Study—Nevada   43

4   Understanding Rollover Crashes 45
An Overview of Rollover Crashes   45
Rollover Dynamics   45
Rollover Speed   47
Improving Rollover Thresholds and Vehicle Stability   50
Rollover Prevention   51
Understanding Electronic Stability Control   52
Active Rollover Protection   54
Rollover Warning Systems   54
Avoiding Rollover Crashes   55

5   The Friction Circle and Skid Control 57
Why Tires Skid   57
The Friction Circle   59
The Friction Circle—Dry Road   60
The Friction Circle—Wet Road   62
Braking and Accelerating in Curves   64
Braking in Curves—Modern Vehicles   66
Case Study—Maryland   67
Oversteer and Understeer   68
Types of Skids   70
Correcting a Skid   72
"Look Where You Want to Go"   73
Skids to Rollovers   73
Wet/Dry Switch   73
Case Study—Connecticut   74
Preventing Skids   75

6   Siren Limitations 77
Outdated Technology   77
How a Siren Works   78
Real Life Scenario   81
Windows Up or Down   82

Localization   82
Civilian Response to a Siren   83
Emergency Response Protocols   84

7   Negotiating Intersections 87
Introduction   87
Notice of Approach   87
Best Practice   88
Account for All Lanes of Traffic   90
Caravanning   90
Case Study—Florida   91
Green Lights   92
Case Study—Pennsylvania   92
Apparatus vs. Apparatus   94
Case Study—Illinois   94
Rolling an Intersection   95
Traffic Preemption Devices   96
Conclusion   99

8   Seat Belts and Occupant Protection Devices 101
Introduction   101
Rollovers and Seatbelts   102
Delta-V   103
Wearing Helmets While Responding   104
Rollovers and Ejection   105
Ambulances and Occupant Restraints   107
NFPA Requirements   108

9   Railroad Crossings 111
Introduction   111
Railroad Crossing Design   111
Best Practice at Railroad Crossings   113
Case Study—Virginia   118

10   Air Brake Operations 121
Introduction   121
Why Use Air Brakes?   121
Spring Brakes—Used for Parking and Emergency Braking   122
Service Brakes—Used for Routine Braking   122
Inspecting the Air Brake System   124
Brake Fade   126

Preventing Brake Fade   128
Downshifting and Auxiliary Braking   135
Snub Braking   137

11 Tire Safety 143
Hydroplaning   143
Handling a Hydroplane   146
Tire Blow Outs   147
Proper Tire Pressure   148
Dual Tires   152
Handling a Tire Blowout   153
Inspecting and Reading a Tire   155
Snow Tires   158
Tire Chains   159

12 Drunk and Impaired Driving 161
Introduction   161
Impaired Driving Defined   161
Understanding Blood Alcohol Concentration   162
NIOSH, IAFC, and NFPA Recommendations   165
Supervisor Considerations   168
Case Study—Wyoming   169

13 Overweight, Oversized, and Modified Vehicles 173
Introduction   173
Overweight Issues   173
Brake Fade   174
Skids and Handling   175
Tank Baffles   175
Tire Blowouts   175
Case Study—Utah   176
Case Study—New Mexico   176
Case Study—Texas   177
Height, Weight, and Width of Vehicles   177
Road Defects   179

14 Retiring Apparatus, Out-of-Service Apparatus, and Annex D .. 183
Introduction   183
Obtaining Funding for New Apparatus   185
Out of Service Criteria   187
Out of Service Procedures   189

15 Tiller Ladders 191

Tractor Trailer Dynamics   191
Preventing Jackknifes   194
Tiller Ladder Skid Control   195
Rollovers   195
Backing Operations   196
Turning a Tiller Ladder   197

16 Fifteen Passenger Vans 199

Introduction   199
Federal Studies   200
Rollovers   200
Implementing Safety Procedures   201
Case Study—Oregon   204

17 Trailer Operations 205

Introduction   205
Selecting a Tow Vehicle   205
Selecting a Hitch   207
Selecting a Trailer   209
Understanding Trailer Brakes   210
Adjusting Trailer Brakes   212
Breakaway Systems   213
Sway Control Systems   213
Trailer Operations   213
Acceleration   215
Braking   215
Turning and Curves   216
Backing   217
Trailer Sway   217
Maintenance and Inspections   218

18 Off-Road Driving 221

Introduction   221
Understanding Four-Wheel Drive   221
Understanding Differentials   225
Off-Road Driving Considerations   226
Off-Road Operations   226
Off-Road Seatbelt Use   230

19 Off-Road Recovery Operations 233

Introduction 233
Tow and Recovery Points 234
Winches 234
Anchor Straps and Recovery Straps 236
Using a Winch for Recovery Operations 237
Anchor Points 238
Calculating Winch Pulls 240
Using a Recovery Strap 240

20 Backing the Apparatus 243

Best Practices 243
Case Study—New Jersey 247

21 Driving in Hazardous Conditions: Nighttime and Weather 249

Driving at Night 249
Seeing at Night 249
Understanding Headlights 250
Two-Lane Highways 253
Expectancy 254
Adaption 255
Roadway Geometry 255
Be Careful Driving at Night 256
Driving in Inclement Weather 257
Weather and Stopping Distance 265
Weather and Critical Curve Speed 267
Case Study—Virginia 267

22 Distracted, Sleepy, and Fatigued Driving 269

Distracted Driving 269
Fatigued and Sleepy Driving 272
Causes of Fatigue 274
Countermeasures to Fatigued or Sleepy Driving 276
Case Study—Michigan 277

23 Hose Loads, Wheel Chocks, and Electronic Recording Devices...279

Loading Fire Hose 280
Securing Hose Loads 281
Understanding Wheel Chocks 281
Types of Wheel Chocks 283
Proper Use of a Wheel Chock 286

Vehicle Data Recorders   287
On-Board Video Systems   289

24 Developing Driver Training Programs and Selecting Drivers...291
National Standards   292
Standard Operating Policies and Procedures   292
Writing a Policy and Procedure Manual   294
Selecting Instructors   295
Lesson Plans   296
Selecting the Fire Apparatus Operator   296
Motor Vehicle Record Checks   297
Medical Surveillance   298
Training Records   298
Training the Driver   299
Understanding Stress   300
On-going Training   301
Driver Training Exercises   302
Special Hazard Training   303
Risk Management Programs   304
Case Study—Texas   305

25 Training Ideas...307
G-Meters   307
Speed   308
Stopping Distance   309
Night Driving and Visibility   309
Blind Spots   310
Inclement Weather Driving   311
Sight Distance and Stopping Distance   312
Intersections and Sight Distance   313
Tire Blowouts   314
Antilock Brakes   315
Siren Audibility   316
Backing Cameras   316
Tire Pressure Charts   318
Cone Courses   318
Explaining Skids   319
Vehicle Inspections   319
Drunk Driving   319
On-Board Cameras and Event Data Recorders   320
Critical Curve Speed and Lateral G-Force   320

Off-Road Driving   321
Winch Operations   322
Backing Maneuvers   322
Field Trips   322
Conclusion   323

26 Emergency Vehicle Crash Investigation 325
The Crash   325
The Scene   326
Evidence   327
Photographs   330
Statements   330
Video and Electronic Data   332
Siren Audibility and Warning Lights   332
Vehicle Records and Inspections   333
Traffic Pre-emption Devices   335
NFPA Crash Reporting Requirements   335
Conclusion   336

Appendix A   NIOSH Firefighter Fatality Reports....................337

Appendix B   Understanding the Math 399

Index   415

Preface

Before we get started, please allow me to introduce myself. My name is Chris Daly and I am currently a 29-year veteran of the fire service. In addition to my fire service career, I have spent the past 22 years as a police officer. Please hold your applause, I already know how much firefighters love cops...

Over the course of my police career I have specialized in crash reconstruction. We are the cops who tie up the road for hours while trying to put a crash back together. We try to determine who was at fault, who was going how fast, and who needs to be criminally charged, if anyone.

As I progressed through my fire and police careers, I couldn't help but notice the number of first responders who were killed or injured in vehicle crashes. While vehicle crashes are a leading cause of death for firefighters, police officers, and emergency medical responders, there isn't much training to address this topic. In fact, few fire departments prioritize driver training. While "bread and butter skills" such as stretching attack lines and venting roofs are often practiced, these skills are used infrequently when compared to driving. Those fire departments that do conduct driver training often address basic skills such as low speed maneuvering and vehicle positioning. While these low speed skills are important, many training programs lack an in-depth discussion of vehicle dynamics. This lack of understanding on the topic of vehicle dynamics creates a dangerous gap in driver training. Fire apparatus operators must understand how a vehicle maneuvers at roadway speed, and more importantly, why does it crash? Just as a doctor cannot heal the human body without a thorough understanding of anatomy and physiology, an emergency vehicle operator cannot safely drive a fire apparatus without an in-depth knowledge of vehicle dynamics.

Based on these observations, I decided to develop a training program on my own. After two years of research and hard work, "Drive to Survive" was born. The "Drive to Survive" training program combined my fire service experience with my years of training in crash reconstruction. The goal of the program was to teach emergency responders that no matter how long they have been driving or how "good" they think they are, at some point Mother Nature will take over and the vehicle will lose control. Truthfully, "Mother Nature" is just an easy way of saying "physics," but if I used the term "physics" most of my students' eyes would glaze over and they would tune me out for several hours.

Little did I know how successful and well received the seminar would become throughout the country. What started as a 3-hour training module has now expanded to cover several days. One thing I always seemed to notice was that at the end of each class, several people would come up and tell me "you should write a book."

For years I had no time to write a book until one day I decided that I had no more excuses. What you are holding in your hands is the end result of over 15 years of training and research. While there may be other "driver training" textbooks out there, few of these textbooks provide an in-depth examination of fire apparatus vehicle dynamics. Driver training textbooks usually focus on pump operations, aerial operations, and preventative maintenance. These books only provide a brief overview of vehicle dynamics and crash causation. However, "Drive to Survive—The Art of Wheeling the Rig" is designed to address issues specific to DRIVING the vehicle, not OPERATING the vehicle. The purpose of this book is to "fill in the gaps" which relate to vehicle dynamics and crash causation. Armed with this advanced knowledge of vehicle dynamics, fire apparatus operators will learn the limits of driving an emergency vehicle and come to understand that no matter how long they have been driving or how good they think they are, at some point physics will take over and the vehicle will lose control.

In addition to the general knowledge that this book will provide for the individual driver, the book is also designed to provide an in-depth knowledge of those topics required in the following NFPA standards:

- NFPA 1002 "Standard for Fire Apparatus Driver/Operator Professional Qualifications"—Section 4-3
- NFPA 1451 "Standard for Fire and Emergency Service Vehicle Operations Training Programs"
- NFPA 1500 "Standard on Fire Department Occupational Health and Safety Program"—Section 6

These topics include:

- The importance of donning passenger restraint devices and ensuring crew safety
- The common causes of fire apparatus accidents

- Recognition that the fire apparatus operator is responsible for the safe and prudent operation of the vehicle under all conditions
- The effects on vehicle control of liquid surge
- Braking reaction time
- Load factors
- Effects of high center-of-gravity on rollover potential
- General steering reactions
- Speed
- Centrifugal force
- Laws and regulations
- Principles of skid avoidance
- Night driving
- Shifting and gear patterns
- Negotiating intersections
- Railroad crossings
- Bridge safety
- Identification of automotive gauges
- Operational limits
- Vehicle dimensions
- Turning characteristics
- Spotter signaling
- Principles of safe vehicle operation
- Manufacturer's specifications
- Rules and regulations of the jurisdiction

By addressing these topics, the book is designed to act as a definitive textbook for emergency vehicle operator classes (EVOC). It is my hope that by addressing these issues and providing EVOC instructors with methods to convey these concepts, we will be able to reduce the number of emergency vehicle crashes throughout the world.

Remember that the concepts you learn in this book relate to more than just driving a fire apparatus. These principles apply to anyone who drives a vehicle, including family members and loved ones. Read this knowledge, grasp this knowledge, and share this knowledge with everyone you know.

I hope you enjoy the book and learn something from it. Should you have questions, concerns, or comments, please don't hesitate to contact me at station56@ aol.com. I look forward to hearing from the readers.

An injury or death on the fireground is a traumatic event. However, the direct effects of the incident are usually contained to the fire personnel who were involved at the scene. Rarely will a mistake on the fire ground cost a civilian their life. The same cannot be said for an emergency vehicle crash. Many emergency vehicle crashes result in injuries and fatalities to civilians.

Firefighters and emergency responders are supposed to save lives, not take lives. When fire apparatus operators drive in a reckless or unsafe manner, they not only risk the lives of firefighters, but also the innocent civilians who share our roads.

Acknowledgments

I want to thank several people who have provided support and assistance with this project. Most importantly: Bill Peters (Jersey City Fire Department, retired) and John Daily (Teton County Sheriff's Office, retired) who took the time to lend their invaluable expertise to review this project for any errors or omissions. Bill has vast expertise in fire apparatus design, operations, maintenance, and NFPA standards. John literally wrote the book on crash reconstruction. Also, thanks to James Sobek (Wolf Technical Services), one of the world's leading experts on nighttime visibility and optics, for reviewing the chapter on nighttime driving and bad weather. Thanks to Dr. Frank Navin, another person in the crash reconstruction world who also literally wrote the book on commercial motor vehicle crashes and rollovers. Frank was kind enough to review the chapters on lateral g-force and rollovers. Finally, I'd like to thank Rand Smith, Ryan Depew, Sreenivasan Ranganathan, Ken Holland, Alan Korn, and Roger Lackore, who answered various questions along the way.

Fire Apparatus Stopping Distance

Introduction

The distance it takes a vehicle to come to a stop is known as *stopping distance*. Many fire apparatus operators believe that the stopping distance of a vehicle is the distance it will take the vehicle to *skid or brake* to a stop. However, the braking distance of a vehicle is just one of four phases. The distance the vehicle travels during each of these four phases must be separately calculated and added together to determine the vehicle's *total stopping distance* (fig. 1–1). These four phases include the following:

1. *Perception Distance:* The distance it takes the driver to perceive a hazard
2. *Reaction Distance:* The distance it takes the driver to react to the hazard
3. *Mechanical Lag Distance:* The distance it takes the vehicle to respond to the driver's input
4. *Braking Distance:* The distance it will take the vehicle to dissipate its energy and come to a stop

> **Total Stopping Distance = Perception Distance + Reaction Distance + Mechanical Lag Distance + Braking Distance**

Author's Note: I would just like to point out that the first five chapters of this book have a lot of numbers in them. I realize this may not be ideal, but it is important for a fire apparatus operator to have a thorough understanding of the concepts discussed in these chapters before they get behind the wheel. I debated putting these chapters later in the book, but in the end, I felt this would be best, as these first few chapters will make the later topics easier to understand. If you read these first few chapters slowly and thoroughly, I promise the numbers won't be any worse than pump school. That said, if you would prefer, feel free to start reading the book at Chapter 6 "Siren Limitations". The book reads more like a novel from that point forward and will allow you to sink your teeth in before you circle back to Chapter 1.

Many factors play a part in determining the total stopping distance of a vehicle, including the reaction time of the driver, the road conditions, the vehicle's braking efficiency, how forcefully the driver applies the brakes, and the speed of the vehicle. Fire apparatus operators must have a thorough understanding of each of these factors and maintain a constant awareness of the time and distance that will be necessary to bring a moving fire apparatus to a safe stop.

Fire apparatus operators must also remember to account for the time and distance it will take a civilian driver to perceive, react, and yield the right-of-way to an approaching fire apparatus. If the fire apparatus operator drives in an aggressive manner, the civilian might not have enough time to perceive, react, and grant the right-of-way to the emergency vehicle.

The distance it will take an emergency vehicle to come to a controlled stop is often an overlooked facet of many driver training programs. The following sections will examine each component of a vehicle's total stopping distance and give the fire apparatus operator a better understanding of the issues involved.

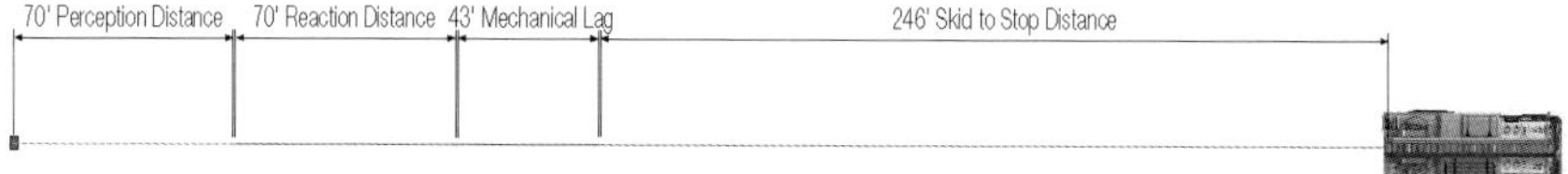

FIGURE 1–1. The total stopping distance of a vehicle includes four pieces. Each piece must be separately calculated and then added together to determine the total stopping distance of the vehicle. Here we see the total stopping distance of a fire apparatus skidding to a stop from 60 mph on dry asphalt ($f = 0.75$) with 65% braking efficiency. Braking efficiency and drag factor will be discussed later in this chapter.

Perception and Reaction Time

When a person uses their senses to gather information, the process is known as *perception*. Perception allows a driver to detect and identify a hazard in the roadway. While most drivers will use sight to perceive a hazard in the roadway, there may be times when a driver uses sound, vibration, or g-force to detect and identify a hazard.

Once the driver perceives a hazard in the roadway, their brain must process the hazard and decide a course of action. When the brain selects a course of action, it will send signals to the body to react. This reaction usually involves turning the wheel or applying the brakes. The time it takes for a person to detect, identify, and react to a hazard is known as *perception and reaction time*.

Many published texts state that the average perception and reaction time for a driver is 1.6 seconds. However, studies have shown that a driver's perception and reaction time will vary, based on the circumstances. Some of these circumstances are discussed next.

Age of the Driver

Many elderly drivers will not perceive and react to a hazard as well as a younger driver. This is because an elderly driver may not see or hear as well as they used to, or their cognitive functions may not be as acute as they once were. Because of eye disease or degeneration, an elderly driver may not see a hazard until it is too close to avoid. A similar problem may arise if an older driver detects the problem but cannot interpret it because they have an underlying cognitive problem, such as dementia.

Location of the Hazard

If the hazard is directly in front of the vehicle, the response time should be quicker than if the problem is coming from the driver's peripheral vision. A hazard emerging from the right or left will require the driver to move his eyes to focus on the problem. The process of moving a person's eyes is known as *saccade*. Average saccade time can be as long as one-third of a second on a clear day.[1]

If the hazard is emerging more than 15 degrees from the front of the vehicle, a driver will likely have to turn their entire head to focus on the problem.[2] It can take a driver as much as 1.7 seconds to move their head all the way in one direction and all the way back.[3] The 1.7 seconds required for a person to move their head back and forth is in addition to the perception and reaction time.

Nature of the Hazard

Many hazards are not unexpected hazards because drivers encounter them on a regular basis. A red light requires a driver to perceive and react, but it is an expected part of driving. Therefore, drivers are better able to anticipate and react when they encounter a red light.

An approaching emergency vehicle, however, or a child who runs into traffic is not as common as a red light. Drivers do not have much experience reacting to these situations because they are not often encountered. While a driver may anticipate a red light, there is usually no warning that a child is about to run into the road. Infrequent encounters may lead to longer response times as the driver must determine how to handle an unusual situation.

Surrounding Circumstances

Is the driver on a straight, open road, traveling 35 mph, with no obstacles as far as the eye can see? Or is the driver stuck between two tractor trailers, trying to decide whether to swerve into a Jersey barrier or hit a baby carriage on the other side of the road? As drivers encounter more complex driving scenarios, their perception and reaction times may increase as they are forced to weigh the consequences of their evasive actions. If a driver finds himself in a lose/lose scenario, he may not react at all. This is why fire apparatus operators often encounter civilian drivers who freeze when an emergency vehicle approaches. When the emergency vehicle approaches, the civilian may be faced with limited options on how to react and get out of the way.

Distractions

If a driver is distracted, he may have a longer perception and reaction time. If the fire apparatus operator is listening to the radio, looking at an approaching scene, or discussing incoming orders with the officer, he may have a delayed perception and reaction time. Despite common belief, a human cannot multitask. Instead, a driver must quickly change his thought process from one task to another as rapidly as possible. If the fire apparatus operator is dealing with a task that relates to driving the fire apparatus or arriving on scene, he may not react appropriately to a hazard in the roadway. Distracted driving will be discussed in more detail later in this text.

Drive Defensively

Fire apparatus operators must understand how each of the aforementioned factors will affect their ability to perceive and react to hazards. Fire apparatus operators must also understand that these issues will affect how a civilian perceives and reacts to an approaching emergency vehicle. When a fire apparatus operator approaches a civilian vehicle, he will not know the age, experience level, or mental and physical condition of the driver. This is why an apparatus operator must assume the worst and drive in a defensive manner. Driving the apparatus too fast, rushing up to an intersection, or tailgating a vehicle will not give a civilian driver enough time to perceive and react to the approaching apparatus. Furthermore, this type of aggressive driving may cause the civilian driver to panic or freeze up, resulting in a failure to give the right-of-way. Worse yet, an aggressive emergency vehicle driver may force a civilian to make the wrong decision and cause a crash.

Perception and Reaction Distance

In the previous section, we discussed the perception and reaction *time* of the driver. While perception and reaction time is the amount of time it takes a driver to perceive and react to a hazard, the distance the vehicle travels during this time is known as the perception and reaction *distance*. The perception and reaction distance will depend on the speed of the vehicle and the amount of time it takes the driver to perceive and react. A distracted driver who is not paying attention will take a longer amount of time to perceive and react to a hazard. Therefore, a distracted driver will have a longer perception and reaction distance. Traveling at a faster speed will also result in a longer perception and reaction distance, as the vehicle will cover more distance during the time it takes the driver to perceive and react to the problem.

As an example, let's compare the perception and reaction **distance** of a vehicle if the driver has two different reaction **times**: 1.6 seconds and 3.0 seconds (table 1–1).

TABLE 1–1.   This table examines the perception and reaction distance of a driver at different speeds. Note the difference in perception and reaction distance if the driver takes more time to respond. As an example, at 60 mph the distracted driver may take approximately 123 extra feet to respond to a hazard. These distances demonstrate the dangers of distracted driving.

Perception and Reaction Distance		
Vehicle Speed	Perception/Reaction Time of 1.6 Seconds	Perception/Reaction Time of 3.0 Seconds
10 mph	23 ft	43 ft
20 mph	46 ft	87 ft
30 mph	70 ft	131 ft
40 mph	93 ft	175 ft
50 mph	117 ft	219 ft
60 mph	140 ft	263 ft
70 mph	164 ft	307 ft

When we examine each speed in table 1–1, we see that if a driver is distracted the perception and reaction distance will increase substantially. Therefore, when a distracted driver encounters a hazard it will take the vehicle a longer distance to come to a stop. Also take note of how the perception and reaction distance increases as the speed of the vehicle increases.

Remember: the perception and reaction time is the time it takes the driver to perceive and react to the problem (in seconds). The perception and reaction distance is the distance (in feet) that the vehicle travels as the driver perceives and reacts to the problem.

Braking Distance: Understanding Energy

Once the driver perceives and reacts to a hazard, they will need to initiate an evasive maneuver. In some cases, the driver will accelerate or turn the wheel in an attempt to avoid a hazard. In many cases, the driver will attempt to bring the vehicle to a stop, often by forcefully applying the brakes. The distance it takes the vehicle to brake to a stop is the next phase of the total stopping process and is commonly referred to as the *braking distance.* If the driver locks the tires and skids, this phase is often called the *skid distance.* For the purpose of this discussion, we will be discussing the skid distance of a vehicle. This assumes that the driver has fully engaged the brakes while attempting to avoid a hazard. The wheels of the vehicle will be locked or the antilock braking system will be fully engaged. However, it is important to remember that in most cases, when the driver is gently depressing the brakes to come to a normal stop, the braking distances will be much longer. This is because the driver is not using the full braking ability of the vehicle. In either case, a fire apparatus operator must understand that the braking distance of the vehicle is directly related to how much energy the vehicle possesses at the time of the braking maneuver. Therefore, before discussing the braking distance of the vehicle, we must first understand the concept of kinetic energy.

A moving fire truck has kinetic energy. Kinetic energy is the energy of motion. The amount of kinetic energy depends on the vehicle's size and speed. The heavier the fire apparatus or the faster it is traveling, the more kinetic energy it will possess. If a vehicle crashes on an interstate, the injuries are usually more severe than if the vehicle had crashed in a parking lot. This is due to the increased kinetic energy that results from the faster speed. Fire apparatus operators must have a thorough understanding of kinetic energy to fully appreciate the dynamics of a moving vehicle.

If a vehicle has more energy, it will take a longer distance to dissipate the energy and come to a stop. This is why the stopping distance of a vehicle increases as the speed of the vehicle increases. The vehicle requires more time and distance to dissipate all the energy. To better understand the concept of kinetic energy, consider a bucket full of water. The fire apparatus is the bucket, and the kinetic energy is the water in the bucket. If we want the vehicle to slow down, we must get rid of some of the kinetic energy, or dump some water out of the bucket. If we want the vehicle to stop, we must get rid of all the kinetic energy, or dump all the water out of the bucket. Unfortunately, kinetic energy does not just disappear. Instead, it must be converted into a different type of energy and dissipated.

In most cases, the driver will apply the brakes and convert the kinetic energy into heat energy so it can be dissipated into the atmosphere. However, there may be instances where the driver does not have enough time and distance to dissipate all of the vehicle's kinetic energy with the braking system. When this happens, the kinetic energy is dissipated by skidding the tires or crushing the vehicle. Let's examine each scenario more in depth.

Brakes

In most cases, the braking system will convert the vehicle's kinetic energy into heat energy and dissipate it into the atmosphere. When a driver applies the brakes, the brake pad will rub against the disc or drum and create friction. The friction between the pad and drum creates heat, which burns off the kinetic energy, slowing and stopping the vehicle. This scenario takes place during a normal braking maneuver.

Skidding Tires

Problems arise when there is not enough time and distance to burn off all the vehicle's kinetic energy with the braking system. When this occurs, the driver is forced to slam on the brakes and lock the wheels. When the wheels lock, the brake pad is no longer rubbing against the disc or the drum to create friction. Instead, the kinetic energy is converted to heat energy by the friction of the tire rubbing across the road surface. Anyone who has seen a vehicle lock its wheels at a high speed can attest to the large amount of smoke that wafts across the road as the vehicle skids to a stop. Keep in mind that modern vehicles are equipped with antilock brakes. Even though the tires don't fully lock, energy is still dissipated as the tires rub against the road surface. Antilock brakes will be discussed in more detail later in this book.

Crush

In some cases, a vehicle is unable to dissipate all its kinetic energy and come to a safe stop by using the brakes or skidding the tires. When this occurs, the vehicle slams into another object and the kinetic energy is dissipated by crushing the structure of the vehicle. The more kinetic energy present when the vehicle strikes the object, the greater the crush to the vehicle. This is why high-speed crashes typically result in severely crushed vehicles.

Braking Distance: Understanding Drag Factor

We have just discussed the three ways that a fire apparatus will dissipate its kinetic energy and come to a stop: applying the brakes, skidding the tires, or crushing the vehicle. It is my hope that crushing the vehicle is not the driver's first choice of action, so instead we will discuss how long it takes a vehicle to skid to a stop by using the brakes. Our discussion will assume that the tires are locked or the antilock braking system (ABS) is fully engaged. In other words, the driver is attempting to use 100% of the vehicle's braking ability. Keep in mind that during a normal deceleration, such as approaching a traffic light or a curve under normal conditions, the driver may only use a small percentage of the vehicle's braking ability. In these cases, the braking distance may be much longer.

When a driver slams on the brakes and attempts to use the full braking ability of the vehicle, the distance it takes the vehicle to skid to a stop will depend on the speed of the vehicle and the drag factor of the roadway. While the speed of a vehicle is straightforward, calculating the drag factor of a roadway is more complex. The drag factor of the roadway will determine how well a vehicle is able to grab the roadway and come to a stop. In order to calculate the drag factor of a roadway, we must know three things: the coefficient of friction of the road surface, the braking efficiency of the vehicle, and the slope of the road. Once these three factors are known, they are entered into the following formula, which will then determine how well the vehicle is able to grab the road and come to a stop:

$$\text{Drag Factor} = (\mu)(n) \pm m$$

Where

μ = the coefficient of friction of the roadway

n = the braking efficiency of the vehicle

m = the slope of the roadway

It is important for a fire apparatus operator to have a thorough understanding of each factor in this formula. This will allow him to better conceptualize how well his vehicle will grab the road and come to a stop. Keep in mind that this concept addresses the worst-case scenario: a fire apparatus that is in a panic stop with the tires skidding or the ABS fully engaged. In many cases, the driver may be able to brake to a stop without actually skidding the tires or engaging the ABS. With that in mind, let's discuss how each factor in this formula will affect a vehicle's ability to grab the road and stop.

Coefficient of Friction

The "stickiness" of the roadway is called the *coefficient of friction.* The higher the coefficient of friction, the more sticky the road is. A roadway with a high coefficient of friction will result in a shorter stopping distance. A roadway with a low coefficient of friction will result in a longer stopping distance.

There are many ways a crash investigator can measure the coefficient of friction of the roadway. Measurements typically range from 0.2 to 0.9, with higher numbers corresponding to stickier roadways and lower numbers corresponding to slicker roadways. While precise measurements are always best, there are general rules of thumb. Table 1–2 provides a general overview of roadway friction coefficients of a skidding vehicle based on actual skid tests.

TABLE 1–2.   This table provides an overview of roadway friction coefficients based on the type of road surface and the weather conditions. Although actual testing of a crash scene is the preferred method of determining these values, this table is useful for demonstrating how the stickiness of the roadway can change with the weather. A lower coefficient of friction will mean a longer stopping distance for the fire apparatus operator. (Reference: R. W. Rivers, Traffic Accident Investigation: A Training and Reference Manual [Jacksonville, FL: Institute of Police Technology and Management, 1988], 409.)

Roadway Material	Type of Material	Coefficient of Friction
Portland Cement	Dry and well-traveled	0.60 to 0.75
Portland Cement	Wet and well-traveled	0.45 to 0.70
Asphalt	Dry and well-traveled	0.55 to 0.80
Asphalt	Wet and well-traveled	0.40 to 0.65
Gravel	Loose	0.40 to 0.70
Gravel	Packed	0.50 to 0.85
Ice	Cold, Frost	0.10 to 0.25
Ice	Warm, Wet	0.05 to 0.10
Snow	Dry, Loose	0.10 to 0.25
Snow	Wet, Loose	0.30 to 0.50
Snow	Dry, Packed	0.25 to 0.55
Snow	Wet, Packed	0.30 to 0.60

Braking Efficiency

The braking efficiency of a vehicle determines how well the braking system will stop a skidding vehicle. In a passenger vehicle equipped with hydraulic brakes and standard automobile tires, it is assumed that the braking efficiency is 100%, provided all of the brakes are working properly and all four wheels lock. Unlike a passenger vehicle, a fire apparatus will not have 100% braking efficiency. The reduced braking efficiency of a fire apparatus is attributed to the rubber composition of the truck tires and the mechanical lag time of the air brake system.

Truck Tires

Truck tires are made of stiff rubber compounds that are designed to carry heavy loads over long distances without wearing out. While a truck tire may be more durable and less likely to wear out, the stiff rubber does not grip the road as well as a standard automobile tire. As a result, truck tires contribute to a longer stopping distance when the vehicle enters a skid or ABS deceleration.

Air Brakes

When the driver of a vehicle with hydraulic brakes presses the brake pedal, the brake system engages almost instantaneously. There is a very small lag time, but this lag time is negligible. In a vehicle equipped with air brakes, it takes longer for the brake system to fully engage. This is because it takes time for air to travel through the brake system, enter complex brake chambers, and fully engage the air brakes. This time delay is known as *mechanical lag time* or *air pressure lag time*. Depending on the age and the condition of the vehicle, this lag time may be over half a second. Depending on the speed of the vehicle, the mechanical lag time of the air brake system may add up to 100 ft of stopping distance.

Slope

The drag factor is also affected by the slope of the road. A vehicle skidding downhill will have a lower drag factor than a vehicle that is skidding uphill. Therefore, a vehicle skidding downhill will have a longer stopping distance than a vehicle skidding uphill.

Braking Distance: The Skid to Stop Formula

After calculating the drag factor of the road and determining the speed of the vehicle, it is possible to determine the vehicle's braking distance. The braking distance of the vehicle is calculated using an equation known as the skid to stop formula. The skid to stop formula will calculate the braking distance of a vehicle based on the conditions that were present at the time of the skid:

$$SD = \frac{S^2}{(30)(f)}$$

Where

SD = stopping distance of the vehicle in feet

S = speed of the vehicle in miles per hour

f = drag factor, which is the coefficient of friction of the
roadway adjusted for the road grade and braking
efficiency of the vehicle

Note that nowhere in the formula does it say to rate yourself on a scale of 1 to 10 and multiply, or to divide by the number of years you have been driving. Nor does it ask you to enter a different variable if you are en route to working house fire instead of an automatic fire alarm. The fire apparatus operator who has been driving for 30 years will skid to a stop in the same distance as the kid who just got hired, regardless of the type of call you're going to.

This is one of the most important points in this book!

You can't beat Mother Nature!

Total Stopping Distance

Once the vehicle's braking distance has been calculated, it can be added to the perception and reaction distance that we previously talked about. Adding the braking distance to the perception and reaction distance will determine the total stopping distance of the vehicle. Keep in mind that this section deals with a worst-case scenario in which the driver is slamming on the brakes under emergency conditions and skidding the vehicle to a stop. In a normal situation, the driver will apply the brakes and gently bring the vehicle to a controlled stop. If the driver gently decreases speed instead of slamming on the brakes, the total stopping distance will be considerably longer.

Many students ask if the skid to stop process is the same if the vehicle is equipped with antilock brakes. The easy answer to this question is yes. While antilock brakes work differently from standard brakes, the stopping distance will be similar. Antilock brakes will be discussed in depth in Chapter 10.

How Changing Conditions Affect Skid Distance

Having examined the skid to stop formula, it is evident that there are several factors that will affect a vehicle's skid distance. These factors include speed, road conditions, the braking efficiency of the vehicle, and the slope of the road. Let's examine how the stopping distance of a fire apparatus will change based on each of these factors.

Speed

The faster a vehicle is traveling, the more kinetic energy it will have. If a vehicle has more kinetic energy, it will require a longer distance to burn off all that energy and bring the vehicle to a stop. Table 1–3 examines the skid distance of a fire apparatus as it skids to a stop at various speeds on a dry road or wet road. Remember that these tables only provide the *skid distance* of the vehicle. To determine the total stopping distance of the vehicle, we would have to *add the perception and reaction distance to the skid to stop distance* for each speed.

Table 1–3 demonstrates that the skid distance of the fire apparatus will increase as the vehicle's speed increases. Also note that the stopping distance of the apparatus will be significantly longer on a wet day than on a dry day. This increase in skid distance is a result of the reduced coefficient of friction

TABLE 1–3.   As the speed of a fire apparatus increases, the amount of kinetic energy will also increase. When it comes time to stop the vehicle, it will take a longer distance to burn off the extra kinetic energy. As a result, an increase in speed means an increase in skid distance. Notice how the skid distance increases on a wet roadway because of the slick conditions.

Skid Distance of a Fire Apparatus as Speed and Road Conditions Change		
Speed	Dry Road (coefficient of friction = 0.75)	Wet Road (coefficient of friction = 0.4)
10 mph	7 ft	13 ft
20 mph	27 ft	51 ft
30 mph	62 ft	115 ft
40 mph	109 ft	205 ft
50 mph	171 ft	321 ft
60 mph	246 ft	462 ft
70 mph	335 ft	628 ft

These distances assume that the fire apparatus has a 65% braking efficiency.

associated with the wet roadway. *Remember that this increased stopping distance is a scientific fact that no amount of skill or experience will be able to change.*

Slope of the Road

A vehicle that is skidding downhill will skid further than a vehicle that is skidding uphill. In fact, there may be times on a snow- or ice-covered road when the vehicle can't stop at all as it skids on a steep downhill grade. Fire apparatus operators should preplan these locations and avoid them during slick weather. Table 1–4 examines the difference in skid to stop distance as the road grade changes on a dry or wet road.

Table 1–4 demonstrates that a fire apparatus traveling *down* a 10% slope on a dry road will take 73 ft longer to stop than a fire apparatus that is traveling *up* a 10% slope. While 73 ft doesn't seem like a lot, it may be the difference between skidding into an intersection instead of stopping safely. Notice how the difference in skid distance is more significant in wet weather. A fire apparatus skidding *down* a 10% slope on a wet road will skid an additional 290 ft as compared to the fire apparatus skidding *up* a 10% slope on a wet road. Extreme caution must be used when traversing a downhill grade, especially during inclement weather.

TABLE 1–4.   The slope of the road will change the skid distance of a vehicle. Notice how a vehicle sliding downhill will take a longer distance to come to a stop. As the grade becomes more severe, the difference in stopping distance increases.

Skid Distance of a Fire Apparatus at 50 mph as Slope and Road Conditions Change		
Slope	Dry Road (coefficient of friction = 0.75)	Wet Road (coefficient of friction = 0.4)
+0.20 uphill	121 ft	181 ft
+0.15 uphill	131ft	203 ft
+0.10 uphill	142 ft	231 ft
0.0 no slope	171 ft	321 ft
−0.10 downhill	215 ft	521 ft
−0.15 downhill	247 ft	758 ft
−0.20 downhill	290 ft	1,389 ft

These distances assume that the fire apparatus has a 65% braking efficiency.

Braking Efficiency

A fire apparatus has a lower braking efficiency than a passenger vehicle. This reduced braking efficiency is a result of the rubber composition of the truck tires and the mechanical lag time of the air brake system. Studies have shown that the braking efficiency of a fire apparatus could be as low as 65% when compared to the braking efficiency of a standard passenger car. Table 1–5 compares the difference in skid distance between a passenger vehicle and a fire apparatus on a dry road, while table 1–6 compares the difference in skid distance on a wet road.

As a result of the reduced braking efficiency, a fire apparatus traveling 50 mph on a dry asphalt roadway will take approximately 60 ft more than a passenger vehicle to come to a stop. Take note that as the speed of the fire apparatus increases, this disparity in stopping distance becomes even larger. If the speed of the vehicle increases from 50 mph to 60 mph, the difference in stopping distance increases to 86 ft.

Because fire apparatus have reduced braking efficiency and longer stopping distances, fire apparatus operators must understand the danger of tailgating the vehicle in front of them. If a civilian vehicle realizes there is a fire truck behind them and slams on the brakes, the civilian vehicle is going to come to a stop much quicker than the fire truck. This will result in the fire truck skidding into the back of the civilian vehicle.

> The total stopping distance of a vehicle includes two parts: the distance the vehicle travels while the driver perceives and reacts to the problem (perception and reaction distance), and the distance it takes the vehicle to stop once the brakes are applied (skid distance). Adding these two distances together provides the total stopping distance of the vehicle.

TABLES 1–5.   Provides the skid distances on a dry road. The skid distance of a fire apparatus is longer than the skid distance of a standard passenger vehicle. This is because the fire apparatus is equipped with air brakes and truck tires. Note the difference in skid distance at each speed between the passenger vehicle and fire apparatus.

Skid Distance on a Dry Road		
Speed	Passenger Vehicle—Braking Efficiency 100%	Fire Apparatus—Braking Efficiency 65%
10 mph	4 ft	7 ft
20 mph	18 ft	27ft
30 mph	40 ft	62 ft
40 mph	71 ft	109 ft
50 mph	111 ft	171 ft
60 mph	160 ft	246 ft
70 mph	218 ft	335 ft

Assumes a drag factor of 0.75

TABLE 1–6.   Provides the skid distances on a wet road.

Skid Distance on a Wet Road		
Speed	Passenger Vehicle—Braking Efficiency 100%	Fire Apparatus—Braking Efficiency 65%
10 mph	8 ft	13 ft
20 mph	33 ft	51ft
30 mph	75 ft	115 ft
40 mph	133ft	205 ft
50 mph	208 ft	321 ft
60 mph	300 ft	462 ft
70 mph	408 ft	628 ft

Assumes a drag factor of 0.40

Notes

1. Marc Green, "Let's Get Real About Perception and Reaction Time," last modified 2013, https://www.visualexpert.com/Resources/realprt.html.
2. Green, "Let's Get Real."
3. G. Long and A. Nitsch, "Effect of Dead Turning on Driver Perception-Reaction Time at Passive Railroad Crossings," Transportation Research Board 2007 Meeting CD, 2008.

Sight Distance

Overview of Sight Distance

A driver must be able to see a hazard far enough away from their vehicle to ensure that there is enough time and distance to safely maneuver their vehicle or take evasive action. The distance the driver can see is known as sight distance. Sight distance is most easily defined as "the length of roadway ahead, or to the sides, visible to the driver." Drivers must ensure there is enough sight distance to see oncoming traffic and then decide if they are able to perform maneuvers such as the following:

1. Stopping suddenly because of a hazard in the road ahead
2. Making left turns against oncoming traffic
3. Looking to the left or right before accelerating onto a highway
4. Passing another vehicle on a multi-lane roadway

A key element to safe driving is the ability to see oncoming traffic and then judge if there is enough time and distance between oncoming vehicles to make a maneuver on a roadway. Drivers who take greater risks tend to use smaller gaps in traffic to make a maneuver, while a safer driver will wait patiently for a larger gap in traffic. Keep in mind that the faster the oncoming traffic, the larger the required gap in traffic.

Stopping Sight Distance

Chapter 1 discussed the time and distance that is required to safely stop a moving fire apparatus. Armed with this understanding, we must now discuss the importance of being able to see far enough ahead of the vehicle to bring the fire apparatus to a safe stop. The minimum sight distance required to perceive, react, and bring the vehicle to a safe stop should a driver encounter a hazard in the road is known as *stopping sight distance.* Stopping sight distance is affected by rain, fog, smoke, hills, or curves in the roadway. If conditions obscure the road ahead, the stopping sight distance is reduced and drivers must slow down.

As an example, let's say you are driving along a straight road on a clear day. Because of the straight road and clear weather, you are able to see a half mile in front of the vehicle. As you are driving, you notice a disabled vehicle in your lane of travel a half mile away. Because you can see the disabled vehicle so far ahead of you, there is plenty of time to perceive the hazard, react to the hazard, and bring the fire apparatus to a gentle and controlled stop without hitting the disabled vehicle.

Now replace the clear weather with fog. Because it's foggy, you can only see 100 ft in front of the fire apparatus (the stopping sight distance is reduced from ½ mi. to 100 ft). Now, instead of being able to detect the disabled vehicle in your lane of travel from a half mile away, the disabled vehicle suddenly emerges from the fog, just 100 ft in front of you. Instead of having enough time to perceive, react, and stop the fire apparatus in a gentle and controlled fashion, you are slamming on the brakes and trying to skid the rig to a stop before slamming into the back of the disabled vehicle.

To determine the forward sight distance a fire apparatus operator will need in order to stop the vehicle before crashing into a hazard, we must first calculate the total stopping distance of the fire apparatus at the given speed. As an example, let's examine the total stopping distance of a fire truck that is traveling 55 mph on a dry asphalt road with an average drag factor of 0.70. We will assume the fire apparatus has a braking efficiency of 65%, which is typical for a commercial motor vehicle. We will also assume that the driver will perceive and react to the hazard with a perception reaction time of 1.6 seconds (table 2–1).

The total stopping distance for the fire apparatus in this scenario is 350 feet. This means that in order to avoid the hazard, the driver must be able to see the hazard from 350 ft away. This will give the driver enough time and distance to perceive the hazard, react to the hazard, and bring the fire truck to a screeching halt before striking the hazard. If the road conditions were wet or slick, this stopping distance would increase dramatically.

Keep in mind that our imaginary scenario also assumes that the driver is slamming on the brakes and skidding the fire truck to a stop. Skidding the fire truck to a stop is not a safe way to drive the rig. Instead, we should drive in such a way that we leave ourselves enough room to decelerate the vehicle in a safe and controlled fashion. If we were to decelerate the vehicle from 55 mph in a firm but controlled fashion, we would need nearly 450 ft of stopping sight

TABLE 2–1.   Calculated total stopping distance at 55 mph. A fire apparatus traveling into this particular curve at 35 mph will have enough time to perceive, react, and skid to stop if they see a hazard 200 ft into the curve. However, if the driver is traveling 55 mph, he will not have enough time to stop the apparatus before slamming into the hazard. Drivers must slow down and account for the available forward sight distance.

Perception/Reaction Distance Calculated*			
Vehicle Speed (mph)	Vehicle Speed (fps)	Perception/ Reaction Time (seconds)	Perception/ Reaction Distance (ft)
55	80.6	1.6	129

*(55 mph x 1.466 = 80.6 fps) x 1.6 seconds = 129 ft

Skid Distance Calculated*			
Vehicle Speed (mph)	Drag Factor of the Roadway	Braking Efficiency of the Fire Truck	Skid Distance (ft)
55	0.70	65%	221

*(55 x 55) = 3025 divided by (30) x (.70 x.65) = 221 ft

Total Stopping Distance Calculated*		
Perception/Reaction Distance (ft)	Skid Distance (ft)	Total Stopping Distance (ft)
129	221	350

*129 ft + 221 ft = 350 ft

distance. This mathematical exercise demonstrates how much forward sight distance is needed to safely decelerate a vehicle should the driver encounter a hazard in the road.

Now let's consider what would happen if the sight distance in front of the vehicle was affected by a sight obstruction, such as a curve in the road. Let's say you are rounding a curve at 55 mph. As we've just calculated, at 55 mph it will require 350 ft to perceive, react, and skid the truck to a stop should you encounter a hazard. However, in our imaginary scenario there is only 200 ft of forward sight distance into the sharp curve. If you round the curve and suddenly see a stopped vehicle just 200 ft in front of you, there will not be enough time and distance to perceive, react, and skid the truck to a stop. Instead, the fire apparatus will crash into the vehicle before it is able to skid to a safe stop (fig. 2–1).

This is why drivers must slow down and account for the available forward sight distance. Situations that reduce forward sight distance include curves, hills, dips in the road, fog, and weather. Anytime the existing conditions restrict

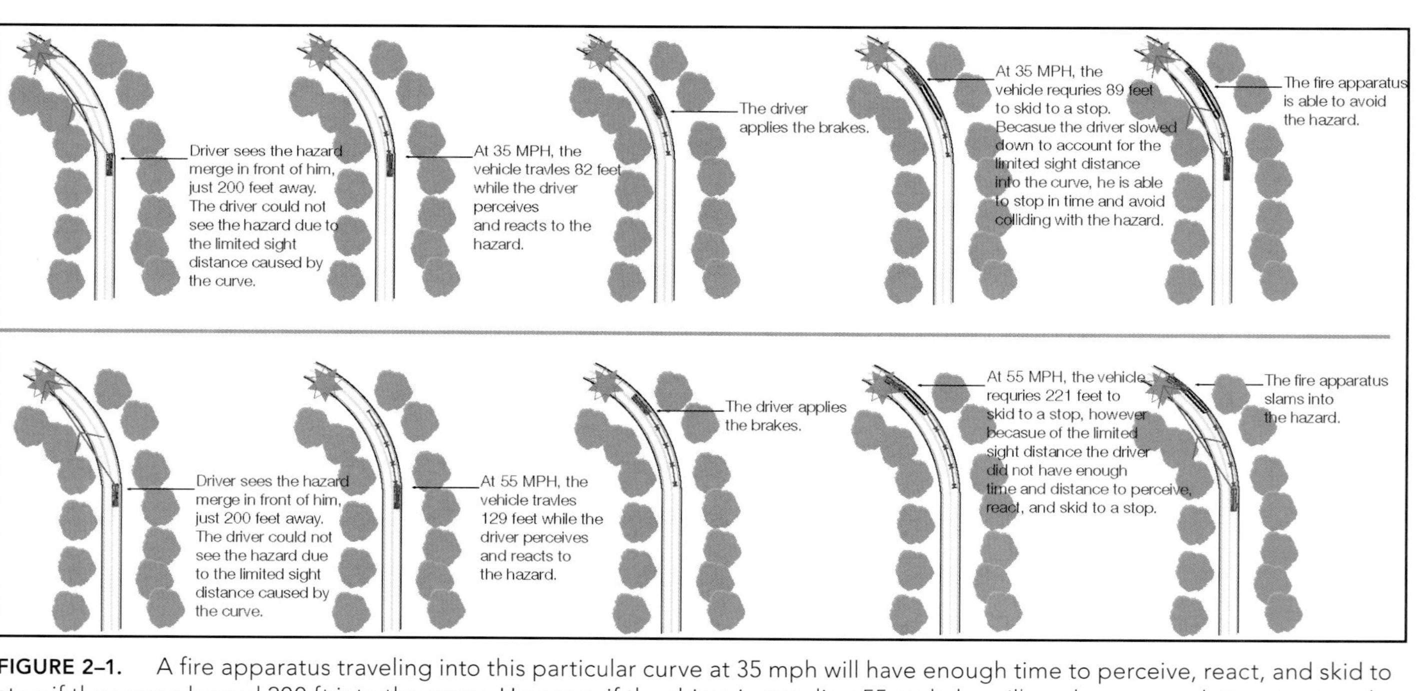

FIGURE 2–1. A fire apparatus traveling into this particular curve at 35 mph will have enough time to perceive, react, and skid to stop if they see a hazard 200 ft into the curve. However, if the driver is traveling 55 mph, he will not have enough time to stop the apparatus before slamming into the hazard. Drivers must slow down and account for the available forward sight distance.

forward sight distance, drivers must slow down and ensure that they give themselves enough time and distance to come to a safe stop if they should encounter a hazard in the road.

As proof of this theory, let's look at what would happen if the apparatus had slowed from 55 mph to 35 mph while approaching the imaginary curve. If the fire apparatus had slowed down and entered the curve at 35 mph instead of 55 mph, the total stopping distance would have dropped from 350 ft to 171 ft (table 2–2). Now when the fire apparatus enters the curve, the driver will have enough time to perceive, react, and skid the fire truck to a stop if they should see a hazard 200 ft into the sharp curve. This is because the driver will see the hazard 200 ft away and it only takes him 171 ft to come to a stop.

TABLE 2–2.   Calculated total stopping distance at 35 mph.

Perception/Reaction Distance Calculated*			
Vehicle Speed (mph)	Vehicle Speed (fps)	Perception/ Reaction Time (seconds)	Perception/ Reaction Distance (ft)
35	51.3	1.6	82

*(35 mph x 1.466 = 51.3 fps) x 1.6 seconds = 82 ft

Skid Distance Calculated*			
Vehicle Speed (mph)	Drag Factor of the Roadway	Braking Efficiency of the Fire Truck	Skid Distance (ft)
35	0.70	65%	89

*(35 x 35) = 1225 divided by (30) x (.70 x.65) = 89 ft

Total Stopping Distance Calculated*		
Perception/Reaction Distance (ft)	Skid Distance (ft)	Total Stopping Distance (ft)
82	89	171

*129 ft + 221 ft = 350 ft

Stopping Sight Distance and Speed Limits

Despite the best efforts of traffic engineers, nearly every fire department has an intersection or road in their district with limited sight distance. The limited sight distance makes it difficult to enter the highway or turn against oncoming traffic. While traffic engineers try to mitigate these issues when they design a road, there

FIGURE 2–2. Notice the difference in forward sight distance in these two scenarios. In this one the driver can see for nearly a mile ahead.

FIGURE 2–3.   In this scenario the driver's forward sight distance is extremely limited because of the curve in the road.

are times when it can be difficult. This is especially true in areas that have older roadways or extreme topography such as large hills and sharp curves (figs. 2–2 and 2–3).

In situations where there is limited sight distance and the issue cannot be resolved through other means, traffic engineers may lower the speed limit. Lowering the speed limit will accomplish two things. First, it will allow a vehicle more time to make a maneuver because approaching vehicles will not cover as much distance during that period of time. Think of how much easier it is to make a left turn when the oncoming traffic is traveling 25 mph instead of 55 mph. Second, a lower speed limit will reduce the stopping distance of an approaching vehicle if it is cut off by another vehicle that is turning or entering the highway. If a vehicle is cut off and has to panic stop, the vehicle will come to a stop more quickly and may be able to avoid a crash. If a crash does occur, a lower speed limit should result in a less severe crash.

Sight Distance and Emergency Response

It is extremely important to consider sight distance issues when driving to an emergency. Remember that the other vehicles on the road require adequate sight distance just as the fire apparatus operator does. This is important to remember when it comes time to pull onto a highway, make a turn, pass another vehicle, or make a U-turn. The fire apparatus operator must be able to see far enough away to safely turn or maneuver on the highway. If a driver is unable to see far enough away, he may pull directly in front of another vehicle and cause a crash (that is, cut them off).

The gap in traffic required to safely make a maneuver depends on the speed of the oncoming traffic, the acceleration rate of the vehicle making the maneuver, and the road conditions at the time. The fire apparatus operator must leave enough gap in oncoming traffic to safely turn or enter a roadway, reducing the chance of cutting off other vehicles. As an example, if a fire apparatus arrives at a four-way intersection and intends to make a left turn, it is important to make sure that oncoming vehicles have enough time and distance to see the fire truck, react to the fire truck, and slow down to grant it the right-of-way. This is especially true in a large, cumbersome emergency apparatus that may take extra time to accelerate. Aggressive fire apparatus operators will pull up to an intersection and turn left directly in front of oncoming traffic. This practice is extremely unsafe, as the fire apparatus operator may not give the oncoming driver enough time to perceive the emergency vehicle as a hazard, react, and yield the right-of-way (fig. 2–4). Remember that oncoming drivers may not immediately recognize the fire truck and what it plans to do. Giving civilian drivers time to recognize the emergency vehicle and make an appropriate driving decision is the key to providing a civilian driver with adequate notice of approach. Notice of approach will be discussed in Chapter 7.

Any maneuver made on the roadway should not affect the behavior of other vehicles and cause them to have to panic stop or take sudden evasive action. For this reason, fire apparatus operators should consider the roads and intersections where there is limited sight distance to safely turn or accelerate onto a highway. There may be intersections or roads in the district that a driver will want to avoid all together, as the fire apparatus does not have the needed acceleration to safely pull onto a road or make a turn due to approaching high speed traffic. Drivers must ensure there is enough sight distance to safely turn from one road to another without cutting off another vehicle. Should the forward sight distance be obstructed by weather, smoke, or the layout of the roadway, drivers must also remember to slow down to leave themselves enough time and distance to safely stop the vehicle should a hazard emerge ahead.

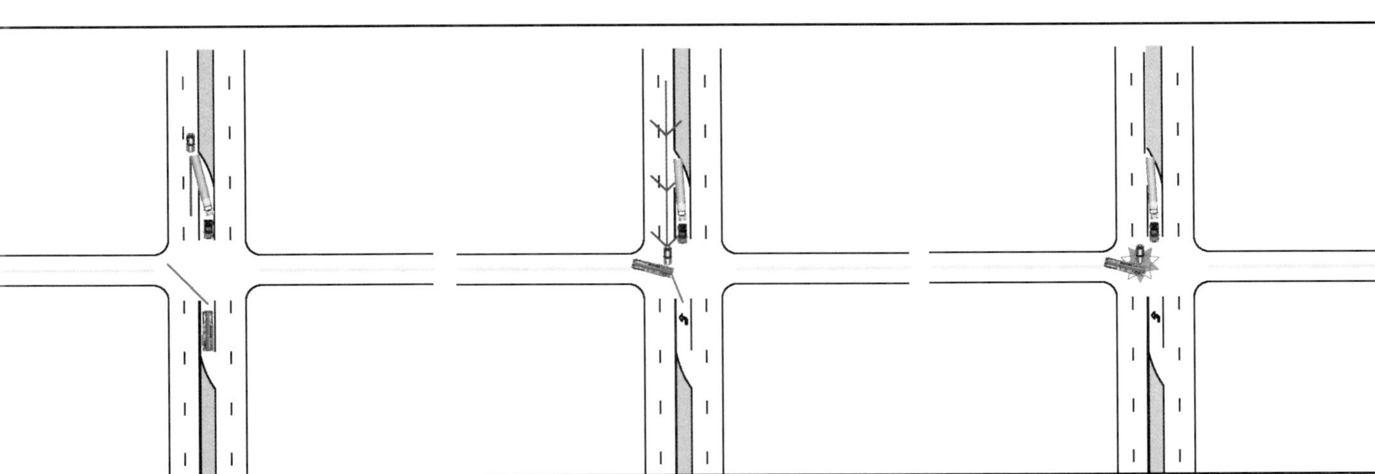

FIGURE 2–4.   A fire apparatus operator is attempting to make a left turn at an intersection. However, the fire apparatus operator's forward sight distance is blocked by stopped traffic in the oncoming left-hand turn lane. Regardless, the fire apparatus operator makes the left turn and turns directly in front of a vehicle traveling in the thru-lane. The oncoming vehicle could not see the fire apparatus either, as his forward sight distance was also blocked by the turning traffic.

G-Force and Curve Dynamics

The Dangers of Curves

Curves in the road are one of the greatest dangers a fire apparatus operator may face. The design characteristics of a large fire apparatus, combined with excess speed and poor directional control, make the vehicle highly susceptible to losing control or rolling over. This is especially true while rounding a curve. To prevent a curve-related crash, every fire apparatus driver training program must include a thorough understanding of g-force and curve dynamics.

G-Force

When most people think of acceleration, they think of a car stopped at a red light and then accelerating when the light turns green, but the term acceleration means more than just accelerating from a stop. Any change in a vehicle's direction of travel results in acceleration.

When a vehicle changes direction, it undergoes *lateral acceleration*. Lateral acceleration pushes the car from side to side and is noticeable to the driver and passengers as *centrifugal force*. The amount of lateral acceleration a vehicle will experience during a steering maneuver will depend on how fast the vehicle is traveling and how sharply the driver turns the wheel. Lateral acceleration is commonly referred to as "gs" or "g-force."

In order to better understand vehicle dynamics, fire apparatus operators must understand the concept of *g-force*. Most readers are familiar with the term "g-force" as it pertains to fighter pilots and their aircraft. A fighter pilot making a tight turn at high speed will experience high g-force.

The same concept applies to fire apparatus operators. Just as a fighter pilot can create g-force on the airframe by moving the control stick and throttle, a fire apparatus operator can create g-force on the apparatus by using the gas pedal and steering wheel. If a driver puts too much g-force on the fire apparatus, the vehicle may roll over or lose control.

When discussing g-force, we must understand that there are two types of g-force. The first is longitudinal g-force, which acts on the fire apparatus in the front-to-back direction. A fire apparatus will experience longitudinal g-force when the vehicle accelerates or brakes. The fire apparatus operator controls the amount of longitudinal g-force with the accelerator and the brake pedal.

The other type of g-force is lateral g-force. Lateral g-force will act on the vehicle from side to side. The amount of lateral g-force a fire apparatus will experience is determined by how fast the vehicle is traveling and how sharply the driver turns the wheel (curve radius).

A fire apparatus operator can also create a combined g-force by accelerating or braking while turning the wheel. An example of this scenario would be a driver who is braking while turning into a parking space or accelerating through a curve. Depending on the circumstances, a combined g-force can be very hazardous.

G- Force and Vehicle Handling

The handling characteristics of a vehicle are often judged by the amount of lateral g-force a vehicle can experience without rolling over or losing control. Automobile manufacturers use a simple process to determine how many gs a vehicle can safely handle before it rolls over or breaks traction with the road. A circle is painted on a large driving pad. The vehicle drives around the circle at ever-increasing speeds. As the vehicle drives around the circle at increasing speed, the vehicle will experience more and more lateral gs. Eventually, the vehicle will reach a point where it rolls over or breaks traction with the road.

This method allows automobile manufacturers to calculate the *cornering power* of the vehicle using the radius of the circle and the speed at which the vehicle broke traction with the road. The higher the cornering power of the vehicle, the better the vehicle will handle in a curve. A vehicle that can handle more gs has better handling abilities than a vehicle that can handle less gs. As we will see, the cornering power of a fire apparatus is quite limited when compared to a standard passenger automobile. In fact, most fire apparatus will roll over before the driver is able to generate enough lateral g-force to break traction with the road.

Rollover or Spinout

For the purpose of crash prevention, fire apparatus operators must understand the safety issues related to excess g-force. This is because a fire apparatus can only absorb so much g-force before things start to go badly for the driver and the crew. If the lateral g-force exceeds the rollover threshold of the vehicle, it will roll over. This is the most likely case for a fire apparatus, especially on a dry road (fig. 3–1). On the other hand, vehicles that have a low center of gravity will rarely be able to generate enough lateral g-force to exceed their rollover threshold and roll over. Instead, when a vehicle with a low center of gravity generates more lateral g-force than the available drag factor of the roadway, the tires will break traction with the road and slide out (fig. 3–2).

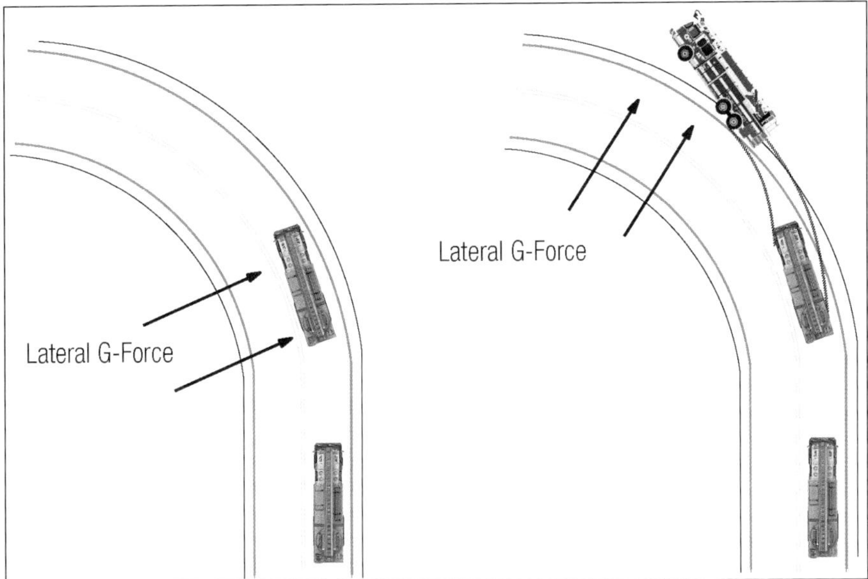

FIGURE 3–1.   As a fire apparatus rounds a curve, lateral g-force is acting on the vehicle. The amount of lateral g-force depends on the speed of the vehicle and the radius of the curve. If the lateral g-force exceeds the rollover threshold of the vehicle, the vehicle will roll over.

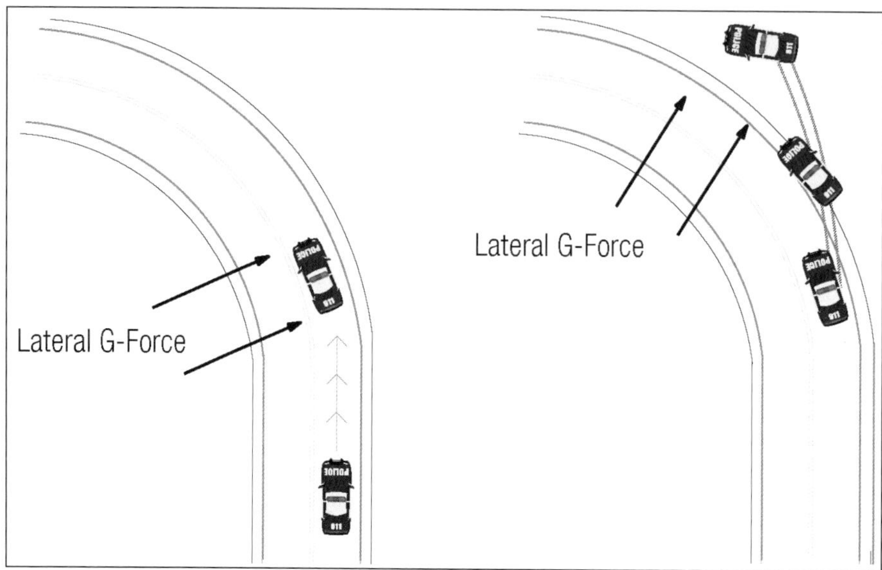

FIGURE 3–2. In this case, a police car is rounding a curve. Because the police car has a low center of gravity, the amount of lateral g-force never exceeds the rollover threshold of the vehicle. However, if the amount of lateral g-force exceeds the drag factor of the roadway (stickiness of the road), the vehicle will break traction and spin off the road.

Understanding Rollover Thresholds

The rollover threshold of a vehicle is a relationship between the height of the center of gravity and the track width, which is the distance between the center of the rear or dual wheels (fig. 3–3). The rollover threshold is calculated using the following formula, which is also known as the static stability factor. The resulting value calculated by this formula is the amount of lateral g-force the vehicle can absorb before it rolls over.

This formula demonstrates that if two vehicles have the same width, the vehicle with the higher center of gravity will have a lower rollover threshold. Therefore, the vehicle with the high center of gravity will tend to be less stable. Because of their design and function, most fire apparatus have a high center of gravity which makes them relatively unstable when compared to other types of vehicles.

NFPA 1901 addresses the issue of fire apparatus rollover thresholds. The standard provides rollover threshold requirements that must be met in order for the

$$\text{Rollover Threshold} = \frac{\text{Track Width of the Vehicle}}{2 \times \text{Height of the Center of Gravity}}$$

The rollover threshold is calculated with this formula, also known as the static stability factor. The track width is defined as "the distance between the centers of the right and left tires along the axle," and "the center of gravity is measured in a laboratory [or by the manufacturer] to determine the average height above the ground of the vehicle's mass."[*]

Don't get confused. As the height of the center of gravity goes up, the rollover threshold goes down. As the height of the center of gravity goes down, the rollover threshold goes up.

*DOT Technical Report—DOT HS 809 868

FIGURE 3–3.   The rollover threshold of a vehicle is a relationship between the height of its center of gravity and the track width of the wheels.

apparatus to be NFPA compliant. If the vehicle is unable to meet these requirements, it must be equipped with a stability control system.[1]

In order to meet the NFPA requirements, the manufacturer can calculate the height of the vehicle's center of gravity and the subsequent static stability factor or use the results of a tilt table test. In a tilt table test, the apparatus is chained to a large platform that rises up and tilts until the vehicle begins to roll over. Tilt table testing will provide the manufacturer with the angle at which a vehicle will begin to roll over. This angle can then be used to calculate the vehicle's rollover threshold.

Using the requirements in the NFPA 1901 standard, the minimum rollover threshold requirements (without an ESC system) can be calculated. The calculated minimum rollover thresholds are provided in table 3–1:

TABLE 3–1.   Calculated rollover thresholds using the information provided in NFPA 1901.

NFPA Requirement	Rollover Threshold[*]
Tilt table result no less than 26.5°	0.49
Height of CG no more than 80% of the track width	0.62

* DOT Technical Report—DOT HS 809 868

This NFPA requirement demonstrates that most modern-day fire apparatus will be able to absorb just 0.50–0.60 lateral gs without rolling over. Keep in mind that these rollover threshold requirements are for newer fire apparatus which meet the requirements set forth in the most recent version of NFPA 1901. Some fire departments may be operating a fire apparatus with an older design. As such, an older vehicle may have a rollover threshold that is lower than current NFPA requirements, which means that it may not be able to absorb as much lateral g-force without rolling over. Also remember that some apparatus may not meet the NFPA rollover threshold requirements and instead be equipped with an electronic stability control system. However, relying on the electronic stability control system to keep the vehicle from rolling over is not a sound idea.

Having calculated the rollover thresholds required by NFPA, we can now compare the minimum rollover thresholds of a fire apparatus to the average rollover thresholds of a civilian vehicle. Table 3–2 demonstrates the relatively low stability that is common to most fire apparatus when compared to passenger automobiles.

TABLE 3–2. Average rollover thresholds of different types of vehicles. Note the relatively low rollover threshold of a fire apparatus as compared to other types of vehicles.

Type of Vehicle	Common Rollover Threshold[*]
Passenger Car	1.30–1.50
Pickup Trucks and Vans	1.00–1.30
Jeeps	0.80–1.00
Fire Apparatus Minimum Requirements	0.49–0.62

* DOT Technical Report—DOT HS 809 868

Fire apparatus operators must understand the difference in stability between their personal vehicle and that of a fire apparatus. An average passenger car can absorb 1.3–1.4 lateral gs before it rolls over. This compares to the average fire apparatus which can typically absorb no more than 0.50–0.60 lateral gs without rolling over. If the fire apparatus experiences more than 0.50–0.60 lateral gs for a sustained period of time, such as during a severe steering maneuver or while rounding a curve, the vehicle will roll over.

This difference in rollover thresholds is an important teaching point for fire apparatus operators. It is not uncommon for a fire apparatus operator to arrive at the firehouse in their personal vehicle, having just driven through a curve in the road at high speed. The driver dons their turnout gear and gets behind the wheel of a much larger fire apparatus. The driver then drives out of the firehouse and begins to traverse the same curve at the same speed that they drove in their personal vehicle. However, the fire apparatus flips onto its side. Why? Because the vehicle dynamics of the vehicle changed considerably when they exited their personal vehicle and got behind the wheel of the much larger fire apparatus. As the rollover threshold tells us, a fire apparatus is considerably less stable than a passenger vehicle. Fire apparatus operators must understand and appreciate this fact when they get behind the wheel of a large fire department vehicle.

G-Force While Driving

Having discussed how much g-force a fire apparatus can absorb before it rolls over, we must now discuss how much g-force a vehicle will experience as it rounds a curve or makes an evasive maneuver. Understanding how to control g-force is another important teaching point for fire apparatus operators.

As we have discussed, a fire apparatus operator can create two types of g-force: longitudinal g-force and lateral g-force. Longitudinal g-force will act on the fire apparatus from front to back. A fire apparatus will experience a longitudinal g-force when it is braking or accelerating. The harder the fire apparatus operator presses down on the brake or accelerator pedal, the more longitudinal g-force the fire apparatus will experience. If the driver creates too much longitudinal g-force, the vehicle's tires will break traction with the road and skid. This concept, known as "the friction circle," will be discussed in Chapter 5.

Lateral g-force will act on the fire apparatus from side to side. A fire apparatus will experience lateral g-force anytime the driver turns the steering wheel, such as while rounding a curve or turning from one road to another. The amount of lateral g-force a fire apparatus will experience as it rounds a curve will depend on how fast the vehicle is traveling and how sharply the driver turns the steering wheel (also referred to as the **radius** of the vehicle's path of travel). The smaller the curve radius, the tighter the vehicle must turn in order to negotiate the curve. Many intersections will have a curve radius at the corner of just 30–60 ft, while country roads and large highway ramps may have sweeping curves with a radius of several hundred feet.

The formula for calculating lateral g-force is shown in Figure 3–4. While some readers may have no interest in examining the formula for g-force, it is important to take a quick look. Take note that the formula does not say to rate your driving ability on a scale of one to ten and multiply, or divide by the number of years you have been driving. More importantly, the formula does not say to subtract 12 if you are going to a working fire versus an automatic fire alarm. The driver who has been driving for 30 years will create the same amount of g-force as a new driver, regardless of the type of call you are responding to.

Remember that lateral g-force will increase as the speed of the vehicle increases or as the driver continues to turn the steering wheel more sharply. A fire apparatus can only absorb so much lateral g-force before it begins to roll over or lose control.

$$\text{G Force} = \frac{\text{Speed}^2}{\text{Curve Radius} \times 15}$$

FIGURE 3–4.   The formula for calculating lateral g-force.

Real Life

Having examined the formula for g-force, let's take a look at a real-life example and examine a curve in the road that has a radius of 320 ft (fig. 3–5). Take a look at table 3–3 and notice that as vehicle speed increases, so does the amount of lateral g-force. Considering the fact that most fire apparatus have an average rollover threshold of 0.50–0.60 gs, the average fire apparatus would roll over if it rounded this curve at a speed any greater than 50 mph. At this speed, the amount of lateral g-force would exceed the rollover threshold of the vehicle and cause it to flip over.

Keep in mind that a speed of 50 mph means that the fire apparatus operator is driving the vehicle at almost 100% of its ability. At this speed, there is no room for error. Just a slight increase in speed or slight turn of the steering wheel would increase the lateral g-force to a point which would cause the vehicle to roll over. This is why fire apparatus operators must drive in a manner that leaves a cushion between the amount of lateral g-force that is pushing on the vehicle and the rollover threshold of the vehicle.

In the case of this curve, the driver should strive to maintain a speed less than 25 mph as it will reduce the amount of lateral g-force to a comfortable level for the driver while also increasing the safety cushion between the actual g-force and the vehicle's rollover threshold. In fact, many highway design engineers post curve advisory speeds which limit the amount of lateral g-force experienced by the driver to no more than 0.15–0.20 gs (fig. 3–6). This is because many drivers will begin to feel uncomfortable when they experience a lateral g-force that is any greater than 0.20 to 0.30 gs. Drivers who begin to feel uncomfortable as they round a curve may panic and make an inappropriate steering or braking maneuver. This inappropriate steering or braking maneuver may lead to a crash. The concept of lateral g-force and how it relates to driver comfort is yet another important teaching point for fire apparatus operators.

FIGURE 3–5.   Consider this real-life example. The radius of this curve is approximately 320 ft, which is not unusual for many areas.

TABLE 3–3.   This table shows how much g-force a vehicle will experience as it rounds the curve depicted in Figure 3–5. Notice the increase in lateral g-force as the speed of the vehicle increases.

Speed	Curve Radius	G-Force
10 mph	320 ft	0.02 g
15 mph	320 ft	0.05 g
20 mph	320 ft	0.08 g
25 mph	320 ft	0.13 g
30 mph	320 ft	0.19 g
35 mph	320 ft	0.26 g
40 mph	320 ft	0.33 g
45 mph	320 ft	0.42 g
50 mph	320 ft	0.52 g
55 mph	320 ft	0.63 g
60 mph	320 ft	0.75 g
65 mph	320 ft	0.88 g

FIGURE 3–6. In the case of our real-life example, the posted advisory speed for the curve is 25 mph. Notice that this speed corresponds to a lateral g-force of 0.13 gs. Highway design experts try to minimize the lateral g-force felt by a driver so the driver does not feel overwhelmed and begin to panic.

Speed is not the only factor that can increase or decrease lateral g-force. Lateral g-force will also increase as the radius of a curve decreases or gets tighter. Table 3–4 demonstrates that if a curve gets sharper, the lateral g-force will increase even when the vehicle maintains a constant speed.

Consider a vehicle that is traveling 35 mph around different curves with different radiuses. Notice the increase in lateral g-force as the radius of the curve gets tighter. This is why so many fire apparatus crashes have occurred while the

TABLE 3–4. In this table, we can see the increase in lateral g-force as the curve gets sharper. Notice that while the speed of the vehicle stays the same, the lateral g-force increases with the sharper curves.

Radius of the Curve	Speed of the Vehicle	Lateral G-Force
450 ft	35 mph	0.18 g
400 ft	35 mph	0.20 g
350 ft	35 mph	0.23 g
300 ft	35 mph	0.27 g
250 ft	35 mph	0.33 g
200 ft	35 mph	0.41 g
150 ft	35 mph	0.55 g

vehicle is turning from one road onto another at an intersection. The radius of a 90° intersection corner could be as low as 30–60 ft. A speed as low as 15 mph may create a lateral g-force that exceeds the rollover threshold of the apparatus and causes it to roll over.

Steering-Induced Rollovers

While lateral g-force is commonly associated with rounding a curve or turning a corner, countless rollover crashes have occurred on a straight road due to a steering input from the driver. Remember that anytime the driver turns the wheel, they create an artificial curve in the road. A driver who turns the wheel too sharply will create a curve in the road which may induce a rollover. This is known as a *steering-induced rollover.*

Steering-induced rollovers often result from the following scenario: the driver rounds a curve too fast and the passenger side tires drop off the roadway and onto the shoulder. Rather than stop the apparatus and bring it back onto the road in a safe fashion, the driver fights to get the apparatus back onto the road. As the driver turns the wheel to get the apparatus back onto the road, the front tires suddenly mount the road edge and the vehicle shoots back onto the road, heading into the oncoming lane. At this point, the driver turns the wheel sharply in the opposite direction to try and get the vehicle back into its original lane of travel. By turning the steering wheel sharply, the driver creates a curve in the otherwise straight road. The artificial curve in the road, combined with the speed of the vehicle, creates a lateral g-force that is greater than the rollover threshold of the vehicle. As a result, the fire apparatus rolls over (fig. 3–7).

Weight Shift

Weight shift is another thing to keep in mind when discussing g-force. As a fire apparatus rounds a curve and lateral g-force begins to push on the side of the apparatus, the side of the vehicle on the outside of the curve will start to sink into the suspension. As the vehicle sinks into its suspension, the vehicle's center of gravity will start to shift towards the outside of the curve. This shift in the vehicle's center of gravity will reduce the track width on that side of the vehicle. Because the rollover threshold of a vehicle depends on the vehicle's track width, this reduction in track width caused by the shifting center of gravity will further reduce the rollover threshold of the vehicle and cause the vehicle to become less stable (fig. 3–8).

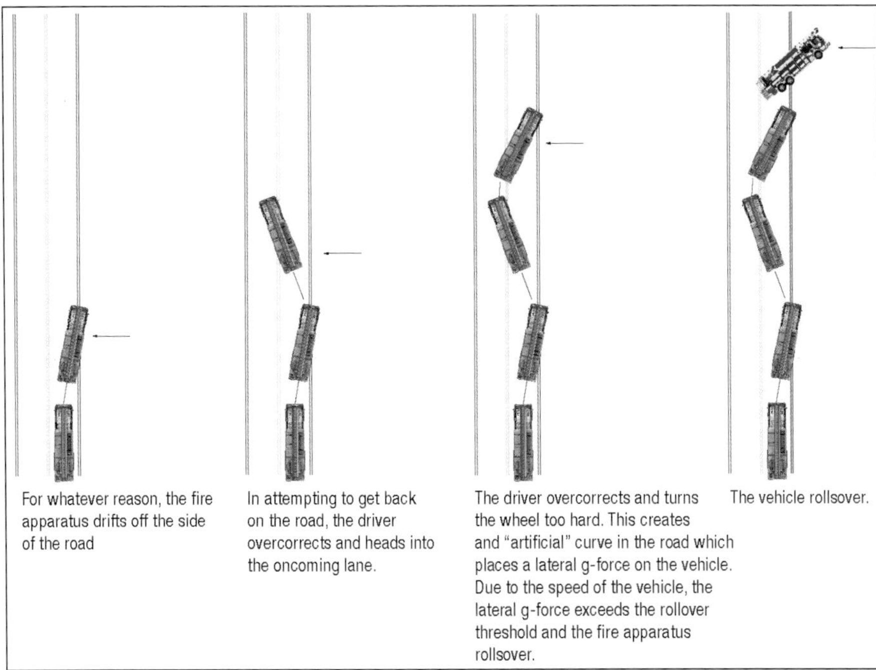

FIGURE 3–7.   A fire apparatus operator may experience a steering-induced rollover. In this case, the driver drifts off the road for some reason. In an attempt to get back on the road, the driver turns the wheel too hard and ends up traveling into the oncoming lane. The driver then attempts to re-correct in order to return to the appropriate lane of travel. When attempting to re-correct, the driver again turns the wheel too hard, which creates an artificial curve in the road. If the driver turns the wheel too hard and creates a curve radius that is too sharp for the speed that the vehicle is currently traveling, the lateral g-force will exceed the rollover threshold of the vehicle and the vehicle will roll over.

This is why lateral g-force can be such a problem. Not only is lateral g-force likely to flip you over, it can also shift the vehicle's center of gravity, reducing the vehicle's rollover threshold and making it easier for the vehicle to roll. One study demonstrated that weight shift reduced a vehicle's rollover threshold from 0.45 to 0.26.[2] This is a considerable reduction in vehicle stability as a result of weight shift, especially considering the fact that it does not take much speed to generate 0.26 lateral gs while steering or rounding a curve.

Weight shift can also have a negative effect on a vehicle in the longitudinal direction (front-to-back). If a vehicle brakes hard, the weight will shift to the front axle. This weigh shift will cause the rear of the vehicle to unload, making the rear tires more likely to lock-up and skid. Reducing the amount of weight shift a vehicle experiences in the front to back direction is an important part of skid control. This is best accomplished with smooth acceleration and braking.

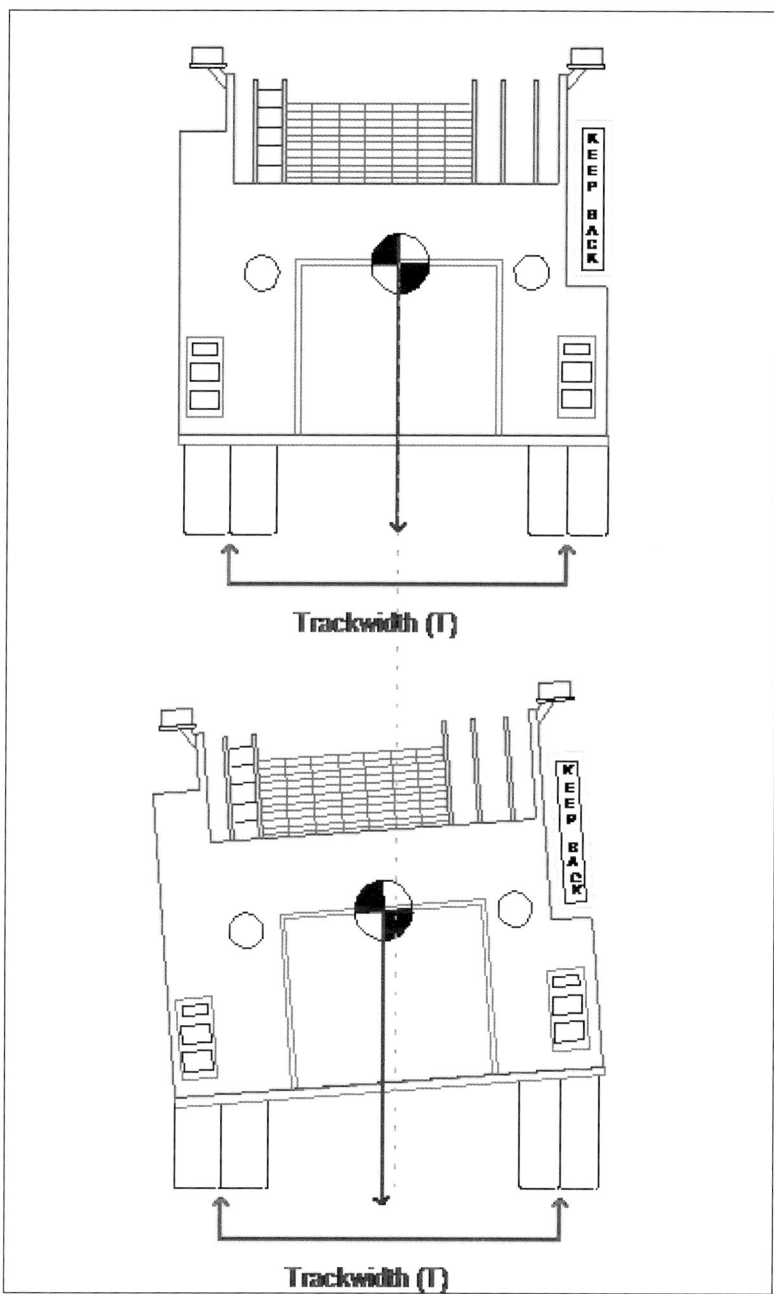

FIGURE 3–8.   As lateral g-force acts on the vehicle, the body of the vehicle will begin to shift on its suspension towards the outside of the curve. When this happens, the center of gravity will shift within the track width, reducing the vehicle's rollover threshold and further degrading the stability of the vehicle.

Air Force Testing

The United States Air Force conducted extensive testing to investigate rollovers that were associated with the P-18 crash trucks commonly used throughout the military. As part of a comprehensive study, the Air Force conducted several performance tests that monitored the amount of lateral g-force the P-18 could absorb before it rolled over or lost control. One of the solutions that came from the study was the use of a special strut that was installed on the vehicle's suspension system. This strut reduced the amount of weight shift experienced by the vehicle and increased the amount of lateral g-force the vehicle could absorb before it rolled over.

If we examine how weight shift can contribute to a vehicle's likeliness to rollover, the use of this suspension strut makes sense. If the vehicle's center of gravity shifts during a maneuver, the track width on that side of the vehicle is reduced, causing the vehicle to become less stable. If a suspension component is in place that will help mitigate how far the vehicle's center of gravity shifts, it will help to prevent a reduction in the vehicle's rollover threshold which keeps the vehicle more stable. Even though the rollover threshold of the P-18 is still low compared to other types of vehicles, it will not be further reduced during a dynamic maneuver by a shifting center of gravity. This is one reason why a stiffer suspension helps to improve vehicle stability.

Another interesting point related to the Air Force P-18 study was the fact that the improvements to the vehicle's suspension system helped to raise the vehicle's rollover threshold to a point where its tires would break traction with the roadway before the vehicle began to rollover. This is because the reduced weight shift during a turn resulted in the rollover threshold of the vehicle remaining at a value that was above the drag factor (or stickiness) of the roadway. Remember that if the amount of lateral g-force required to roll the vehicle is greater than the amount of grip between the tires and the road surface, the tires will tend to slide out before the vehicle generates enough lateral g-force to flip over. This important teaching point is a concept known as "the friction circle" and will be discussed in Chapter 5.

Slosh

Fire apparatus operators must remember to enter a curve at a reduced speed to help reduce the g-force that is acting on the vehicle. Reducing g-force will reduce weight shift and help maintain vehicle stability. This is especially important for fire apparatus that carry a water load because of a phenomenon known as *liquid surge* or *slosh*. Slosh is especially dangerous in fire apparatus that operate with a partially loaded water tank, such as a wildfire tender that pumps and rolls as it fights a wildfire.

Water slosh inside a fire apparatus water or foam tank is the direct result of inertia. Inertia is defined as the resistance of an object to any change in its speed or direction of motion. As a fire apparatus changes direction, slows down, accelerates, or makes an evasive maneuver, inertia will cause the water inside the tank to resist this change and continue at the vehicle's original speed or in the vehicle's original direction of travel. The inertia of the water load will cause the water to push on the inside of the tank and create a force that will push the entire vehicle towards its original direction of travel, causing the phenomenon we call "slosh."

If there is enough liquid inertia, it can cause the vehicle to lose steering control, increase stopping distance, or roll over as it rounds a curve. The same can be said for a sudden evasive maneuver in which the driver needs to quickly change the vehicle's direction of travel. If there is enough inertia from the liquid load, the vehicle may not complete the evasive maneuver effectively, or it may roll over in the process.

Liquid surge is even more of a problem in a partially loaded water tank where water is more free to slosh back and forth and push against the walls of the tank. This motion can cause the vehicle's center of gravity to shift as the vehicle attempts to absorb the energy of the sloshing load by sinking or shifting on its suspension. When the center of gravity shifts, the rollover threshold of the vehicle will be reduced, and the vehicle will be more prone to rolling over.

In addition to a loss of lateral control, the energy of the water wave sloshing back and forth against the inside of the tank can contribute to an increased stopping distance. This is why water tanks should be kept completely full or completely empty. A completely full or completely empty water tank will help reduce the effects of liquid surge caused by the sloshing water in a partially filled tank (fig. 3–9).

Baffles

Another method of reducing liquid surge is to install baffles inside the water tank. Baffles are partitions that help reduce the energy of a surging liquid load. Baffles act as a shock absorber to help prevent a large wall of water from striking the inside of the tank. The NFPA 1901 standard discusses two types of tank baffling methods:[3]

Containment method

This method uses a series of swash plates to divide the tank into a series of smaller, interconnected compartments.

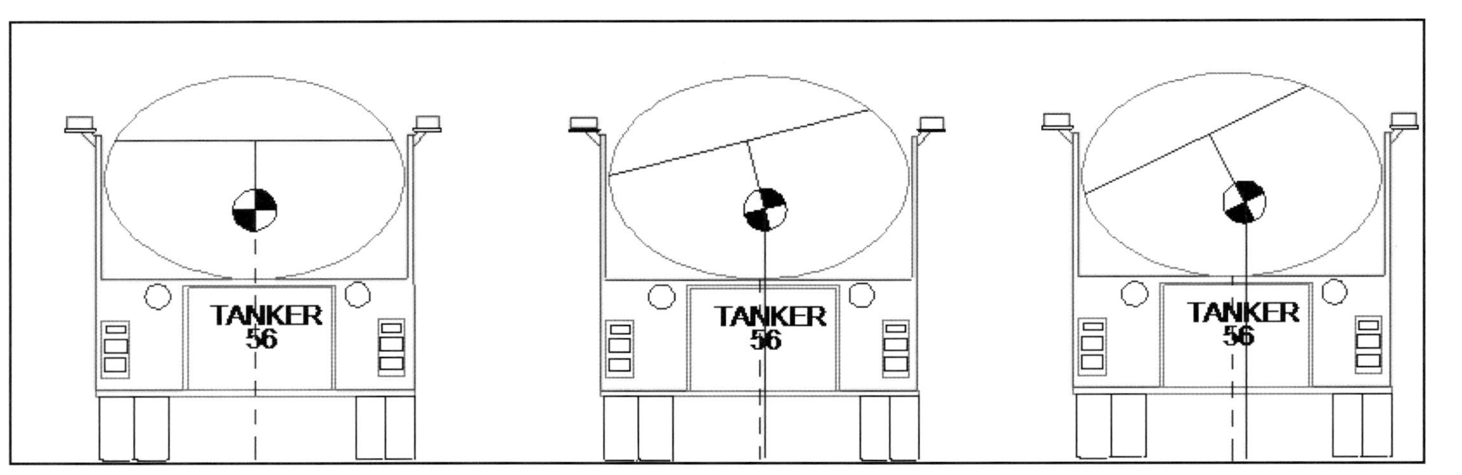

FIGURE 3–9. In a partially loaded water tank, lateral acceleration will cause the liquid load to shift as the apparatus rounds a curve. Notice how the center of gravity shifts to the outside of the curve as the apparatus rounds the curve. If the center of gravity shifts far enough, the apparatus will roll over.

Dynamic method

This method uses a series of baffles to disrupt the movement of water by changing its direction of travel. These baffles are often staggered so that the changing direction of the water creates a turbulent motion that results in the water absorbing most of its own energy.

Regardless of what type of water baffles the fire apparatus is equipped with, it is important to understand the dangers associated with carrying a liquid water load. At no time should the water tank be operated when it is partially full. To do so will result in more liquid surge because the water in the tank is less constrained and able to slosh about freely. This excess slosh can lead to a loss of stability, even at low speeds.

However, there are areas of the country where pump-and-roll operations are common. In these cases, fire apparatus operators must use extreme caution and thoroughly understand how sloshing water in the tank can have a significant impact on vehicle dynamics. This is also true for those that may operate a modified or aftermarket vehicle that is not equipped with tank baffles. The issue of modified vehicles will be discussed in Chapter 13.

Case Study—Nevada

On August 14, 2006, a Nevada Bureau of Land Management engine company was en route to a wildfire. The engine was being driven with a half-empty water tank when the operator attempted to negotiate a curve in the roadway. While negotiating the curve, the engine rolled over and came to rest on its passenger side.

Fortunately, all members were wearing seat belts at the time of the crash. Four of the members were able to self-extricate themselves via the left-rear window. The driver was trapped between the driver's seat and the steering wheel and had to be extricated from the vehicle. A postcrash inspection revealed that, in addition to poor road conditions and a speed that was too fast for those conditions, a contributing factor to the crash was the partially filled water tank, which resulted in a sudden shift in the vehicle's center of gravity.

Notes

1. NFPA 1901 (4.13)—Vehicle Stability
2. Chris Winkler, "Rollover of Heavy Commercial Vehicles," *UMTRI Research Review 31*, no. 4. (December 2000).
3. NFPA 1901, Section A.18.2.6

Understanding Rollover Crashes

An Overview of Rollover Crashes

When a fire apparatus operator creates too much g-force using the accelerator pedal and the steering wheel, the result is often a rollover crash. An examination of fire apparatus rollover crashes reveals these three common scenarios:

1. An apparatus that rounds a curve too fast and rolls over.
2. An apparatus that rounds a corner at an intersection too fast and rolls over.
3. An apparatus that rolls over as a result of the vehicle leaving the road and the driver overcorrecting in attempt to bring the vehicle back onto the road. In these cases, the vehicle usually leaves the road because the driver enters a curve too fast, the driver is distracted, or the road collapses under the weight of the apparatus.

Rollover Dynamics

Anytime a vehicle moves from side to side, such as rounding a curve or making an evasive maneuver, the vehicle will experience *lateral acceleration*. The amount of lateral acceleration will depend on how fast the vehicle is traveling and how tightly the driver turns the steering wheel. Lateral acceleration is more easily understood as *lateral g-force*. If a vehicle experiences high lateral g-force, it may roll over or spin out.

Most passenger vehicles have a low center of gravity. Because of this low center of gravity, high lateral g-force will likely cause a passenger vehicle to break traction with the road and spin out. Most passenger vehicles will not roll over unless they are tripped by an obstruction in the road, such as a curb or pothole.

However, most fire apparatus have a high center of gravity and are inherently unstable vehicles. Excess speed and severe steering inputs may cause a fire apparatus to roll over while rounding a curve or during an evasive maneuver. Whether a fire apparatus will roll over or spin out will depend on the vehicle's rollover threshold and the drag factor of the roadway. If the road is very slippery, the apparatus will likely break traction with the road and spin out. If the road is dry, the fire apparatus will likely roll over.

To put this in perspective, imagine pushing a mannequin on an ice-skating rink. If pushed properly, the mannequin will stay upright and slide across the surface of the ice. This is because there is very little friction between the mannequin's feet and the ice rink. If a lateral force is applied to the mannequin's torso, its feet will break traction with the icy surface, and the mannequin will slide along the rink.

However, if the mannequin were to be pushed in an asphalt parking lot, it would fall over. The asphalt surface of the parking lot will grip the mannequin's feet more than the icy surface of the skating rink. When pushed, the mannequin's feet will stick to the asphalt road surface. While the mannequin's feet are stuck to the roadway, the top of the mannequin tips over due to the lateral force that was applied. As a result, the mannequin falls over.

This same scenario holds true for a vehicle. As the vehicle rounds a curve or makes an evasive maneuver, lateral forces will begin to push on the vehicle. If the vehicle is traveling too fast and enough lateral force is generated, the vehicle will lose control. On a wet road, the fire apparatus tires will likely break traction with the roadway, and the vehicle will start to skid. On a dry road, the tires will likely maintain their grip on the road while the top of the fire apparatus begins to tip over from the lateral g-force. In this case, the fire apparatus will most likely roll over.

There are two types of rollover crashes: *tripped* and *untripped*.

Tripped rollovers

Tripped rollovers are caused when an object in the vehicle's path blocks the movement of the tires, causing the vehicle to "trip" into a rollover. A tripped rollover can occur when the vehicle tires strike a curb, a ditch, a pothole, or a collapsing soft shoulder. Trying to determine the speed at which a fire apparatus will experience a tripped rollover is extremely complex. There are many factors that must be considered when examining a tripped rollover, such as the speed when the apparatus struck the object, the height of the object, and the suspension and tire deflection of the fire apparatus.

The most common scenario for a tripped rollover occurs when a fire apparatus rounds a curve too fast and the vehicle's tires begin to lose their grip on the roadway. When the tires break traction and lose their grip, one of the axles slides off the road. When the tires leave the roadway, they may sink into the soft shoulder or strike an object on the roadside. This tripping action results in the fire apparatus tipping over.

Untripped rollovers

Untripped rollovers are caused by the frictional force between the tire and the road surface. These types of rollovers are also called *maneuver-induced rollovers*.[1] Imagine driving on a wide-open parking lot and suddenly turning the wheel as hard as you can, causing the vehicle to roll over—this is an untripped rollover.

Untripped rollovers are usually associated with a steering maneuver at a relatively high speed. This turn of the steering wheel creates a lateral g-force that causes the vehicle's weight to shift. When this weight shift occurs, the vehicle's center of gravity will shift towards the side of the vehicle on the outside of the turn. If the center of gravity is high enough and shifts far enough, the vehicle will roll over. Quantifying the speed at which an untripped rollover will occur is more straightforward than calculating the speed of a tripped rollover.

Rollover Speed

The first step in determining the speed at which a vehicle will experience an untripped rollover is calculating the vehicle's *rollover threshold*. The rollover threshold of a vehicle is a relationship between the track width and the height of the center of gravity. The formula for calculating the rollover threshold of a vehicle is discussed in Chapter 3.

If two vehicles share the same width, the vehicle with the higher center of gravity will typically be less stable than the vehicle with the lower center of gravity. To better understand this concept, consider a firefighter who wants to lift a heavy object off a victim. If the firefighter wants to lift the object and save the victim, he will most likely use a lifting bar. The longer the lifting bar, the easier it is to lift the object due to the increase in mechanical advantage. This same concept holds true for rollover thresholds. A vehicle with a higher center of gravity will have a longer lever acting on the vehicle. The longer the lever acting on the vehicle, the easier it is to flip the vehicle onto its side.

Having determined the rollover threshold of a vehicle, we must now determine how much lateral g-force will be generated as the driver turns the steering wheel. The amount of lateral g-force generated by the driver will depend on the speed of the vehicle and how sharply the driver turns the steering wheel (see Chapter 3). If the driver creates a lateral g-force that exceeds the rollover threshold of the vehicle, it will roll over.

Most fire apparatus have a rollover threshold that falls between 0.50–0.60 lateral gs. This means that if the driver creates a lateral g-force that exceeds 0.50–0.60 lateral gs for a sustained period of time, the fire apparatus will roll over. This compares to passenger cars which have rollover thresholds that typically average around 1.2–1.5 lateral gs. This significant difference in rollover thresholds explains why fire apparatus are so unstable when compared to the typical passenger car. A fire apparatus operator who just drove to the firehouse in a small sedan must realize that the dynamics of the vehicle changed dramatically when he got behind the wheel of an emergency vehicle with a high center of gravity.

Understanding the relative instability of a fire apparatus is important. This is especially true when we consider that a vehicle will routinely experience 0.25 gs of lateral acceleration during a normal drive. If a vehicle routinely undergoes 0.25 gs of lateral acceleration and the average fire apparatus rollover threshold is only 0.50, we can see that there is little room for error. Depending on the vehicle speed and the situation, one slight turn of the steering wheel or an increase in speed of just a few miles per hour may increase the amount of lateral g-force significantly, creating more lateral g-force than the vehicle is able to absorb.

This fact is evident when we examine the g-force a vehicle would experience while rounding the curve depicted in Figure 4–1. The radius of this curve is 80 ft. Table 4–1 demonstrates that a typical fire apparatus would rollover at a speed of just 25–30 mph, which most fire apparatus operators would not consider very fast.

Fire apparatus operators must understand how important it is to maintain a safe speed. This is especially true while rounding a tight corner or sharp curve in the roadway. Tight curves and sharp corners will result in a large lateral g-force, even if the fire apparatus is traveling at a relatively low speed.

TABLE 4–1. This table provides the amount of lateral g-force a vehicle will experience while rounding the 80-ft curve depicted in Figure 4–1.

Speed	Radius	G-Force
10 mph	80 ft	0.08 g
15 mph	80 ft	0.18 g
20 mph	80 ft	0.33 g
25 mph	80 ft	0.52 g
30 mph	80 ft	0.75 g
35 mph	80 ft	1.02 g
40 mph	80 ft	1.33 g

FIGURE 4–1.   The radius of this curve is 80 ft. When traversing a curve like this, it will not take much speed to generate enough lateral g-force to exceed the vehicle's rollover threshold and cause it to roll over.

Reducing the speed of the vehicle and using steady, controlled turns of the steering wheel will help the fire apparatus operator minimize how much lateral g-force is placed upon the vehicle. Reducing lateral g-force will increase the driver's margin for error while rounding a curve or making an evasive maneuver and reduce the chance of a rollover crash.

Improving Rollover Thresholds and Vehicle Stability

Improving a vehicle's rollover threshold will help make the vehicle more stable and better able to resist a rollover crash. There are several ways to increase the rollover threshold of a vehicle. Some of these methods include the following:

1. Increasing the track width and making the vehicle wider
2. Lowering the center of gravity and making the vehicle lower
3. Stiffening the suspension components of the vehicle

However, these methods have practical limitations. Any increase in the track width of the vehicle is limited by the width of the roadway, and any decrease in the height of the vehicle is limited by the type of vehicle. The height of a ladder truck can only be lowered so much. Often, the only option is to stiffen the suspension of the vehicle.

The issue of stiffer suspension components was examined during the United States Air Force testing of the P-18 crash truck that was discussed in Chapter 3. During this testing, it was discovered that as the vehicle's suspension system was stiffened, it became more resistant to the lateral (side-to-side) forces working against it. These stiffer suspension components resulted in an increased amount of lateral g-force required to roll or yaw the vehicle (spin around its center of gravity). This was because the stiffer suspension system increased the vehicle's resistance to lateral forces and reduced the amount of weight shift. If a vehicle's center-of-gravity does not shift as far, the track width of the vehicle will not be reduced during a dynamic maneuver which results in the rollover threshold of the vehicle remaining relatively stable.

An interesting side effect was also discovered during this testing. It was noted that as stiffer suspension components made the vehicle more stable, the vehicle gave better feedback to the driver as it was driven closer and closer to the point of impending rollover. This feedback was characterized by the front tires of the vehicle starting to lose traction and slide out, instead of the vehicle starting to tip over. Because a driver's natural reaction to skidding front tires is to slow down, the driver usually slowed the vehicle to a point that allowed the tires to regain traction and resume the vehicle's intended path of travel before it rolled over.[2] The skidding tires provided a warning that the vehicle was close to tipping over and gave the driver time to make corrections.

The sliding tires on the P-18 acted as an improvised warning device for the driver, warning him to slow down and avoid the rollover before it happened. Unfortunately, an untrained driver will not recognize this warning sign about the approaching rollover. Drivers must also realize that if a vehicle with a low rollover threshold is driven on a roadway with a high coefficient of friction (dry road), there will be little warning before the vehicle rolls over. The vehicle will roll over well before the tires start to break traction and make noise.

Rollover Prevention

When a large lateral g-force is combined with a high-center-of-gravity vehicle and a sloshing liquid load, it is easy to see why driving a fire apparatus is so dangerous. Too much g-force in the lateral and longitudinal directions will make it more difficult to stop and turn the vehicle, especially during an evasive maneuver or panic stop. Excessive g-force can also cause a vehicle to tip over or lose control.

While there are many ways to prevent a vehicle from rolling over, the best way is to reduce the speed of the vehicle. When a driver slows down, he will reduce the lateral and longitudinal g-forces acting on the vehicle. Ensuring that the fire apparatus is kept at a safe speed will also reduce the amount of kinetic energy that must be kept under control, making it easier to handle and stop the vehicle. In addition to controlling vehicle speed, these other steps can be taken to reduce the risk of rollover:

1. Ensure that the vehicle is properly weighed each year so it is not overweight or overloaded.
2. Ensure that the weight is properly distributed across the vehicle.
3. Ensure that the vehicle is equipped with the proper tank baffles.
4. Ensure that the vehicle's tires are well-maintained and properly inflated.
5. Ensure that apparatus operators have a thorough knowledge of their district and of their neighboring departments. Knowing the location of curves, intersections, and tight turns will allow the fire apparatus operator to slow down and drive through the hazard at a safe speed. Midway through a curve is not the time to realize the vehicle is traveling too fast. By then it will be too late.
6. Do not approach an intersection too fast or suddenly slam on the brakes. This can lead to a loss of control, especially in a vehicle that is carrying a liquid load.
7. Avoid drifting off the road. When a vehicle drifts off the road, the shoulder of the road may give way or the driver may overcorrect, both of which may lead to a rollover.
8. If the vehicle drifts off the road, come to a complete stop. Once the situation has stabilized, the driver can bring the vehicle back onto

the road slowly and safely. A driver who attempts to bring the vehicle back onto the road at a highway speed may overcorrect, enter the oncoming lane, and cause a head-on crash. Or while overcorrecting and traveling into the oncoming lane, the driver may attempt to turn back into the appropriate lane, creating sudden steering input that may cause the vehicle to trip and roll over.

There are also ways that apparatus manufacturers can help reduce the risk of a vehicle rollover. In order to make a fire truck more stable and less likely to roll, manufacturers can either shorten the lever (lower the center of gravity) or increase the distance between the rear wheels (increase the track width). Another option would be to stiffen the vehicle's suspension components to reduce the lateral shifting of the vehicle's center of gravity.[3]

The US Air Force P-18 study provided five recommendations to increase rollover safety among crash trucks. These recommendations are applicable to all fire apparatus and are listed as follows:[4]

1. Speed Notification Device: An audible device or heads-up display on the windshield would relay the speed of the vehicle to the operator without taking attention away from the road.
2. Governor: A governor would limit the operating speed of the vehicle and prevent the operator from exceeding the stability limits. Because this vehicle is not a primary firefighting vehicle, a few seconds delay in arrival to the scene will not compromise the capabilities or responsiveness of the firefighters.
3. Rollover Warning Device: A device should be installed to warn the operator when the vehicle is approaching the roll angle or lateral force required to roll over or slide out.
4. Black Box: A device similar to an aircraft black box would provide data on the status of the vehicle and information on the response of the driver throughout the duration of vehicle operation.
5. Dual Tires on the Rear Axle: The addition of dual tires on the rear axle of the P-18 would enhance and complement the stability of the retrofitted vehicle by widening the wheelbase.

Understanding Electronic Stability Control

Most fire apparatus manufacturers and automobile manufacturers are now equipping their vehicles with electronic stability control (ESC). There are many types of ESC systems. Most of these systems go by proprietary names that are assigned by specific vehicle manufacturers. In a generic sense, the Insurance Institute for Highway Safety (IIHS) defines an electronic stability control system as follows:

> *ESC is a vehicle control system compromised of sensors and a microcomputer that continuously monitors how well a vehicle responds to a driver's steering input, selectively applies the vehicle brakes and modulates engine power to keep the vehicle traveling along the path indicated by the steering wheel position. This technology helps prevent the sideways skidding and loss of control that can lead to rollovers. It can help drivers maintain control during emergency maneuvers when their vehicles otherwise might spin out. The systems are marketed under various names, including dynamic stability control, vehicle stability control, and dynamic stability and traction control, among others.*

For lack of a better term, ESC "takes over" when the vehicle begins to understeer, oversteer, or rollover (understeer and oversteer will be explained in Chapter 5). A vehicle equipped with electronic stability control has sensors which detect how the vehicle is moving or sliding around its center of mass. These sensors are able to determine the vehicle's heading in relation to the position of the steering wheel. If the vehicle is not heading where it's supposed to be going, the ESC will activate one or more of the brakes and decrease the throttle in an attempt to correct the heading of the vehicle (figs. 4–2 and 4–3). By correcting the vehicle's path of travel, the ESC reduces the likelihood that the vehicle will lose control or roll over.

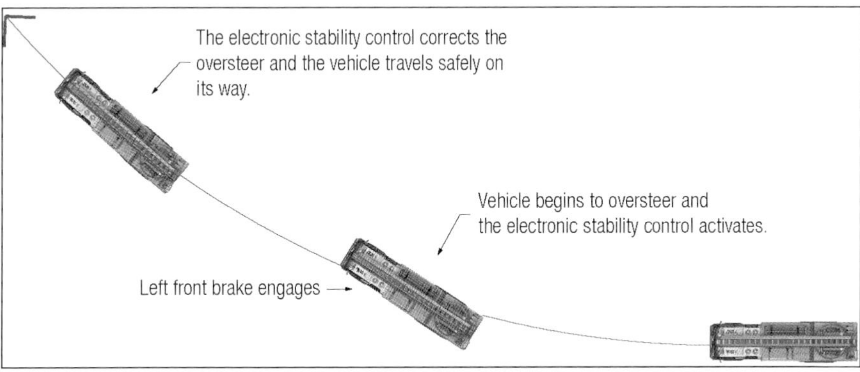

FIGURE 4–2.   Electronic stability control activation scenario 1.

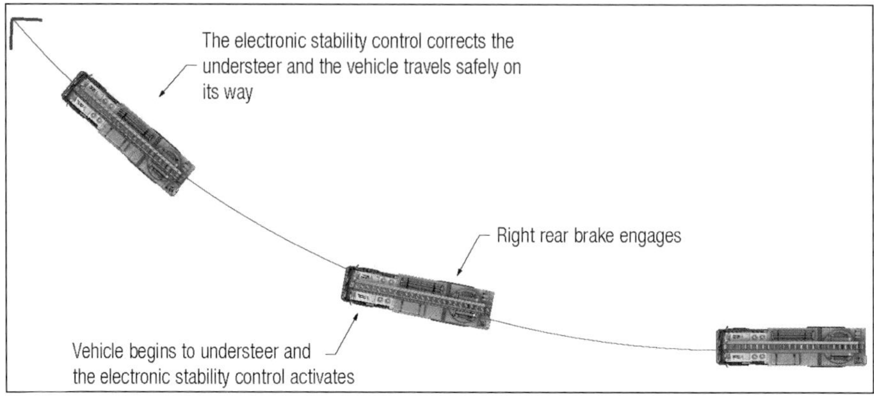

FIGURE 4–3. Electronic stability control activation scenario 2.

Active Rollover Protection

While electronic stability control systems measure and correct a vehicle that is in an understeer or oversteer situation, active rollover protection (ARP) will attempt to actively interrupt a vehicle that is about to roll over. ARP will monitor the amount of force being applied about the vehicle's lateral axis. When the ARP detects a potential rollover, it will apply the brakes and decrease engine torque. This type of ARP deployment usually activates during an untripped rollover caused by excessive speed or a sudden steering input.

In the case of a tripped rollover, where the vehicle strikes an object and begins to roll around its lateral axis, the ARP will engage an *active suspension system*. When the active suspension system deploys, it will produce a force in the opposite direction of the rolling force in an attempt to push the vehicle back onto its wheels.

Rollover Warning Systems

Many manufacturers now provide rollover warning systems which will alert the driver when the lateral g-force force acting on the vehicle is too dangerous. While there are many types of rollover warning systems, most resemble a turn-bank indicator on an airplane. As the vehicle begins to experience a lateral force from side-to-side, the force is represented on a dashboard display. If the lateral force is acceptable, the lights remain green. As the force become more dangerous, the lights will change to yellow or red. If your apparatus is equipped with this type of warning system, ensure that you know what it means and how to use it.

Avoiding Rollover Crashes

Rollover crashes are extremely dangerous to the occupants of a heavy vehicle. While only 4.4% of tractor trailer accidents are rollovers, 58% of truck driver fatalities occur in a rollover crash.[5] Additional studies indicate that while rollovers only accounted for 2% of all crashes in 2002, rollovers accounted for nearly 1/3 of all fatal crashes.[6] Rollover crashes are highly lethal and are often caused by careless or inattentive driving and excess speed.

Professional fire apparatus operators must understand the dangers associated with a high-center-of-gravity vehicle, liquid surge, and excess speed. Operators must be familiar with their first-due districts and mutual aid districts so they can anticipate curves, intersections, and tight corners and slow down before traversing them. When driving through these hazardous situations, fire apparatus operators must use extreme caution. The high severity rate of these rollover crashes are also one more reason why firefighters should be wearing a seat belt.

Notes

1. Rollover Prevention Docket No. NHTSA-2000-6859 RIN 2127-AC64. Accessed July 1, 2015
2. Kieth Bagot, "Evaluation of Retrofit ARFF Vehicle Suspension Enhancement to Reduce Vehicle Rollovers." DOT/FAA/AR-02/14. Office of Aviation Research, Washington DC.
3. "P-18 Suspension Roll Stability Test," Jennifer Kalberer and Leo Davis, October 1999.
4. "P-18 Suspension Roll Stability Test," Kalberer and Davis.
5. Chris Winkler, "Rollover of Heavy Commercial Vehicles," *UMTRI Research Review* 31, no. 4, (December 2000).
6. Liebemann, E.K. et al, "Safety and Performance Enhancement: The Bosch Electronic Stability Control (ESC)."

The Friction Circle and Skid Control

Why Tires Skid

A vehicle can only absorb so much g-force before things start to go badly for the vehicle and the driver. If the vehicle experiences a g-force which exceeds the rollover threshold of the vehicle, the vehicle will roll over. However, in the case of a vehicle with a low center of gravity or a fire apparatus on a slick road, too much g-force may cause the tires of a vehicle to break traction with the road **before** enough g-force is generated to cause the vehicle to roll over. This is because there is only so much grip available between the tires and the road surface. If the vehicle experiences a g-force which is greater than the amount of grip between the tires and the road surface, but less than the rollover threshold of the vehicle, the vehicle will start to skid before it rolls over. There are many different scenarios that can cause a vehicle's tires to lose their grip on the road. Let's examine each of these scenarios more thoroughly.

Over-Braking

Using the brakes is the best way to burn off a vehicle's kinetic energy and bring the vehicle to a safe stop. However, there may be times when this is not possible due to a sudden emergency which leaves the driver no time to slow down. When this happens, the driver may be forced to fully apply the brakes. This forceful braking action will create a longitudinal g-force which is greater than the available friction of the road. When this occurs, the wheel will lock up and stop turning. When the wheel stops turning, the tire will stop rotating and will drag or skid along the road. This scenario is more common in wet and inclement weather because it will take less braking force to overcome the frictional force between the tire and the slick road surface. Hence, a vehicle is more susceptible to skidding on a wet or slippery road.

Over braking is also an issue with an unloaded fire apparatus. Keep in mind that the braking system on a fire truck is designed to handle a fully loaded truck with a full tank of water. If the vehicle is driven with an empty water tank, the vehicle will be over-braked. An over-braked vehicle is more susceptible to skidding because there is an excess braking force on the wheels. This situation is especially hazardous for drivers of water tankers who may be driving with an empty tank during a tanker shuttle or to and from a maintenance facility.

Drivers must understand that smooth braking and acceleration is the best method for preventing a skid in the longitudinal direction. Looking far ahead to anticipate hazards will allow the fire apparatus operator to gently decrease speed instead of having to forcefully apply the brakes. Should the vehicle start to skid, decreasing the pressure on the brake pedal may reduce the braking force that caused the skid and allow the tires to regain traction with the road.

Over-Accelerating

If a driver applies too much acceleration, the motor may turn the wheels hard enough to break traction with the road. A classic example of over-acceleration is a burnout, in which a driver at a red light will floor the accelerator, squeal the tires, and create a large cloud of smoke. When this occurs, the rear end of the vehicle tends to fishtail as the tires break traction with the road.

When rounding a curve or executing an evasive maneuver, drivers must be very aware of acceleration. Too much acceleration may cause the tires to spin and break traction with the road. Smooth acceleration and smooth use of the steering wheel will help reduce the chance of a skid caused by over-acceleration.

Auxiliary Braking Systems

If used on a wet, snow covered, or icy road, an engine retarder or driveline retarder may cause a vehicle to skid. This is because the retarder creates a braking force on the wheels through the drive train. If the retarder provides more braking force on the wheels than the available friction between the tire contact patch and the road, the wheels may lock up and cause the tires to skid. Fire departments must implement a written procedure regarding the proper use of a retarder in inclement weather. Fire apparatus operators must remember to shut off certain types of retarders when driving in wet weather. Check with the manufacturer of the auxiliary braking device to determine the best course of action on slick roads.

Turning or Cornering

A vehicle can also skid in the lateral direction when the driver turns the steering wheel to round a curve or make a turn. If the speed and turning radius of the vehicle combine to create a lateral g-force that is greater than the available friction of the roadway, the tires will lose their grip on the road. This scenario is a common cause of curve-related vehicle crashes, especially in wet or slippery weather.

The Friction Circle

To better understand what causes a tire to skid, the fire apparatus operator must understand the friction that exists between the tire contact patch and the road. As a vehicle drives on a road, the tire is rotating around the axle and gripping the road at the contact patch. When the contact patch grips the road, it will push the vehicle forwards, backwards, or side to side, depending on the driver's input. If the tire contact patch is unable to grip the road, it will start to slide, and the vehicle will begin to skid.

A tire has a limited amount of friction to work with. The amount of friction a road can provide will depend on weather conditions, the slope of the road, and the condition of the tires. Because there is only so much friction available to work with, it is important for a driver to understand how the friction is used. Friction can be used to do the following:

1. Decelerate and stop the vehicle
2. Accelerate and drive the vehicle forward
3. Turn the vehicle in one direction or the other
4. Any combination of the above three

A vehicle will enter a skid when the frictional capabilities of the tires have become saturated. When this occurs, the tire will start to slide rather than maintain a firm grip on the road. When teaching fire apparatus operators how to understand the concept of tire grip, it is useful to discuss the concept of a *friction circle*. A s will provide a visual representation of how much grip is available to the driver to accelerate, brake, and turn the wheel. While the friction circle of each apparatus will differ based on several factors, including speed, tire stiffness, air pressure, and axle weights, let's discuss the concept in general terms.

If the driver is using friction to move forwards or backwards, he is using *longitudinal friction*. If the driver is using friction to move side to side, he is using *lateral friction*. Longitudinal friction refers to the braking and accelerating capabilities of the tire, while lateral friction refers to the cornering ability of the tire. There is only so much friction available for a drive to use in the lateral direction, the longitudinal direction, or a combination of the two directions. The amount of friction available for the fire apparatus operator to use is equal to the drag factor (or stickiness) of the roadway. Therefore, the radius of our friction circle will be equal to the drag factor of the road.

As an example, consider the fact that the drag factor of a dry, flat asphalt roadway usually falls around 0.75, though it may vary for several reasons. However, for the purpose of this discussion, we will assume that the drag factor of the road is 0.75 which will make the radius of our friction circle 0.75. As long as the vehicle does not generate more than 0.75 gs in the lateral, longitudinal, or a combination of both directions, the tires will be able to hold their grip on the roadway.

For the purpose of this training discussion, we will assume that our friction circle is round. However, in real life this may not be the case. As some studies have demonstrated, a roadway will not provide as much grip in the lateral (side-to-side) direction, especially on a wet road and at higher speeds. Therefore, a vehicle's friction circle may not be truly round.

Another issue we must consider when drawing our friction circle is the fact that a fire apparatus is equipped with truck tires. Truck tires do not grip the road as well as a passenger car tire. Due to differences in design and rubber composition, the grip of an average truck tire is usually around 80–85% the grip of a passenger car tire. This will reduce the radius of a fire apparatus friction circle when compared to that of a passenger car.

Once we have examined the road conditions and vehicle limitations to determine the size and shape of our friction circle, we can examine how much g-force a vehicle can absorb before it will break traction with the roadway. First, we will consider a vehicle that is rounding a curve at a steady speed with no acceleration or braking. This scenario would be considered a "steady state curve" because all of the grip between the tire and the roadway is available to use in the lateral (side-to-side) direction.

The Friction Circle—Dry Road

Let's consider a curve with a radius of 320 ft. Using this curve radius, we can create a table that will tell us the lateral g-force the vehicle will experience as it traverses this curve at different speeds. These speeds and the corresponding g-force are provided in table 5–1. Table 5–1 tells us that a vehicle traveling through this curve at 35 mph will experience 0.26 lateral gs. When we plot 0.26 lateral gs on our friction circle, we see that there is still plenty of g-force left to work with. This concept is demonstrated in Figure 5–1.

Now let's examine the friction circle if the vehicle was traveling 55 mph as it rounded the curve. In this case, the vehicle is experiencing 0.63 lateral gs as it rounds the curve. At this speed, there is not much g-force left to work with (fig. 5–2).

TABLE 5–1.   This table shows how much g-force a vehicle will experience as it rounds a curve with a radius of 320 ft

Speed	Curve Radius	G-Force
10 mph	320 ft	0.02 g
15 mph	320 ft	0.05 g
20 mph	320 ft	0.08 g
25 mph	320 ft	0.13 g
30 mph	320 ft	0.19 g
35 mph	320 ft	0.26 g
40 mph	320 ft	0.33 g
45 mph	320 ft	0.42 g
50 mph	320 ft	0.52 g
55 mph	320 ft	0.63 g
60 mph	320 ft	0.75 g
65 mph	320 ft	0.88 g

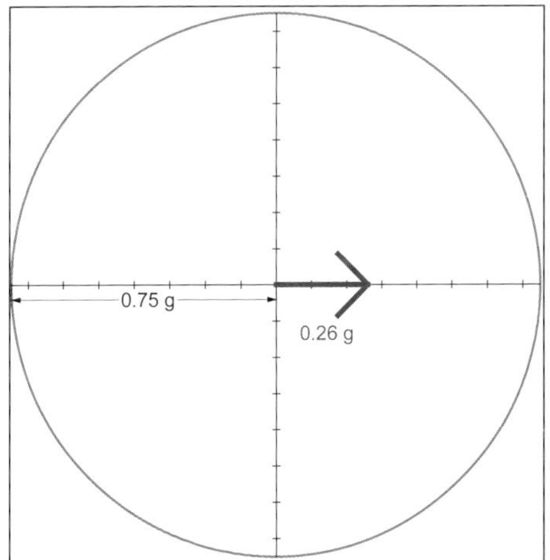

FIGURE 5–1.   This is an example of a friction circle. Notice that the radius of the circle is equal to the drag factor of the roadway (0.75). As long as the fire apparatus operator does not create a g-force greater than 0.75 and stays inside the circle, the tires will be able to hold their grip on the road. In this case, as the operator rounds a 320-ft curve at 35 mph, they create 0.26 lateral gs. The friction circle shows that there is still plenty of g-force left to work with. However, we must remember that if a fire apparatus exceeds 0.50–0.60 lateral gs on a dry road, the vehicle may roll over.

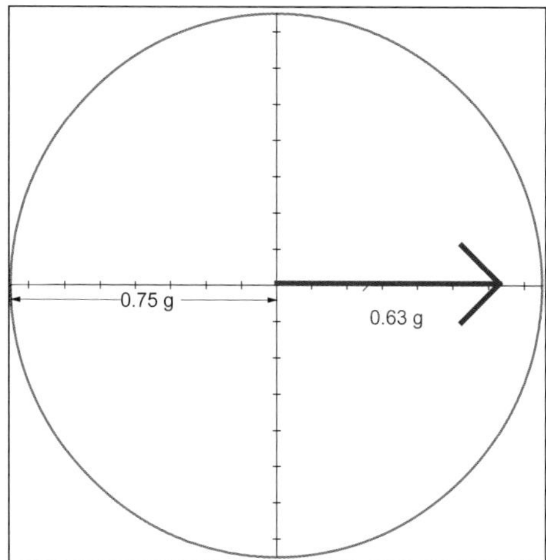

FIGURE 5–2.   In this case, a vehicle is rounding the 320-ft curve at 55 mph. At this speed, the driver creates 0.63 lateral gs. This does not leave much g-force left to work with, and the friction arrow is growing very close to the edge of our friction circle. On a dry road, a fire apparatus would have probably rolled over before reaching this point.

Just a slight increase in speed, or a slight turn of the steering wheel, will increase the amount of lateral g-force to a point which will grow outside the friction circle and cause the vehicle's tires to lose their grip on the roadway.

Finally, what happens if the vehicle rounds the curve at 65 mph? At this speed, the vehicle will experience 0.88 lateral gs as it rounds the curve. This amount of lateral g-force will put the vehicle outside of the friction circle. The vehicle will have used up all of the available traction and the tires will have lost their grip on the road. This is shown in Figure 5–3.

The Friction Circle—Wet Road

Driving in inclement weather requires significant adjustments to speed and strategy. This is because a wet road does not have as much grip to offer as a dry road. To better understand this issue, let's redraw our friction circle and assume that the road is wet or snow-covered. Because of the inclement weather, the drag factor of the roadway has dropped to 0.30 gs. Therefore, the radius of our friction circle has shrunk considerably (fig. 5–4).

Using the same 320-ft curve that we discussed before, let's look at what happens if our apparatus travels through the curve at 40 mph. At 40 mph, the apparatus will experience a lateral g-force of 0.33 gs. Notice that the amount of g-force

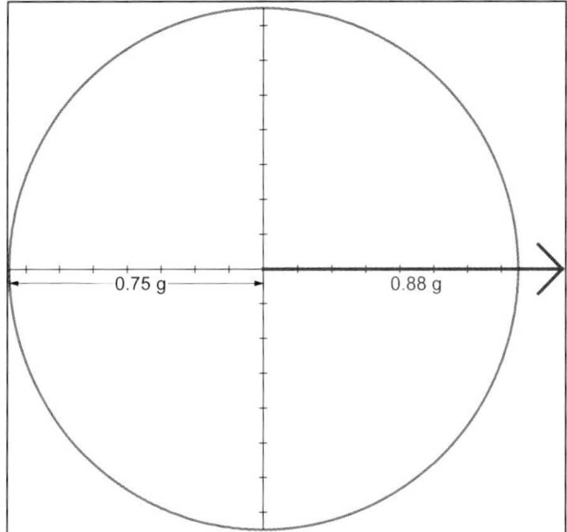

FIGURE 5–3.   At 65 mph, the driver has created 0.88 lateral gs as they round the 320-ft curve. At this speed, the driver has created a lateral g-force that is greater than the radius of our friction circle. The vehicle has used up all of the available friction, and the tires will break traction with the road.

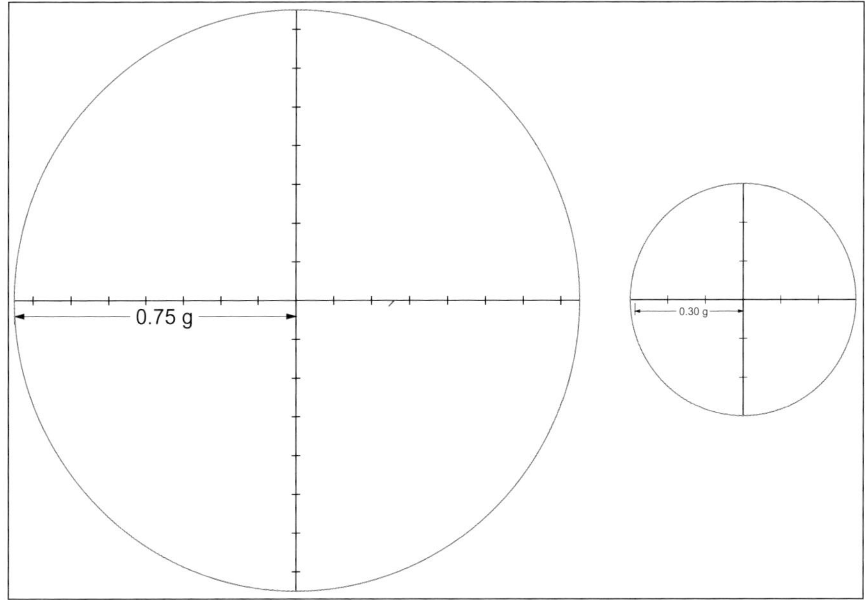

FIGURE 5–4.   In inclement weather the drag factor of a roadway will be reduced. In this case, the slick road has reduced the drag factor from 0.75 to 0.30. Notice how much this reduction in drag factor has caused the friction circle to shrink. This means that the fire apparatus operator does not have as much g-force to work with and the vehicle will be much more susceptible to a skid.

the apparatus will experience is the same regardless of the road conditions. However, on a wet road, this amount of lateral g-force will use up all of the available grip between the tires and the road surface, causing the vehicle to break traction with the roadway. This scenario is shown in Figure 5–5.

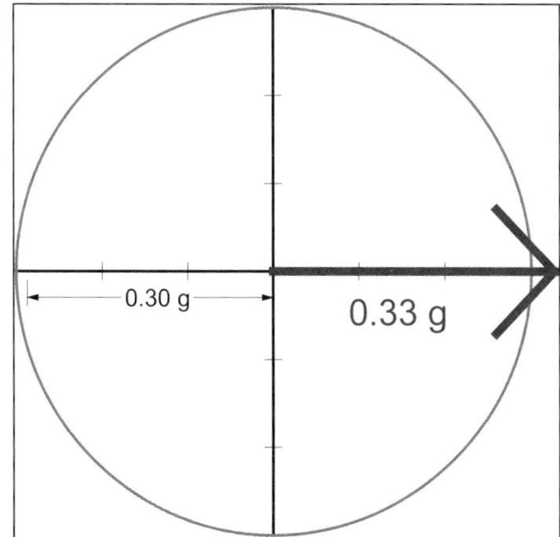

FIGURE 5–5. In this case, the driver drove through the 320-ft curve at 40 mph. This creates a lateral g-force of 0.33 gs. On a slick road, where the radius of the friction circle is only 0.30 gs, the driver has created a lateral g-force that is greater than the available grip between the tires in the road. The vehicle breaks traction and loses control.

The wet road scenario we have just discussed demonstrates why it is so important to slow down in wet weather. This is especially true while rounding a curve. In the case of this 320-ft curve, a vehicle could not drive any faster than 39 mph before the vehicle used up all of the available grip and slid off the road. Considering the fact that the speed limit for this real-life roadway is 45 mph, a vehicle would lose control even if it traveled at a speed that was less than the posted speed limit. Keep in mind that at 39 mph, the driver is just at the edge of the friction circle and about to break traction. Just a slight increase in speed or a small turn of the steering wheel would increase the amount of lateral g-force to a point that causes the vehicle to lose control.

Braking and Accelerating in Curves

Keep in mind that the previous curve-related scenarios depict a vehicle that is traversing a curve at a steady speed with no braking or acceleration. What would happen if the driver decides to apply the brakes or accelerate while rounding a curve? As an example, consider a driver who is rounding the 320-ft curve at just over 59 mph. At this speed, the driver is using 0.74 lateral gs and is just at the edge of the friction circle.

Now imagine that the driver feels overwhelmed by the g-force, panics, and applies the brakes. Or the driver comes around the curve and is suddenly faced with a slow-moving vehicle in the apparatus's lane of travel. When the driver forcefully applies the brakes to avoid this hazard, they create a braking force of 0.4gs in the longitudinal direction. When these lateral and longitudinal g-forces are combined, the driver creates a combined g-force of 0.84 gs, requiring more grip than the roadway has to offer (fig. 5–6). The combined g-force will cause the tires to break traction with the roadway, and the driver will lose control. This explains the old adage that you should break **before** the curve instead of **in** the curve.

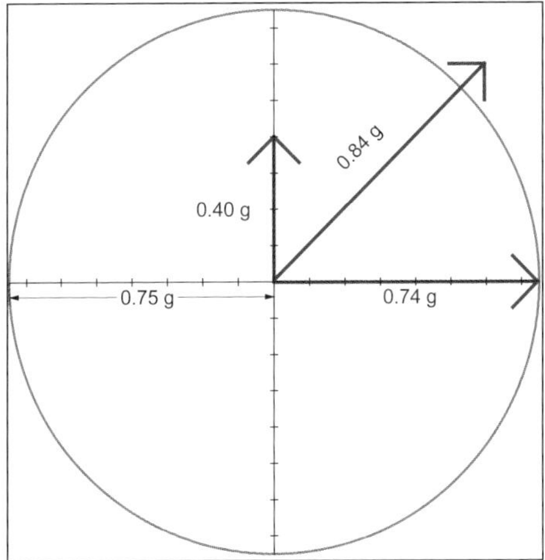

FIGURE 5–6. If fire apparatus operators use braking or acceleration while rounding a curve, they will create a combined g-force in both the longitudinal and lateral direction. In this case, the driver was rounding the curve at a speed that put the vehicle just at the edge of the friction circle. Suddenly applying the brakes created a combined g-force that put the vehicle outside of the friction circle. As a result the tires will break traction and the vehicle will lose control.

A similar scenario would play out if an auxiliary braking system engaged while rounding the curve on a wet roadway. If the driver is traveling at a speed that is close to the edge of the friction circle, a sudden activation of an engine or driveline retarder would be the same as if the driver applied the brakes. If this braking action is great enough, it may create a combined g-force in the lateral and longitudinal direction that exceeds the available grip of the roadway (fig. 5–7). This is why drivers have always been told to shut off an engine or driveline retarder on a wet road. Because a wet road does not have as much friction available to work with, failing to shut off the auxiliary brake could cause the driver to lose control and crash the fire apparatus.

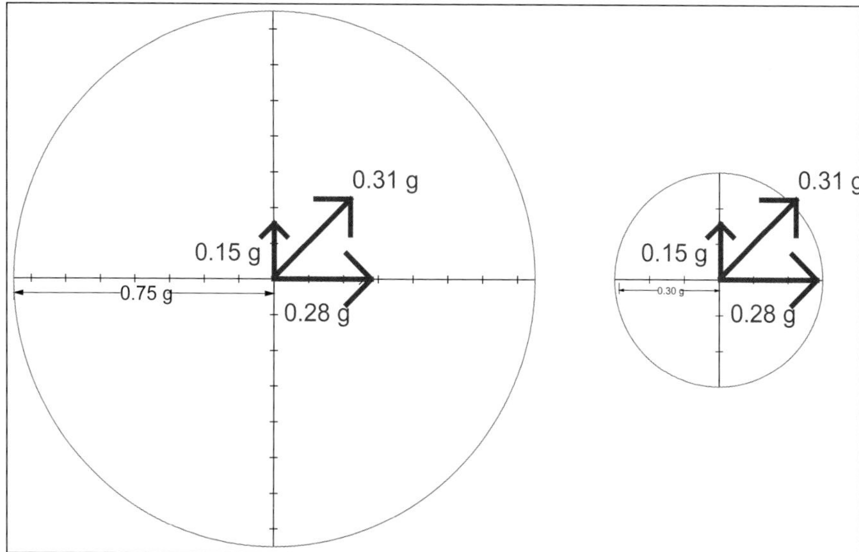

FIGURE 5–7. This figure depicts a dry road with a drag factor of 0.75 and a wet road with a drag factor of 0.30. On the dry road, the fire apparatus operator is rounding a curve at a speed that creates 0.28 lateral gs. When the driver disengages the accelerator pedal, the auxiliary brake activates and begins to decelerate the vehicle in the longitudinal direction at 0.15 gs. The combined g-force is 0.31 gs, which is well within the friction circle of the dry road. However, on the wet road, this combined g-force of 0.31 gs is outside the friction circle. As a result, the activation of the auxiliary brake creates a braking action that causes the apparatus to lose control. This is why it is so important to deactivate some types of auxiliary brakes on a wet or slick road.

Braking in Curves—Modern Vehicles

Remember that a modern fire apparatus may be equipped with electronic stability control. If this is the case in your fire department, ensure that the fire apparatus operator knows how to react if they find themselves losing control while rounding a curve. While the old adage was to not apply the brakes in the curve, an electronic stability control system may actually allow the driver to apply the brakes since the system will activate and assist the driver in regaining control.

This is something that every fire department should research and include in the driver training program. Keep in mind though—relying on the electronic stability control system to engage and save the day is not a safe way to do business. Drivers should still strive to slow down **before** the curve and round the curve at a speed that is safe for conditions so that the assistance of the electronic stability control system is not required.

Also be sure to research the type of auxiliary braking system the apparatus is equipped with. Fire apparatus operators must know if they should disengage the system in wet weather. Most manufacturers will probably recommend that the system is disengaged before driving on a wet or slippery road. If this is the case with your fire apparatus, make sure that the driver training program includes drive time with the engine or driveline brake turned **off**. Many fire apparatus operators conduct all of their driver training time on dry roads with the auxiliary braking system turned on. Driving a fire apparatus for the first time without the system engaged can come as a startling experience as the stopping distance of the vehicle may increase dramatically.

Case Study—Maryland

On June 9, 1990, a pumper-tanker from a Maryland fire department was responding to an emergency call. As the driver turned at an intersection, the driver lost control, causing the vehicle to rotate 180° and overturn into a ditch. Fortunately, all five firefighters were wearing seat belts and only received minor injuries.

At the time of the crash, the roads were wet from an earlier rain shower. The driver stated that as he made the turn, he was traveling 25–30 mph. The driver further stated that he "took his foot off the gas to slow the truck" and he "counted on the engaged engine retarder to slow him down." When the driver took his foot of the gas, he stated "the rear end went very fast, slipped around 180 degrees till I hit a ditch and flopped over." The driver stated that the engine retarder was left on at all times and never turned off.[1]

More than likely, the primary causative factor in this crash was the use of the engine retarder on a wet road. When an engine retarder engages, it slows the driveshaft and the rear drive axle of the vehicle. This action is no different from applying the brakes. If the braking force of the engine retarder overcomes the amount of available friction between the tire and the road surface, the wheels will lock and the tires will start to skid. If the rear tires of a vehicle skid, the rear of the vehicle will have a tendency to fishtail and come around towards the front of the vehicle. The engine retarder should be shut off when driving on wet or slippery roads.

Oversteer and Understeer

When a driver creates a lateral g-force that exceeds the friction circle, the vehicle will break traction with the roadway. When this happens, the vehicle will behave differently based on which tires have lost their grip on the road. If the front tires of a vehicle start to break traction while rounding a curve, the vehicle will *understeer*. In an understeer, the vehicle will lose steering control and will continue to travel in a straight line. While the tires of the vehicle will point one direction, the vehicle will travel in another. Professional racers refer to this phenomenon as a "push" (fig. 5–8).

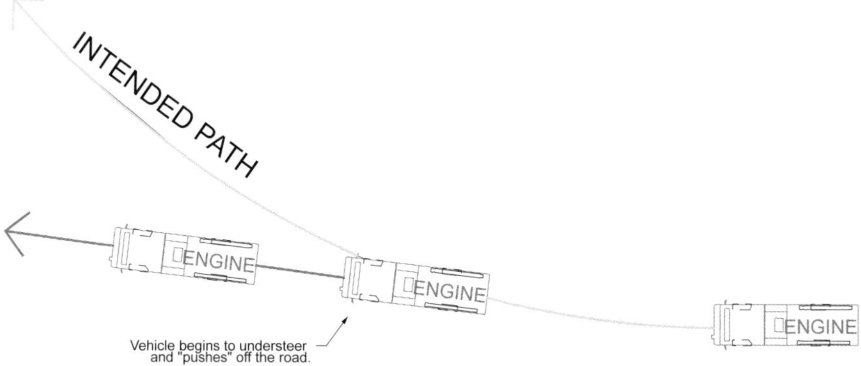

FIGURE 5–8.   A vehicle that is understeering will tend to "push" off the road.

Oversteer is another condition encountered when rounding a curve too fast. If the apparatus oversteers, the rear tires will break traction and the rear end of the vehicle will start to spin around its center of gravity. This is known as a *yaw*. If the rear end of the vehicle swings around far enough, the vehicle will spin out (fig. 5–9).

In real-world driving, several factors will affect whether a vehicle will understeer or oversteer. The main factors include weight distribution, weight shift, suspension geometry, brake adjustments, tire stiffness, and tire pressures. While both scenarios can be disastrous, an oversteer situation is considered to be more dangerous. This is because a vehicle that is understeering is still dynamically stable. While the driver has lost the ability to steer, the vehicle will simply continue in a straight line as it travels off the road. On the other hand, a vehicle that is oversteering will become unstable. When the rear end of the vehicle breaks traction with the road, the vehicle will begin to spin around its center of mass. A spinning vehicle is more difficult to control than a vehicle that is simply pushing

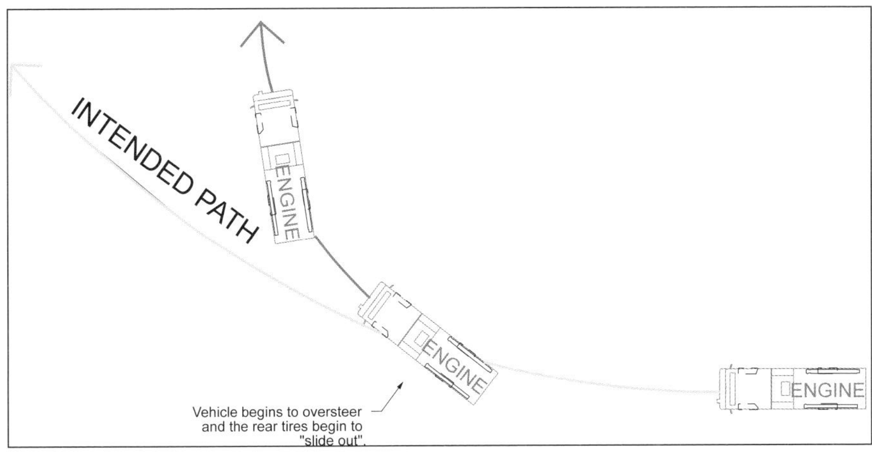

FIGURE 5–9. A vehicle that is oversteering will tend to have a loose rear end.

off the road. This is one of the reasons why most vehicle manufacturers will strive to design a vehicle that understeers instead of oversteers. The thought process is that when a vehicle begins to understeer, the natural tendency for a driver is to slow down and turn the wheel harder, which is the actual solution to the problem. In the case of an oversteering vehicle, the driver would need to turn the wheel in the opposite direction of the spin, which is known as *counter-steering*. Counter-steering is a more complicated solution to the problem and one that not many drivers are trained to do.

When we consider this issue from a firefighting standpoint, a fire apparatus that understeers is better than a fire apparatus that flips over. Granted, firefighters want to avoid **both** scenarios; however, when given the choice, it is relatively safer to push off the road than to flip on your side (keeping in mind that pushing off the road can still lead to a tripped rollover).

You may recall that during the Air Force P-18 rollover tests discussed in Chapter 4, a suspension strut was added to the chassis that increased the stiffness of the suspension components. This stiffening of the suspension reduced the amount of weight shift that the vehicle experienced during lateral maneuvers. By reducing the weight shift, the vehicle's center of mass did not move as far, which helped prevent a reduction of the vehicle's rollover threshold during a dynamic maneuver. As a result of this improvement, the vehicle's rollover threshold remained relatively stable at a value that was higher than the drag factor between the vehicle's front tires and the roadway. Now, as lateral g-force increased during a dynamic maneuver, the g-force would cause the front tires to break traction **before** it rose to a level that exceeded the vehicle's rollover threshold and caused it to rollover. When the front tires began to skid, the driver would usually recognize the fact that they were driving too fast for the given curve radius and start to slow down before the lateral g-force rose to a point that exceeded the rollover threshold of the vehicle. The tires breaking traction provided an improvised rollover warning device that told the driver to slow down.

Keep in mind that the Air Force testing was done on a large open track with plenty of asphalt upon which to maneuver the vehicle. On a real-life 12-foot travel lane, the front tires will quickly leave the roadway if they begin to skid. Once the front tires leave the roadway, the vehicle may trip itself and rollover, or the driver may overcorrect and cause a crash. Therefore, fire apparatus operators must still remember to slow down and drive through a curve at a safe speed.

Types of Skids

Problems will arise when the vehicle is operated at the upper limit of the available friction. If the vehicle is rounding a curve at the upper limit of available cornering friction, there is no friction left for braking. If the brakes are applied while rounding a curve too fast, the braking force will take away from the available cornering friction, causing the tires to slide sideways and the vehicle to lose control.

The opposite effect can be seen when a vehicle is skidding to a stop. In this case, the vehicle is skidding forward, using all of the available friction in the longitudinal direction. As all of the available friction is being used to skid the vehicle to a stop, there is no cornering force left to steer the vehicle from side-to-side. This is why a skidding vehicle loses steering control and skids in a straight line.

To counter the loss of steering control caused by a skidding tire, automobile manufacturers developed antilock brakes. Antilock brakes work by preventing the wheels from locking and skidding. By preventing the tires from skidding and using all of the friction in the longitudinal direction, there is still lateral friction left for steering. This lateral friction allows the driver to maintain steering control, even when the brakes are forcefully applied during a panic stop.

However, despite the use of antilock brakes and electronic stability control, there may still be times when a fire apparatus or emergency vehicle enters a skid. During a skid, the vehicle will behave differently depending on which tires are skidding on the roadway (figs. 5–10 and 5–11).

Front Tire Skids

If the tires on the front axle should lock up and skid, the vehicle will have a tendency to skid in a straight line because all of the friction is being used in the longitudinal direction and there is no friction left to steer the vehicle from side to side. Instead, the vehicle will plow forward until it stops or crashes.

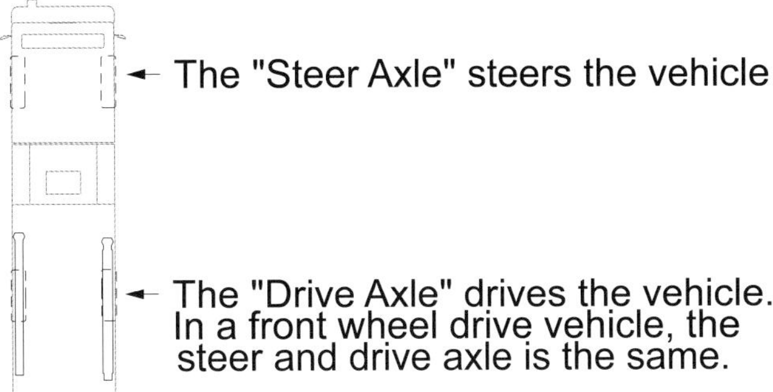

FIGURE 5–10.   The front axle is known as the steer axle because it is the axle which is attached to the steering wheel. The steer axle controls the direction of the vehicle. In a rear-wheel drive vehicle, the rear axles is called the drive axle because this is the axle that is connected to the power train which drives the vehicle forward or backward.

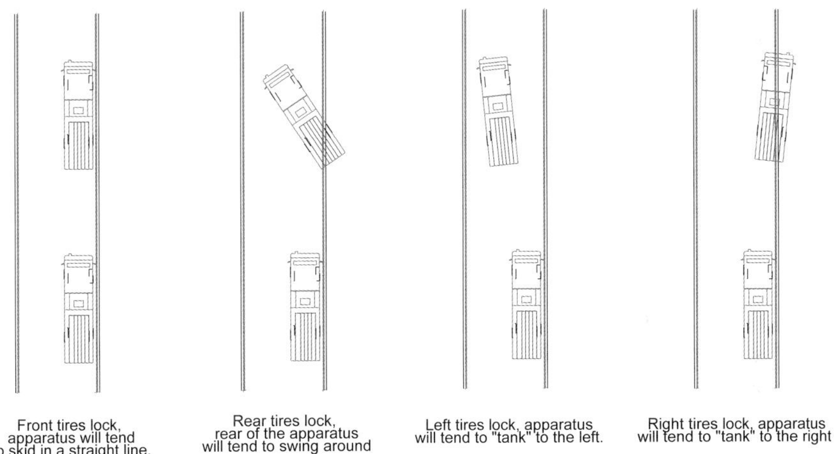

FIGURE 5–11.   A vehicle will react differently depending on which tires lose traction and begin to skid.

Left or Right Tire Skids

If the tires on the left side of the vehicle should lock, the vehicle will start to pull towards the left. If the tires on the right side of the vehicle should lock, the vehicle will start to pull towards the right. This is known as *tanking* because it is similar to the way in which an army tank maneuvers by slowing one tank track or the other.

Full Tire Skid

If all four (or six, or eight) tires skid, the vehicle will usually travel in a straight line. However, the vehicle may have a tendency to drift to the right or to the left.

This is because most roadways are crowned for drainage. If the vehicle skids far enough, and the road crown is great enough, the vehicle may follow the direction of the road crown as it skids to a stop.

Correcting a Skid

Fire apparatus operators must learn to recognize the signs of an impending skid. The most obvious indication of an understeer is that the driver turns the steering wheel but the vehicle still goes straight. Another indication may be a chattering of the front tires as they try to keep contact with the road. When the vehicle enters an understeer situation, the natural inclination for the driver is to apply more steering input by turning the wheel harder. In a severe understeer, this usually won't solve the problem. It is also important to remember that turning the wheel harder will decrease the turning radius of the vehicle. If the vehicle slows down enough and the tires suddenly catch, it could create a lateral g-force that exceeds the rollover threshold of the vehicle and cause the vehicle to roll over.

If a vehicle enters an understeer, the driver should ease off the offending pedal. If the driver is accelerating through the turn and the vehicle starts to understeer, the driver should let off the accelerator. If the driver is applying the brakes when the vehicle starts to understeer, the driver should let off the brakes. The driver may also have to straighten the steering wheel to allow the tires to begin to rotate again. When the tires begin to rotate again, they will be able to regain traction with the road. Just keep in mind that once the tires regain traction, the vehicle will start traveling in the direction that the tires were facing when they started to rotate. If the tires were aimed at the oncoming lane, or off the roadway, this is where the vehicle will want to go. Considering the fact that most fire apparatus are operated on a 10 to 12-ft travel lane, there will not be much room to recover before the vehicle drives off the road.

In order to correct an oversteer situation, the driver must recognize the fact that the rear end of the vehicle is breaking traction and the vehicle is beginning to yaw. If the tires on the rear axle should skid, the rear of the vehicle may begin to spin around. When this occurs, the driver must ease off the offending pedal and turn the wheel in the opposite direction of the spin. Remember that if the rear wheels have broken traction, the front wheels may still have enough grip to allow the drive to steer the vehicle. Turning the wheel in the opposite direction of the spin is known as *counter-steering*. Depending on the circumstances, light acceleration may help bring the rear of the vehicle back in line and correct the yaw.

"Look Where You Want to Go"

When learning methods of skid recovery, most drivers have been told to "turn into the skid" or "counter-steer." While these concepts are useful methods of skid correction, it will be difficult for a driver to think and act in this fashion in an emergency situation. Instead, the easiest method is to "look where you want to go." By focusing on a safe escape route, the driver's brain and arms will work together to get the vehicle where it needs to be. The human brain will act instinctively and make the necessary steering corrections to get the vehicle where the driver wants it to go.

Skids to Rollovers

Keep in mind that a fire apparatus is not driven on a large open racetrack with plenty of room to recover from an understeer or oversteer situation. Fire apparatus are driven on roads with a travel lane usually no wider than 10–12 ft. If the apparatus begins to understeer or oversteer, there will be little room to regain steering control and recover. More than likely, the vehicle will travel off the road and strike a fixed object, or it will drive onto a soft surface such as dirt or grass.

Once the fire apparatus leaves the road and strikes an object or soft surface, the chances are very high that the vehicle will experience a tripped rollover. Once this tripping action occurs, there is nearly nothing that can be done to recover. The apparatus will roll over.

Wet/Dry Switch

Older fire apparatus may be equipped with a limiting valve or *wet/dry switch*. This switch is located in the cab of the truck and allows the driver to reduce the pressure on the front brakes by up to half. These switches were installed because truck drivers believed that reducing the air pressure to the brakes on the steer axle would prevent the front wheels from locking on a wet or slippery road. In other words, the drivers believed that by reducing the air pressure to the steer axle, the wheels would not lock up and they could maintain steering control. Unfortunately, this comes at the cost of severely reduced braking efficiency.

As a result of several serious fire apparatus crashes, the National Transportation Safety Board (NTSB) conducted a special investigation. One of the recommendations that came from this investigation referred to the use of limiting valves (wet/dry switches). The NTSB report stated the following:

> *According to a published NHTSA report, a two-axle vehicle that weighs 27,300 pounds consistently performs better with the front axle limiting valve in the "dry road" position, even on a wet road surface. Use of a limiting valve on this [type of] vehicle appears unwise; it degrades performance.*
>
> *Because fire apparatus often stop suddenly, because they are frequently operated at higher speeds than are conventional vehicles, and because they are operated under hazardous conditions, the Safety Board concludes that the use of manual brake limiting valves can diminish the apparatus stopping capability and, therefore, their use should be discontinued.[2]*

Fire departments that still own and operate older vehicles equipped with limiting valves should adopt policies that require the switch to be left in the "dry road" position at all times. This should also be addressed in the department's driver training program.

Case Study—Connecticut

On May 10, 1990, a Connecticut fire department was dispatched to a fire alarm. The driver of the fire apparatus started the vehicle and noted that the air pressure in the brake system was 120 psi—well within normal operating limits.[3]

The fire apparatus operator drove the vehicle to the fire alarm, downshifting the automatic transmission and then applying the brakes to stop for an intersection. The driver stated that the brakes seemed to work "okay" and reported no issues stopping for the intersection. The driver continued on, negotiating a sharp curve and then traveling down a 10–13% grade. While descending the steep grade, he downshifted and applied the brakes, but "did not feel any braking."

The driver downshifted to "drive -1" and applied the brakes and the parking brake, but the parking brake button kept popping back out. The driver stated that the only deceleration he could detect was from the transmission. The fire apparatus reached another intersection, at which point the driver made a right turn. The driver observed stopped traffic ahead and made a left turn into a parking lot to avoid hitting the stopped traffic. At this point, the apparatus mounted a small curb and struck a tree. As a result of the crash, the driver and left jump seat firefighter received minor injuries. The firefighter in the right front seat received moderate injuries. The firefighter in the right jump seat was fatally injured, as was a second firefighter who was standing in the right jump seat area.

A review of the vehicle's maintenance records was included as part of the investigation into this fatal crash. These records revealed a lengthy history of requests to fix the brakes. One repair request stated "The maxi-brake doesn't hold on hills and the regular brakes have a hard time stopping the engine on emergency runs." On the morning of this crash, the vehicle had been in the maintenance shop while the firefighters exchanged a piece of equipment. While at the shop, the driver spoke to a mechanic about issues related to the brakes. The mechanic checked the air pressure of the braking system and made several brake applications. The mechanic advised the driver to come back after lunch and someone would adjust the brakes. Shortly after this conversation, the call was received that resulted in the fatal crash.

Investigators conducted a postcrash inspection of the vehicle's air brakes. This inspection revealed that while the front brakes appeared to be working and were properly adjusted, three of the four rear axle brake shoes were not making contact with the drum when the brakes were applied. Brake force calculations conducted by crash investigators indicated that at the time of the crash, the vehicle only had 58% of its original braking capability. Compounding this issue was the fact that the limiting valve was in the "wet" position. Because of this, the front steer axle was only operating at 50% of its braking capacity. When combined with the rear brake issues, the fire apparatus only had 36% of its original braking capability at the time of the crash.

The use of the limiting valve, in conjunction with issues related to poor vehicle maintenance, was a contributing factor to this crash.

Following the incident, the Connecticut Department of Motor Vehicles inspected the condition of the apparatus brakes and indicated that the vehicle would have failed a roadside truck inspection if the vehicle had been subject to federal highway standards. Further inspection of all 14 fire department vehicles revealed that 9 of 14 vehicles had to be placed out of service for repairs. The Connecticut Department of Motor Vehicles then inspected 559 fire apparatus from 64 cities and towns. These inspections revealed that 35% of the fire apparatus would have failed a roadside inspection and been placed out of service. Of all out-of-service violations, 50% involved brakes, 18% involved steering, and the rest involved tires, suspension, and fuel leaks.

Preventing Skids

The key to preventing a skid is to slow down and not skid in the first place. Driving a fire apparatus at the upper limits of its handling capabilities leaves no margin for error. This is especially true when navigating curves, approaching intersections, or driving in heavy traffic where a driver may need to take evasive action to avoid a crash. Drivers must remember to look down the road and anticipate hazards before they become a problem. By anticipating the actions of other drivers, slowing down for approaching hazards, and driving in a defensive fashion, a professional driver will negate the need to slam on the brakes. Drivers must remember to use smooth braking, acceleration, and steering techniques to avoid exceeding the frictional capabilities of the tires and inducing a skid. Engine and driveline retarders should be shut off during inclement road conditions, and drivers of unloaded vehicles must be especially aware of the potential for tire lock-up. These methods are especially crucial when driving on wet roads or in inclement weather.

Notes

1. National Transportation Safety Board. "Special Investigation Report Emergency Fire Apparatus". March, 1991
2. National Transportation Safety Board. "Special Investigation Report Emergency Fire Apparatus". March, 1991
3. National Transportation Safety Board. "Special Investigation Report Emergency Fire Apparatus". March, 1991

Siren Limitations

An emergency vehicle is equipped with a siren to warn other motorists of its approach. While sirens are a common tool in the fire service, few fire apparatus operators have a thorough understanding of how a siren works. Fire apparatus operators must understand the need for complete stops at negative right-of-way intersections due to the limited effective range of the warning siren.

Outdated Technology

Unlike their predecessors, modern cars have roofs, doors, and significant sound insulating materials. Modern day vehicles are built to encapsulate the occupants and insulate them from outside sound. The evolution of the modern day passenger vehicle has resulted in a well-insulated vehicle that is designed to keep sound from penetrating into the passenger compartment.

While cars have evolved over time, sirens have not. The fire service is still using the same mechanical sirens and flashing red lights that were used in the days of the Model-T Ford. It is nearly impossible for an outdated emergency vehicle siren to effectively penetrate the passenger compartment of a modern vehicle and warn a civilian driver of the emergency vehicle's approach.

The limited range of a siren seems counter-intuitive to everyone on the fire truck, as the noise of the siren is almost deafening inside the cab of the fire apparatus. While the siren noise in the cab is deafening, a civilian approaching the apparatus will have little chance of hearing the same siren. As a result, many fire apparatus operators overestimate the effective range of their siren.

Countless studies have been conducted to determine the effective range of an emergency vehicle siren. In most cases, the effective range of a siren is limited to 80 ft at a 90-degree intersection. Depending on the circumstances, this distance could be much lower.

How a Siren Works

In order to understand the limitations of a siren as a warning device, fire apparatus operators must understand how a siren works in relation to the sound receiver. The sound receiver is a civilian driver who must perceive and react to a siren that is approaching from a distance. In order for a siren to be effectively heard by a civilian motorist, the siren must penetrate the civilian vehicle at a level that is loud enough to break the civilian's concentration. Unfortunately, sirens are only so loud, and they lose volume as the siren gets further from the siren speaker. This limits the effective range of the siren. Compounding the issue is the fact that modern day vehicles block much of the siren sound from entering the passenger compartment of the vehicle. And once inside the vehicle, the siren must rise above the noise inside the passenger compartment that is created by the motor, the road, stereos, cell phones, and voice conversation. To better understand this process, it is important to examine each of these issues in-depth.

Ambient Noise

The noise inside the passenger compartment of a civilian vehicle is known as *ambient noise.* The amount of ambient noise inside the vehicle depends on the vehicle's speed, the volume of the radio, the fan setting, and any other noise inside the vehicle. For a siren to be heard by a civilian driver, it must rise above the ambient noise in the passenger compartment to a point which is readily discernible by the driver.

Several studies have examined the amount of siren sound required to break a person's concentration while driving. One study demonstrated that in order to detect an emergency siren, a person using a driving simulator required the siren sound pressure level to be 10 dB higher than a person not using a driving simulator.[1] Additional studies concluded that "signal levels 6–10 dB above the masked thresholds will ensure 100% detectability, and that signal levels approximately 15 dB above the masked thresholds are recommended for ensuring rapid response from the listener."[2]

Based on these studies, a siren must rise to a level at least 10 dB higher than the ambient noise inside the passenger compartment to be effectively heard by the driver. This brings up the next point: what is the typical ambient noise level inside a moving vehicle?

The ambient noise level inside a moving vehicle depends on several factors. These factors include how loud the stereo is, whether or not the heater or air conditioning is turned on, what the HVAC fan setting is, and whether the windows are rolled up or down. When all of these factors are put together, they create a significant amount of ambient noise inside the passenger compartment of a vehicle.

Vehicle speed will also affect the ambient noise inside the vehicle because of the noise created by the tires rolling on the roadway, as well as the noise of the vehicle's motor. On average, there will be an increase of 1.2 dB inside the passenger compartment of a vehicle for every 6.2 mph increase in speed.[3] Manufacturer comparison charts reveal that the ambient noise inside the passenger compartment of an idling vehicle ranges from 32.4 dB to 66.7 dB. If the vehicle is traveling at 55 mph, the ambient noise in the passenger compartment ranges from 53.4 dB to 76.3 dB.[4]

Attenuation

Modern-day vehicles are designed to keep noise out. The process of keeping noise out is known as *attenuation*. Attenuation is the difference between the siren sound pressure level outside of the car versus the siren sound pressure level inside of the car.

On average, the structure of a vehicle will reduce the sound pressure level of the siren by approximately 35 dB. If a siren is 110 dB directly outside the driver's window, the sound pressure level of the siren will be reduced to 75 dB inside the car. The amount of siren sound blocked by the structure of the vehicle will depend on several factors, the most important of which is the make, model, and sound insulation properties of the vehicle, as well as the frequency composition of the penetrating sound.

Sound Frequency of the Siren

Sirens are louder at some sound frequencies and quieter at others. Studies have shown that the structure of a civilian vehicle is very effective at blocking the sound frequencies commonly used by sirens. To combat this problem, some siren manufacturers have developed low frequency sirens that are better able to penetrate the body of a civilian vehicle. While these sirens may better penetrate the vehicle, they are often not as loud to begin with. Therefore, the range of a low frequency siren is still limited.

Siren Volume and the Inverse Square Law

The effective range of a siren is directly related to how loud the siren is. Siren manufacturers must make a siren loud enough to be heard, but not loud enough to be a danger and nuisance to society. On average, sirens produce a maximum of 124 dB 10 ft in front of the siren speaker (table 6–1).

Keep in mind that a siren will sweep up and down, and will not always be at maximum volume. As an example, it will take a Wail siren approximately 4 seconds to cycle up and down. The siren is only at its maximum volume for part of the 4 second cycle. As the siren winds down, it is not as loud.

TABLE 6–1.   Sound, or in more technical terms "Sound Pressure Levels," are measured in decibels (dB). To understand how decibels work, examine the average decibel levels of everyday sounds, as compared to an average fire apparatus siren.

Jet Aircraft—150 away	140 dB
Threshold of pain	130 dB
Mechanical Fire Apparatus Siren	124 dB (at 10 feet)
Chainsaw—3 feet away	110 dB
Vacuum Cleaner—3 feet away	80 dB
Rustling leaves	10 dB
Threshold of human hearing	0 dB

The farther you are from the siren, the quieter the sound of the siren. This concept is known as *the inverse square law.* According to the inverse square law, every time the distance from the siren is doubled, the sound pressure level of the siren will decrease 6dB. Most sirens average 124 dB when measured 10 ft in front of the siren. As a result of the inverse square law, most sirens will drop to 106 dB just 80 ft away from the siren (table 6–2).

Remember that the inverse square law is a theoretical concept. In real-life, the sound pressure level of the siren may decrease by more than 6 dB. This usually occurs when there are obstacles blocking or shielding the sound of the siren.

TABLE 6–2.   Assuming a best case scenario for an emergency vehicle, the sound pressure level of the siren will decrease 6 dB every time the distance from the siren is doubled. This table provides the theoretical reduction in sound pressure levels as the distance from the sound source is doubled. As most modern vehicles will require approximately 106–110 dB of sound just outside the driver's window to penetrate into the passenger compartment, we can see why sirens have a limited effective range.

Distance from Siren	10 feet	20 feet	40 feet	80 feet	160 feet	320 feet
Theoretical Sound Level	124 dB	118 dB	112dB	106 dB	100 dB	94 dB

There may also be times when the sound pressure level of the siren does not drop 6 dB when the distance from the siren is doubled. This usually occurs in urban environments which have asphalt roads and concrete buildings that reflect the siren sound. While the sound pressure level does not drop as dramatically, the reflection of the siren off the concrete and asphalt surfaces make it more difficult for a civilian to determine where the siren is coming from.

It is also important to remember that the inverse square law relates to the distance directly in front of the siren. When measurements are taken at an angle from the siren speaker, the sound pressure level will decrease by more than 6 dB every time the distance is doubled. A study conducted in 1977 by the United States Department of Transportation showed that in certain circumstances, the effective range of a siren at a 90 degree intersection was limited to 26–39 feet.[5] The increased reduction in siren sound pressure levels at a 90 degree intersection is mainly attributed to the fact that the siren speaker is aimed towards the front of the vehicle and not towards the sides.

Real Life Scenario

Having discussed how a siren is supposed to work as a warning device, let's now examine these concepts with a real-life example. Let's examine a vehicle that has an average ambient noise level inside the passenger compartment of 65 dB while traveling at 45 mph which is about average for today's vehicles. In order for a siren to effectively break the concentration of a driver, the sound pressure level of the siren would have to reach 75 dB inside the passenger compartment of the vehicle.

If it requires 75 dB of siren sound pressure level inside the passenger compartment to effectively break a civilian driver's concentration and warn of an approaching fire apparatus, and the structure of the civilian vehicle is reducing the siren sound pressure level by 35 dB, the siren would have to be 110 dB directly outside the driver's window (fig. 6–1).

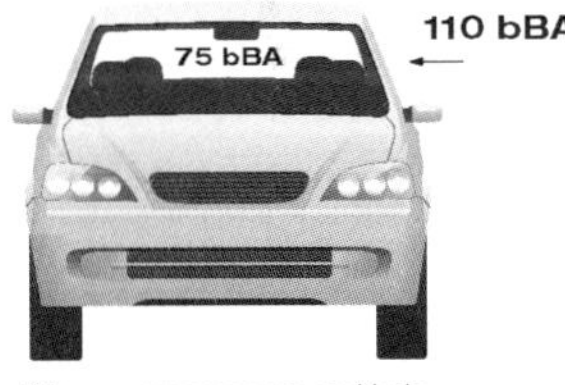

The average ambient noise inside the passenger compartment of a civilian vehicle is approximatley 65 decibels.	Studies indicate that the sound pressure level of the siren must rise to at least 10 decibles above ambient noise. Therefore, at least 75 decibles of siren noise must reach the driver's ear.	If the average passenger car blocks approximately 35 decibles of siren noise from penetrating the passenger compartment, the siren must arrive at the driver's window at 110 decibles.

FIGURE 6–1.

The question now becomes "how far from the emergency vehicle will the siren be louder than 110 decibels?" Most emergency warning sirens are rated at 124 dB, 10 feet in front of the siren. Using the inverse square law as shown in Figure 6–2, it is obvious that the effective range of a siren can be no more than 80 ft. This example demonstrates the science behind a siren's limited effective range and should be a key point for any driver training program.

> The fact that the effective range of a siren can be as little as 40–80 ft **must** be a key point for any emergency vehicle driver training program.

Windows Up or Down

Studies have shown that the ambient noise inside the passenger compartment may increase when a vehicle is traveling at a high speed with the windows down. This increase in ambient noise is caused by the sound of the wind blowing in the window. Driving with the windows down can increase the interior noise level of the vehicle by 10 to 15 dB.[6] While the insertion loss of the vehicle may decrease if the windows are down, the ambient noise inside the vehicle may go up.

If a vehicle is stopped with the windows down, the structure of the vehicle would not block as much of the siren sound. Studies indicate that the attenuation of the vehicle structure would be reduced approximately 10 dB if the driver was sitting still with the window down. However, consider this scenario in real-life: if a civilian is stopped at a red light with the windows down, the fire apparatus has a green light. Therefore, the civilian hearing the siren and yielding the right-of-way is not as great of an issue.

Localization

Once the siren is heard, the civilian driver must then figure out where it is coming from. Extensive studies have been conducted to determine how well a driver can determine where a siren is coming from. One study came to the following conclusions:[7]

1. A control group sitting in an open room could pick out the 45-degree sector where the siren originated 91% of the time.
2. When the same group was seated in an automobile, the identification of the correct 45-degree sector fell to 37.6% with the passenger window open and 26% with the passenger window closed.
3. The study concluded that it is not practicable for a civilian driver to identify the location of an approaching siren.

Once a person sits in a vehicle, it becomes very difficult to determine where the siren is coming from. One explanation for this difficulty is explained by Howard et al, who explain that "Drivers have a greater difficulty localizing the siren because a vehicle enclosure obstructs the direct path of siren noise and redistributes the acoustic energy over the surface of the vehicle, re-radiating into the enclosed space, which has the effect of altering the apparent direction of the external sound source."[8] In other words, as the siren reaches the civilian car, the sound of the siren strikes the body of the vehicle. The body of the vehicle blocks and redistributes the sound over the surface of the car, making it very difficult for a civilian driver to figure out where the siren is coming from.

In addition to the vehicle structure masking and redirecting the sound, there are other environmental factors that may affect a person's ability to determine where the siren sound is coming from. This is especially true in an urban environment where the siren will reflect off of buildings, making it even more difficult for a civilian driver to identify the location of an approaching siren.

This issue is compounded by the fact that the same buildings which are blocking and reflecting the siren are also blocking the sight distance at the intersection. Most urban environments have buildings on the corners of the intersection which make it very difficult for a civilian driver to see an approaching emergency vehicle on a side street. Special care must be taken in these environments.

Civilian Response to a Siren

Anyone who has driven an emergency vehicle with lights and sirens activated knows that it is next to impossible to predict how a civilian will respond to the approach of an emergency vehicle. Every state directs civilian motorists to stop and pull over to the right when an emergency vehicle approaches. However, fire apparatus operators should never rely on the fact that a civilian driver will do as instructed.

Fire apparatus operators must remember that the response of a civilian driver to an approaching emergency vehicle is a complex process that includes many steps. These steps include hearing the siren, localizing the siren (figuring out where it is) and then reacting to the approaching emergency vehicle. This process takes time. All the while, the fire apparatus is closing the distance with the civilian vehicle. If the fire apparatus closes the distance too quickly and does not give the civilian enough time to react accordingly, a crash will occur. As discussed in Chapter 1, the average perception and reaction time for a driver is around 1.6 seconds. In the case of an approaching emergency vehicle, a civilian's perception and reaction time may be higher due to the complex and unusual scenario: a giant truck barreling down with a loud siren and flashing lights.

Fire apparatus operators must also remember that this entire discussion has revolved around a civilian driver with normal hearing. If the driver of the civilian vehicle is hard-of-hearing, it may take even longer for the civilian to react to the approaching emergency vehicle. This is one more reason why emergency vehicle operators should always drive in a defensive manner and never assume that a civilian will immediately yield the right-of-way.

Emergency Response Protocols

Due to the fact that many incidents are not true emergencies, fire departments across the country are implementing "on-the-quiet" or reduced speed responses. These types of responses prohibit the use of lights and sirens by responding apparatus or only allow the first-due company to use lights and sirens. While the list of incidents and apparatus that respond nonemergency should be left to the individual agency, NFPA does state that cover companies relocating to other fire stations must respond nonemergency and obey all traffic laws. Fire departments must identify the types of calls that warrant a nonemergency response and implement policies and procedures to ensure that all fire apparatus respond appropriately.

When responding or driving nonemergency, the fire apparatus operator must remember to obey all traffic control signals, traffic signs, and rules of the road. The emergency warning lights should be turned off so as not to cause confusion and the vehicle should not use its traffic preemption device. When operated in nonemergency mode, the fire apparatus will have no special privileges and should be driven like every other vehicle.

One of the earliest known uses of sound as a warning device can be found in the Bible. In Biblical times, people affected by leprosy were made to warn others of their approach.

—Eldred and Sharp "Are present horns, whistles, and sirens necessary for communications?"

Notes

1. Carl Q. Howard, et al., "Acoustic Characteristics for Effective Ambulance Sirens." *Acoustics Australia 39*, no. 2–45, (August 2011).
2. Howard, "Acoustic Characteristics for Effective Ambulance Sirens."
3. P. Mioduszewski, "Speed Influence on Tire/Road Noise Inside a Passenger Car." Inter.Noise2000, The 29th International Congress and Exhibition on Noise Control Engineering, 27–30 August 2000, Nice France.
4. http://www.auto-decibel-db.com/
5. R. C. Potter, et al., "Effectiveness of Audible Warning Devices on Emergency Vehicles," Report No. DOT-TSC-OST-77-38, August 1977.
6. K. Eldred and B. Sharp, "Are present horns, whistles and sirens necessary for communications?" SAE Paper #720640. 1977.
7. Richard C. Potter and Hsien-Sheng Pei, "How well can audible warning devices of emergency vehicles be localized by the driver of an automobile?" *The Journal of the Acoustical Society of America* 59, no. S16, (1976).
8. Howard, et al. "Acoustic Characteristics for Effective Ambulance Sirens."

Negotiating Intersections

Introduction

Crossing an intersection in an emergency vehicle is dangerous. This is especially true at a negative right-of-way intersection. A negative right-of-way intersection is an intersection where the fire apparatus faces a red light, stop sign, or other traffic control device which indicates that the fire apparatus does not have the right-of-way.

Notice of Approach

Fire apparatus operators often believe that turning on the emergency lights and sirens automatically give the vehicle the right-of-way. This is not the case. An emergency vehicle does not have the right-of-way until a civilian vehicle perceives the approach of the emergency vehicle, reacts to the emergency vehicle, and **grants** the emergency vehicle the right-of-way.

A civilian driver who doesn't know an emergency vehicle is approaching cannot give the right-of-way. Fire apparatus operators must be sure to give a civilian enough time and distance to see or hear the fire apparatus, localize the apparatus, react to the fire apparatus, and take the appropriate action to yield the right-of-way. The amount of time and distance required for the civilian to complete this process is known as *notice of approach.*

Fire apparatus operators must ensure that adequate notice of approach is given to civilian drivers prior to driving the fire apparatus into a negative right-of-way intersection. If the emergency vehicle driver simply slows or pauses at a negative right-of-way intersection, civilian vehicles may not have enough time and distance to perceive and react to the emergency vehicle. The civilian will not be able to come to a stop before the fire apparatus pulls out in front of him. Instead, the civilian vehicle will crash into the fire apparatus as the apparatus proceeds through the intersection against the red light or stop sign.

By coming to a complete stop at a negative right-of-way intersection, the fire apparatus operator will give approaching civilian vehicles time to perceive and react to the fire apparatus. The apparatus operator is giving proper notice of approach. Failure to give proper notice of approach and simply driving into the intersection can result in a crash and subsequent legal ramifications.

Best Practice

Several organizations have published best practices that relate to intersection safety. All of these best practices revolve around common sense. The key to navigating a negative right-of-way intersection is to ensure that the emergency vehicle comes to a **complete stop.** NFPA 1500 requires a complete stop under the following circumstances:[1]

1. When directed by a law enforcement officer
2. At red traffic lights
3. At stop signs
4. At negative right-of-way intersections
5. At blind intersections
6. When the driver cannot account for all lanes of traffic in an intersection
7. When other intersection hazards are present
8. When encountering a stopped school bus with flashing warning lights

The following steps should always be taken when crossing a negative right-of-way intersection:

1. Come to a complete stop.
2. Be wary of creeping too far into the intersection to a point where the front bumper enters the travel lane on the cross street. If an approaching civilian vehicle does not have time to perceive and react to the fire apparatus, it may drive into the bumper. This is especially true with the large, protruding bumpers which are common on modern fire apparatus. Many of these apparatus bumpers are poorly lit, if lit at all. Instead, the emergency lighting on a bumper usually consists of a square strobe at the foremost portion of the bumper. This light is low to the ground and cannot be readily seen from far away. If the fire

apparatus operator is sticking the vehicle bumper into a high speed travel lane, there will not be sufficient warning for the civilian driver to perceive, react, and skid to a stop before striking the emergency vehicle. In other words, there was not enough notice of approach.

3. Look both ways. Ensure that traffic is slowing down.

4. Once civilian traffic on the cross street comes to a stop and yields the right-of-way, make eye contact with the civilian drivers. This will help confirm their intention to stop and allow the apparatus to cross in front of them. A wave of thanks also helps confirm their intentions.

5. Cross the intersection slowly, keeping an eye out in both directions. This is especially true for multilane intersections.

6. If there is more than one lane to cross, make sure to stop at each lane and repeat this process. Many fire apparatus crashes have occurred when the fire apparatus operator stopped on one side of the intersection and then shot across. Even if the civilian vehicles in the closest lane yield the right-of-way, drivers in the other lanes of traffic may not be able to see or hear you (fig. 7–1).

7. Once you have crossed the intersection, continue safely on your way.

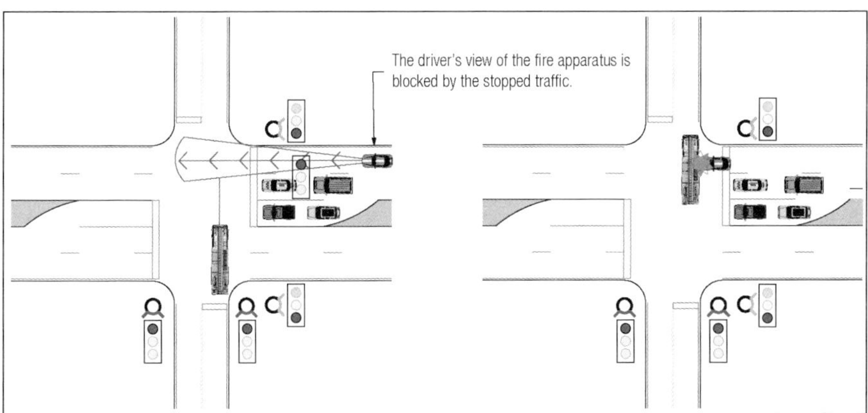

FIGURE 7–1.   When crossing a multilane intersection, the fire apparatus operator must treat each lane as a separate intersection and come to a complete stop for each lane. A civilian driver, who is traveling with a green light, may not be able to see the crossing fire apparatus due to stopped traffic.

When approaching an intersection, many emergency vehicle drivers assume that if civilian drivers cannot see the emergency warning lights, they will hear the siren. As discussed in the previous chapter, fire apparatus operators must remember that sirens have limited effective range at a 90-degree intersection.

Once the civilian hears the siren, they are often too close to the intersection to yield the right-of-way to the crossing fire apparatus. Sirens cannot be relied upon to give proper notice of approach in instances where the fire truck cannot be seen.

Account for All Lanes of Traffic

It is important that the emergency vehicle driver can see and clear **all** lanes of traffic before traversing a negative right-of-way intersection. If the driver cannot see a lane of traffic because there is an obstruction in the way, the driver **must** wait for the obstruction to clear, or for the red light to turn green. This issue is also addressed in the NFPA 1500 standard.[2]

Many drivers argue: "Do you really expect us to sit and wait for the light to turn green?" Yes. If a driver can't account for every lane of traffic, they should not guess. Guessing would be like playing a game of Russian Roulette. Are you willing to risk your career, your livelihood, and possibly your freedom by sticking the bumper of your fire truck into a high speed travel lane and hope that no one is coming?

Remember, if the apparatus driver can't see the civilian vehicle, the civilian driver can't see the fire truck. Unless there is a way for the sight obstruction to move out of the way, as in the case of a large vehicle or tractor trailer, the apparatus driver must wait to ensure control of **all** lanes before beginning to cross.

Fire apparatus operators must remember to stop and clear each lane of traffic on a multilane road. If the vehicle is crossing a four-lane highway, it will have to stop four times and clear each lane before crossing. Stopping at one side of the intersection, clearing only one lane, and then blindly driving across three lanes of traffic could result in a crash.

Caravanning

When multiple emergency vehicles drive directly behind each other to the same emergency, it is called *caravanning*. Caravans are usually the result of multiple units responding from the same fire station, or multiple agencies responding to the same incident (such as police and EMS). Caravanning can be dangerous. This is because a civilian driver will be unable to distinguish the fact that more than one siren, and thus more than one emergency vehicle, is approaching.

The drivers of the second and subsequent apparatus in a caravan must be wary of several issues. The first issue is the fact that the apparatus operator's forward sight distance will be obstructed by the emergency vehicle driving directly in front of him. The only warning the second apparatus operator may have of a hazard ahead will be the sudden activation of the first apparatus's brake lights. Drivers of caravanning apparatus must leave plenty of distance between each vehicle to open their field of vision and increase forward sight distance.

Increasing the distance between responding emergency vehicles will also give the driver of the second vehicle more time and distance to react should the vehicle in the front of the caravan need to stop or slow suddenly. If the first vehicle in the caravan has to take evasive action, the second vehicle may lose control and crash if it is following too closely. Added distance between emergency vehicles will give the second driver more time and distance to react should the vehicle in front of him make a sudden evasive maneuver.

The driver of the second apparatus may also be blinded by the glare of the first vehicle's emergency lights. When following another emergency vehicle on a routine basis, it may be helpful to dim or reduce the amount of rearward facing light on the first apparatus. Installing a dimmer switch will help reduce the amount of glare experienced by other fire apparatus operators in the caravan.

Fire apparatus operators in a caravan must use extreme caution when approaching a civilian vehicle that has yielded the right-of-way. When faced with a caravan of emergency vehicles, a civilian driver may pull over to the side of the road to allow the first emergency vehicle to pass. If the civilian does not realize there is more than one emergency vehicle approaching, the driver may pull back onto the highway, directly in front of the second emergency vehicle. When the civilian vehicle pulls back onto the highway, it is often struck by the second emergency vehicle, or it may force the second emergency vehicle to take an evasive action which results in a crash.

When driving in a caravan to an emergency, fire apparatus operators must leave plenty of distance between each vehicle. This extra distance will give a civilian driver additional time and distance to perceive and react to the approach of the second and subsequent emergency vehicles. Increasing the following distance between each emergency vehicle provides a greater margin of error for the fire apparatus operators driving in the caravan.

Case Study—Florida

Caravanning was a causative factor in a crash in Tallahassee, Florida, on March 24, 2005. At 13:56 hours, an ambulance and a water tanker were responding to a large brush fire. While en route to the brush fire, the ambulance slowed to make a turn. As the ambulance made the turn, the tanker driver took his eyes off the road and looked at some controls on the dashboard. The driver of the tanker looked up and saw the ambulance slowing down for the turn and realized he would not be able to slow down in time before he drove into the back of the ambulance. The tanker driver took evasive action and swerved into the left-hand side of the roadway. As the driver swerved to the left, he observed a civilian vehicle that had pulled over on the left-hand side of the roadway. The driver swerved back to the right to avoid the stopped vehicle, causing the tanker to lose control and roll over. Distracted driving and following too closely were causative factors in this crash.

Green Lights

Negative right-of-way intersections are not the only concern for an emergency vehicle operator. Approaching a steady green light can also be dangerous, especially if the light has been green for an extended period of time. This is known as a stale green light. When approaching a steady green light, apparatus operators must use extreme caution. Drivers should take their foot off of the accelerator, cover the brake pedal with their foot, and scan the intersection for hazards.

Be especially wary if there are no vehicles stopped on the cross street, waiting for the red light to change to green. This is because a stopped vehicle on the cross street will provide a natural barrier that warns other drivers on the cross street of the red light. If there are no vehicles stopped at the red light, a distracted driver may not realize they are approaching an intersection. A distracted driver may drive into the intersection against the steady red light because they never realized they were approaching the intersection. When no cars are stopped on the cross street, be especially wary, as seen in Figures 7–2 and 7–3.

Case Study—Pennsylvania

In 2013, my personal vehicle was destroyed as I approached a stale green light. As I was approaching the green light, another vehicle was approaching on the cross street. The driver of the vehicle traveling on the cross street was looking at her GPS and did not notice the red light ahead of her. The driver drove through the red light and totaled my car. Had a vehicle been stopped at the red light, the driver would have driven into the back of that vehicle, instead of entering the intersection and hitting me. Or, she may have noticed the vehicles stopped for the red light and realized she needed to stop as well. Approaching a green light with no cars stopped on the cross street requires extra caution, especially when driving a large fire apparatus.

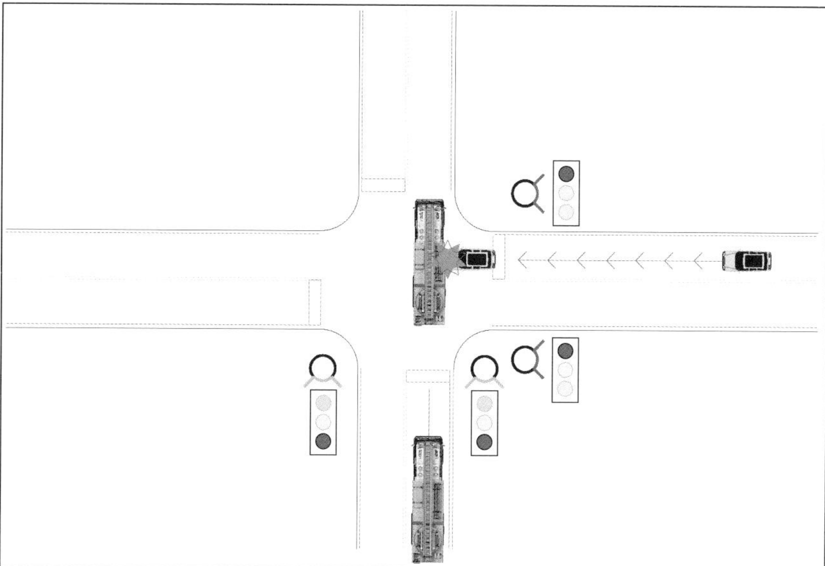

FIGURE 7–2.   If the fire apparatus is approaching a steady green light and no one is stopped on the cross street, use caution. An inattentive driver may not realize the light is red in their direction and drive into the intersection.

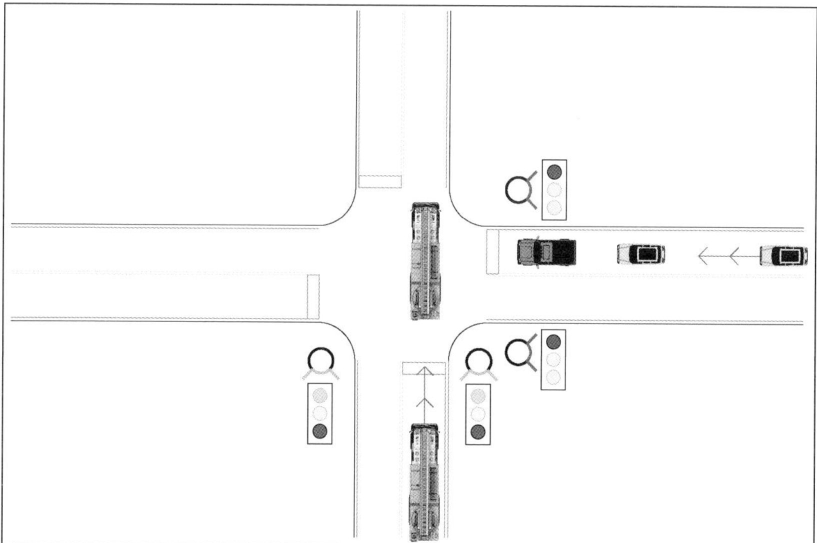

FIGURE 7–3.   If the fire apparatus is approaching a steady green light and someone is stopped on the cross street, it will provide a natural barrier for an inattentive driver who is approaching on the cross street. The driver will notice the stopped vehicle, or drive into the back of it, instead of entering the intersection and striking the fire apparatus.

Apparatus vs. Apparatus

There have been numerous intersection crashes involving two emergency vehicles responding to the same incident. Fire apparatus operators must be keenly aware of the response routes taken by other emergency vehicles that are responding to the same incident. Other emergency vehicles could include police and EMS vehicles, in addition to fire apparatus.

An awareness of other emergency vehicles is especially important for an apparatus operator who is approaching a steady green light. Fire apparatus operators cannot assume that the driver of the emergency vehicle on the cross street will come to a complete stop at a red light. When approaching a green light, fire apparatus operators must always consider who else is responding and where they are coming from. If equipped with a traffic preemption system, drivers must know what type of warning the system will provide if another emergency vehicle is approaching from a different direction.

When approaching an intersection where there may be a potential conflict, attempt to warn other emergency vehicles using the mobile radio. Keep in mind that even if a warning is given over the radio, vehicles from other agencies may not hear the warning if they are on a different radio channel. This is especially true when responding to an incident with agencies such as police or EMS, as well as mutual aid companies from other jurisdictions.

Remember that it will be impossible for fire apparatus operators to hear another emergency vehicle siren over the noise of their own siren. Apparatus operators may have no advance warning that another emergency vehicle is in the vicinity.

There are times when a crew of firefighters may get caught in the rush to arrive first at a working fire or emergency. This attitude may cause the driver and the officer in charge of the truck to suffer tunnel vision which will impair their ability to think about anything other than arriving first. Firefighters can be their own worst enemies if they find themselves racing to a scene to beat another company. These scenarios can lead to tragedy. Emergency agencies should meet ahead of time and preplan response routes that minimize travel times while also reducing the chance of two emergency vehicles colliding at an intersection.

Case Study—Illinois

On April 27, 2004, at 17:35 hours, three fire departments were dispatched to a garage fire. While responding to the fire, an apparatus equipped with a traffic preemption system was approaching an intersection. Investigation revealed that the apparatus had captured the preemption system and the light had turned green. As the apparatus entered the intersection with the steady green light, it was struck just behind the enclosed cab by another engine which was en route to the same fire. As a result of the collision, the striking apparatus rotated approximately 180 degrees and rolled onto its left side. During the course of the collision, the officer of the apparatus was ejected from the vehicle and sustained fatal injuries. The victim firefighter was not wearing a seatbelt.

Rolling an Intersection

Some fire apparatus operators do not come to a full stop at negative right-of-way intersections. Instead, they slow down or *roll* the intersection. Rolling or coasting through a red light may still result in a serious crash, especially in areas where there is limited sight distance which may prevent a fire apparatus operator from seeing approaching traffic.

Consider a fire apparatus that is rolling a red light at 10 mph. The fire apparatus operator enters the intersection and suddenly sees a civilian vehicle approaching the cross street. As the civilian driver did not see the apparatus due to sight obstructions on the corner, the apparatus operator needs to quickly stop the apparatus to avoid a collision. If the driver is rolling the intersection at 10 mph, it will take nearly 30 ft to bring the apparatus to a stop.[3] Considering the design of most intersections, it is doubtful that the apparatus operator will be able to bring the apparatus to a stop before it drives into the path of the approaching civilian vehicle.

It is a far better practice to approach an intersection with the intention of coming to a full stop. This will allow the fire apparatus operator time to look both ways and evaluate the situation. As the fire truck will be at a stop, there is no need to worry about stopping in time if a civilian vehicle is approaching on the cross street. Once the intersection has been deemed clear, the apparatus operator can cross the intersection safely.

I grow tired of hearing emergency vehicle operators claim that they "are trying to save someone." I can assure you that almost 99.99% of the time, the few seconds saved by not stopping at a negative right-of-way intersection will not make a difference in saving a life. I can assure you that the odds are much greater that you will find yourself in a serious crash with a civilian vehicle if you do not come to a complete stop at negative right-of-way intersections.

Keep in mind, our job is to **save** lives, not **trade** lives. By running red lights and stop signs, you are giving yourself permission to act as judge, jury and sometimes executioner for other innocent civilians that are driving on the road.

Traffic Preemption Devices

An emergency vehicle equipped with a traffic preemption device will be able to take control of an intersection as it approaches. The preemption device will change the traffic light facing the emergency vehicle to green while changing the other traffic lights to red. The device can also control pedestrian signals to prevent pedestrians from crossing in front of the emergency vehicle.

In 2006, the Federal Highway Administration conducted an extensive study on the benefits of traffic preemption devices. This study revealed the following:[4]

1. Emergency vehicle preemption has allowed Fairfax County, Virginia, to reduce its response times. The system permits emergency vehicles traveling along U.S. 1 to pass through high volume intersections more quickly and with fewer conflicts, saving 30 to 45 seconds per intersection.
2. Emergency vehicle preemption in the City of Plano, Texas, has dramatically reduced the number of emergency vehicle crashes—from an average of 2.3 intersection crashes per year to less than one intersection crash every five years.

It is clear from these statistics that traffic preemption is a useful tool. Unfortunately, fire departments are often left to the mercy of state highway administrators who control when and where these devices will be installed.

There are several types of traffic preemption systems. Some systems are activated by strobe lights, some by radio signals or sirens, and others that are manually activated by push buttons in the firehouse. Regardless of the type of system in use, fire apparatus operators must have a thorough understanding of these devices, including their limitations.

Light-Emitting/Infrared-Activated Traffic Preemption

In a light-emitting traffic preemption system, a strobe light is mounted on the emergency vehicle. When the strobe light is activated, a detector head on the traffic signal will sense the light and activate the preemption system. A large flood light, usually mounted on the traffic signal cross bar, will activate to notify the driver of the emergency vehicle that they have captured the intersection. When the apparatus operator sees the flood light activate, they know that they have control of the intersection. Light-emitting systems have an average range of 2500 ft.

With an infrared system, there is no flashing strobe light. Instead, there is an infrared transmitter mounted on the emergency vehicle. This type of system is much more discrete and is often used on unmarked police cars. Infrared systems have a slightly shorter range of around 1500 ft.

There can be limitations with light emitting traffic preemption systems, especially if the equipment is not properly maintained. Dirty emitter or receiver heads can disrupt the signal and degrade the effectiveness of the system. If the emitter or receiver heads are struck by objects and knocked out of alignment, there may be issues with the system due to improper aiming. Line of sight obstructions, such as large trucks, low hanging trees, or street signs can block the strobe light from reaching the receiver head. Fog, smoke, or inclement weather can also interfere with transmission of the signal.

Sound-Activated Traffic Preemption Systems

In a sound-activated preemption system, a directional microphone is mounted on the traffic signal arm. When the microphone hears an approaching siren, the system will activate and give the approaching emergency vehicle a steady green light. An advantage to this system is that it does not require a strobe light or infrared transmitter to activate. Most emergency vehicles are equipped with sirens and will be using them during an emergency response. Disadvantages of a sound-activated system include issues with the system tripping unintentionally due to loud horns or other sirens in the area. There may also be issues with the siren reflecting off of buildings. Siren reflection may cause the system to misread the direction of the approaching emergency vehicle and trigger the wrong traffic light.

Global Positioning Traffic Preemption Systems

A global positioning preemption system tracks the emergency vehicle's route of travel, speed, and direction of approach using GPS. The system will then relay this information to the traffic signal so it can control the intersection accordingly. These systems have the same disadvantages of any GPS system, such as problems with cloud cover, buildings, and satellite interference.

Radio-Activated Traffic Preemption Systems

Radio-activated preemption systems rely on a radio transmitter inside the emergency vehicle. The radio transmitter activates the preemption system as the emergency vehicle approaches. An advantage of this system is that it is not affected by line of sight issues, visual obstructions, or weather. The downside to the system is the possibility of interference from other devices that use the same radio frequency. However, recent changes in technology have helped alleviate this issue.

It is said that these systems are also capable of warning other emergency vehicles of possible intersection conflicts using the same transmitting equipment used to activate the preemption system.

The advantages and disadvantages of each preemption system are provided in table 7–1. Fire apparatus operators must be familiar with the proper use and limitations of their traffic preemption system. While each preemption system has advantages and disadvantages, they all have similar safety issues. First and foremost, drivers cannot rely upon these systems or assume that they are always working. Fire apparatus operators must still drive with due regard, especially when approaching an intersection.

TABLE 7–1.   Advantages and disadvantages of various preemption systems.[5]

	Strobe Activated	Siren Activated	Radio Activated
Dedicated Vehicle Emitter Required	YES	NO	YES
Susceptible to Electronic Noise Interference	NO	NO	YES
Clear Line of Sight Required	YES	NO	NO
Affected by Weather	YES	NO	NO
Possible Preemption of Other Approaches	NO	YES	YES

Drivers must also remember that a traffic preemption system will not resolve conflicts with an emergency vehicle that is approaching the same intersection. Drivers must know how the preemption system will give warning if there is a conflict with another emergency vehicle. Some systems will flash the white flood light on the traffic signal arm to notify the fire apparatus operator that another vehicle has captured, or is attempting to capture, the traffic signal. Understanding these warning signs will help the emergency vehicle driver to recognize a potential conflict and slow or stop accordingly.

Another issue to remember is the interoperability of the preemption system in surrounding jurisdictions. This is especially true in areas where the local government, instead of the state highway department, is allowed to select and install traffic preemption devices. If you are using a sound-activated system in your town, but respond to a neighboring jurisdiction that uses a strobe-based system, you must remember that you no longer have the ability to preempt a traffic light.

Fire apparatus operators must also have a thorough understanding of how much time it will take for the traffic preemption system to capture and change the traffic signal. The capture time of the preemption system will have a direct effect on how fast the driver can approach an intersection. There may be intersections where the emergency vehicle driver must slow down or stop while they wait for the preemption system to activate. If the range of the preemption system is limited by sight distance, the emergency vehicle may have to approach the intersection slowly, activate the system and then proceed carefully through the intersection. As a case in point, I assisted in the investigation of an emergency vehicle crash in which an emergency vehicle turned a corner and rushed down a short street. Because the strobe light only had approximately 3 seconds to activate the preemption system before the emergency vehicle arrived at the intersection, the preemption system did not have enough time to capture the traffic signal. The emergency vehicle operator did not consider this fact, and by the time he realized the traffic light had not changed, he entered the intersection and struck a civilian car.

Conclusion

Traffic preemption systems are useful tools that can increase safety and reduce response times. However, every driver training program must include an in-depth discussion on the use and limitations of these systems. NFPA 1451 (7.2.3) requires initial training and annual retraining on traffic preemption systems. In addition, drivers must never assume that the system is working until they receive confirmation that the traffic signal has been captured. Drivers must also remember to never approach an intersection so fast as to overdrive the system. Fire apparatus operators must give the system enough time and distance to receive the signal from the approaching emergency vehicle transmitter and cycle the traffic signal to give the emergency vehicle the right-of-way.

Notes

1. NFPA 1500—Section 6.2.8
2. NFPA 1500—Section 6.2.9
3. Perception and reaction distance at 10 mph, assuming a 1.6 second perception and reaction time is 23.4 feet. Skid distance, assuming a dry road with a drag factor of 0.8 and a braking efficiency of 0.65 is 6.4 feet. Total stopping distance is 29.8 feet.
4. Federal Highway Administration, "Traffic Signal Preemption for Emergency Vehicles—A Cross Cutting Study. Federal Highway Administration," January 2006.
5. J. Collura and E. W. Willhaus. "Traffic Signal Preemption and Priority: Technologies, Past Deployments, and System Requirements." Paper published in the conference proceedings of the ITS America 11th Annual Meeting, Miami Beach, Florida, (June 2001).

Seat Belts and Occupant Protection Devices

Introduction

The culture of the fire service demands a quick response to an emergency. While this belief is admirable, it is often at the expense of firefighter safety. Many firefighters believe they can reduce response times by donning turnout gear and SCBAs inside the fire apparatus instead of wearing their seatbelt.

A common complaint heard throughout the fire service is that a firefighter cannot fit into a jump seat and don a seatbelt while wearing bunker gear, gloves, and an SCBA. Admittedly, it can be difficult to put on a seatbelt while wearing personal protective equipment. Because of these difficulties, apparatus manufacturers have sought to redesign the seats to provide more room and better access to the seatbelt. By improving the design of apparatus seating, it is the hope of manufacturers that the use of seatbelts will increase. While redesigning the seats in new apparatus is an excellent idea, firefighters riding in an apparatus with older seats may use the seat as a continued excuse to not wear seatbelts.

One of the nation's largest fire departments solved this problem by removing all SCBAs from the crew cab. While some may find this a drastic solution, it resulted in some surprising advantages. First, it helps ensure seatbelt use. Second, the extra moments it takes the crew to mask up allows the officer a chance to size-up the situation. Taking a few extra seconds to look around, rather than run up to a burning building while wearing a fogged face mask, will increase situational awareness and go a long way for firefighter safety during the firefight.

In addition to increased situational awareness, removing the SCBAs from the passenger cab reduced leg and knee injuries. This reduction in injuries is attributed to the fact that the bulk and weight of an SCBA makes it more difficult to navigate through the cabin door and exit safely to the ground. Removing the SCBA and only wearing bunker gear made it easier to exit the rig.

Removing SCBAs from the crew cab would be a controversial culture shock to the fire service. However, wearing an SCBA while en route to a call can have a significant impact on the effectiveness of seatbelts and restraint devices. Donning an SCBA inside the crew cab often requires the firefighter to loosen the seatbelt. Loosening of the seatbelt will reduce its effectiveness and contribute to additional injuries in the event of a crash.

A seatbelt is supposed to fit tight against the body, with no gaps. When a seatbelt is placed over the shoulder straps and regulator of an SCBA, it will not fit as it is supposed to. In the event of a crash, the firefighter will have a tendency to move forward against the loose seatbelt, which will compress the SCBA straps, buckles, and regulator against the firefighter's chest. This could result in additional injuries.

Should the seatbelt be loose enough, the firefighter may leave the seat and become a dangerous projectile inside the cab. To compound this matter, the firefighter will be wearing a heavy bottle full of compressed air. This combination could result in disaster.

Admittedly, the issue of transporting and donning SCBAs inside the crew cab is a controversial topic. Fire departments must balance speed and effectiveness with firefighter safety. If a fire department is having trouble ensuring that the members are wearing seatbelts, it is time to examine the issue and determine the root cause. If the matter can be solved through discipline, so be it. If the matter is caused by poor apparatus design, it may be time to examine other methods of mounting the SCBA.

Rollovers and Seatbelts

During a crash, unrestrained firefighters are often ejected from the apparatus, resulting in serious injuries or death. Countless line-of-duty deaths could have been prevented if the firefighter had been wearing a seat belt. According to the United States Fire Administration's "Emergency Vehicle Safety Initiative":

Three out of four people who are ejected from a vehicle will die.

1. *Eight out of 10 fatalities in rollover crashes involve occupant ejection from the vehicle.*

2. *Occupants are 22 times more likely to be thrown from the vehicle in a rollover crash when they are not wearing their seatbelts.*

3. *80% of firefighters killed in vehicle crashes were not wearing seat belts.*

> *4. Only 3% of vehicle crashes in the United States are rollovers, yet rollover deaths represent one-third of all occupant fatalities (over 10,000 people each year).*[1]

Many firefighters believe that there is no need to wear a seatbelt because the size of the apparatus will provide a safety cocoon. This fact could not be farther from the truth. Due to the high risk of a fire apparatus rolling over, the use of seatbelts is imperative. In a rollover crash the purpose of a seatbelt is to prevent a person's head, arms, and other body parts from being ejected, or flailing about inside the vehicle.[2] A seatbelt will keep the occupant's body properly restrained in the seat, where the seat's protective design can help reduce serious injuries. While front and side air bags provide protection during frontal and angular crashes, they do not compensate for a lack of seatbelt use during a rollover.

If the apparatus is built properly, and the roof is strong enough to resist collapse, a rollover crash should be survivable. This is because the collision forces are spread out over a much longer time period as compared to a front or side collision.[3] One need only watch a professional race car driver walk away from a crash in which his vehicle flew into the air, rolled over several times, and came to rest on the infield of the race track. Because the race car is built to withstand such crash forces, and the forces the driver experienced were spread out over a relatively long period of time, the driver is usually able to walk away with minor injuries.

While a rollover crash may be survivable, the occupants must stay within the protective envelope of the vehicle to increase their chance of surviving. Studies have shown that half of the unbelted occupants who died in rollover crashes were totally ejected from the vehicle, and another 10% were partially ejected. According to a NHTSA study, a rollover is the greatest contributing factor to an unbelted occupant being ejected from the vehicle.[4]

In addition to large apparatus, fire departments often operate smaller vehicles that were purchased from the civilian market. Such vehicles include sport utility vehicles, four-door sedans, and pick-up trucks. While smaller vehicles do not rollover as often, their small size may provide less protection during a crash. This is especially true during a high-speed event. Firefighters must wear seatbelts regardless of the size of the vehicle.

Delta-V

While the severity of a rollover crash depends on how many times the vehicle rolled over, the severity of a nonrollover crash is gauged by *delta-V.* Delta-V is the change in a vehicle's velocity over a period of time. As an example, an airline passenger will undergo a change in velocity of 500 mph, but this change takes place over several minutes as the airplane completes its final approach. Because the change in velocity was gradual, the airline passenger does not sustain any injuries. On the other hand, a motor vehicle occupant may undergo a delta-V of 40 mph in just a few milliseconds if the vehicle strikes a wall. This sudden change in velocity, combined with the mechanical damage caused by the crash, will result in serious injuries.

Automobile manufacturers have sought to reduce the severity of a crash by increasing the amount of time it takes the occupants to ride down the crash and come to a stop. By installing crumple zones and air bags, manufacturers have found ways to increase the time it takes for the crash to stabilize, thus decreasing the severity of the crash for those inside the vehicle. Consider jumping out of a building and into an airbag. Because it takes so long for a person's body to fall to the ground through the air bag, the delta-V is not as severe as if the person fell and hit the ground directly. The same idea applies to occupant protection systems.

Even if a vehicle is equipped with airbags, the passengers must still wear a seatbelt. A seatbelt will ensure that the occupant is properly positioned in the seat when the airbags deploy. Proper positioning of the occupant will allow the airbags and other supplemental restraint systems to do their job. Seatbelts also help reduce the risk of ejection.

It is important to note that seatbelts should include three-point harnesses, not lap belt systems. Many older fire apparatus are equipped with lap belts, especially in jump seat areas. While a lap belt may reduce the risk of ejection during a crash, it may contribute to significant injuries as well. Because the upper torso of a firefighter's body is not restrained, it will continue to move forward during a sudden deceleration, while the hips and lower body are still restrained in the seat. This sudden jack-knife can cause head, spinal, and abdominal injuries as the firefighter's body is flung against the lap belt. Any vehicle equipped with a lap belt system should be immediately retrofitted with a three point system.

Wearing Helmets While Responding

NFPA 1500 requires firefighters to wear helmets and eye protection when riding in an open cab or an open tiller seat. NFPA also requires the use of hearing protection if the noise levels inside the passenger cabin exceeds 90 dB. However, NFPA also states that firefighters are **not** to wear helmets while riding inside an enclosed passenger cab. Many firefighters wonder why.

If a vehicle is rear-ended, the seat headrest is designed to absorb the crash energy and prevent whiplash or neck related injuries. If a firefighter is wearing a fire helmet, the brim of the helmet will cause an unnatural position for the firefighter's head as it rests against the headrest. This unnatural position may contribute to more serious injuries if a crash occurs.

Fire helmets can also cause safety issues if there is a rollover. During a rollover crash, head and spinal injuries can occur when the occupant's head drops down and strikes the roof. If a firefighter is wearing a helmet during a rollover crash, the height of the fire helmet will decrease the safe distance between the firefighter's head and the roof of the vehicle. If the vehicle rolls over and the roof begins to crush down, the added height of the fire helmet may cause injury as the roof pushes down on the helmet, and thus the firefighter's head and neck.

The question becomes "where should firefighters put their helmets?" NFPA 1901 addresses this issue in Section 14.1.7.4.1, which states that a location for the storage of helmets inside the vehicle shall be provided. If the helmets are to be carried in the driver or passenger compartment, they must be in an enclosed and latched compartment, or a bracket, that is able to contain the contents when subject to a 9 g force along the longitudinal axis and a 3 g force in any other direction. The purpose of this requirement is to prevent a helmet from becoming a dangerous projectile in the event of a crash. However, practical experience shows that these brackets and compartments are rarely used. If the brackets and compartments aren't going to be used, is it time for the fire service to redesign the shape and design of a fire helmet and make them more crashworthy? Good luck with that.

Remember that fire helmets are not rated as crash helmets. They are only designed to protect firefighters from some of the hazards they will encounter while fighting a fire, such as water running down their neck or debris falling on their head. The design of a fire helmet may increase the risk of injury to the firefighter who is wearing a helmet when the fire apparatus is involved in a crash.

Rollovers and Ejection

When a vehicle begins to rollover, many things begin to happen. This fact is especially true in fire apparatus, which contain multiple passengers, equipment, and other loose items inside the cab.

Kinematics

Whenever a vehicle is involved in a crash, everyone inside the vehicle is subject to *kinematics*. Kinematics describes the motion of bodies or objects inside the vehicle. The simplest example of kinematics is when a vehicle operator slams on the brakes and the occupants keep moving forward. If the occupants are wearing seatbelts, they will feel their bodies press against the seatbelt as it holds them in place. If there is a box of tissues sitting on the seat, it will probably fly forward and land on the floor.

In the case of a rollover crash, the kinematics inside the vehicle are very complex. As the vehicle begins to rollover, the occupants will be flung in a circular manner, similar to a baseball pitcher throwing a ball. If the occupant is on the side nearest to the ground, they will rotate downwards towards the road surface. Occupants sitting furthest from the ground will be rotated and flung even further, most likely striking the roof of the vehicle and then falling down onto the other occupants who were sitting on the near side of the rollover. Anyone inside the vehicle who is not restrained will become a dangerous missile, flying about the passenger compartment. A person wearing a seatbelt may find themselves seriously injured or killed by the person next to them who was too lazy to put a seatbelt on. This explains the importance of ensuring that everyone in the vehicle is wearing a seatbelt.

In addition to the occupants getting thrown about the passenger compartment, loose items inside the crew cab will become dangerous missiles. If not properly restrained, hand tools, helmets, and any other items inside the cab will continue moving. This is why NFPA 1500 and 1901 requires all equipment inside the cab to be properly mounted by "secure mechanical means," or stored in a compartment with a positive latching door.[5] Mounting the equipment securely will ensure that it doesn't fly through the air during a crash and injure a member inside the cab.

Roof crush

One of the most dangerous issues related to a rollover is roof crush. Because the fire apparatus is so heavy, the weight of the vehicle will cause the roof of the crew cab to crush down. This is especially true for older apparatus and some commercial chassis. This is bad for several reasons.

As the roof crushes down towards the occupants, there is a high probability that the roof will come in contact with the heads and necks of the firefighters, resulting in serious, if not fatal, spinal injuries. Keep in mind that if the vehicle's roof is crushing, the vehicle is probably upside down. If the vehicle is upside down, the firefighters will be falling towards the roof as the roof crushes upwards. This combination will result in devastating injuries as the firefighter's head strikes the roof.

These types of injuries are of special concern for firefighters due to the nature of the fire apparatus. Many fire apparatus have raised roofs. These raised roofs will provide a greater distance for the firefighter to travel as the vehicle rolls upside down. The greater travel distance will result in greater force as the firefighter falls towards the roof and strikes the collapsing roof. Greater force on the head and neck results in greater injuries. However, if a firefighter is properly belted inside an apparatus which meets national safety standards for roof crush, there should be enough headspace to prevent the firefighter's head from striking the roof.

After the vehicle rolls and the roof strikes the road, the cab will begin to compress, distorting window openings, breaking window glass, and creating ejection ports that an unrestrained occupant could be ejected through. The rolling motion of the vehicle will create a rotational force that could fling an unrestrained occupant out of an open door or window. Being ejected is bad enough, but if a firefighter is ejected into the path of the rolling vehicle, they could be further injured as the vehicle rolls on top and crushes them.

Ambulances and Occupant Restraints

Patient care in an ambulance has traditionally been an issue when it comes to wearing a seatbelt. It is sometimes difficult for emergency medical personnel to perform necessary medical procedures while properly wearing a seatbelt. While this has admittedly been an issue in the past, new technologies are making their way to the fire service that will allow an emergency medical responder to perform critical medical tasks while still properly restrained.

In order to facilitate the use of a seatbelt by emergency medical responders, fire departments must first identify the primary patient care position in the back of the ambulance. Once this position is identified, the fire department must then train emergency responders how to use the position to provide patient care while allowing the responder to remain seated and belted. The training program must focus on maximizing the amount of time the emergency responder remains belted, as well as emphasizing the importance of using a seatbelt to minimize the risk of death and injury. This training must also include the proper procedures for restraining a patient to a cot for transport.

These training programs should address issues such as the following:

1. Performing as much care as possible while at the scene. Minimizing the amount of treatment required in the back of the ambulance will reduce the amount of time the medical provider must unbuckle themselves.
2. If patient care is required in the back of the ambulance, attempt to time the care when the vehicle is stopped at a red light or stop sign. If there is no urgency to arrive at the hospital, and it is safe to do so, have the driver pull over and stop for a brief period of time.

3.   Design the ambulance so that equipment is within arm's reach. This will prevent the medical provider from having to get up and retrieve equipment from a cabinet. All cabinets and equipment should be securely mounted so they do not break free during a crash.
4.   When purchasing a new ambulance, research medical provider restraint systems and specify a system in the new ambulance. If it is possible to retrofit existing ambulances with a system, do so.

There may be times when an ambulance has to transport a child. Members who ride the ambulance must be trained in the proper method to secure a child in a car seat, as well as the most appropriate place in the transport compartment to secure the child safety seat. Infants should not be positioned in a side-facing orientation, as this may expose them to flying equipment or air bag deployments during a crash. Fire departments should consult with certified child-safety-seat technicians to identify the best method for securing small children for transport.

NFPA Requirements

The NFPA requirements regarding seatbelt use are straight-forward and self-explanatory. The apparatus should never move unless everyone inside the apparatus is wearing a seatbelt. There are three exceptions to this rule: urgent patient care in an ambulance, hose loading operations, and tiller training. Keep in mind that for hose loading and tiller training, NFPA requires strict SOPs related to how these activities take place. Failure to adhere to these requirements could result in legal sanctions should a tragedy occur.

While NFPA provides procedures and equipment requirements to increase seatbelt use among firefighters, nothing will change until the fire service undergoes a complete culture shift. Having ridden in the back of fire trucks for the past 27 years, I understand the thought process of those inside the rig. Firefighters are aggressive, type-A personalities, who believe they are immortals put on this planet to fight a fiery beast. Because of this attitude, firefighters would rather take a risk and save a few seconds on the fire ground by masking up with no seatbelt on.

Each time a new technology is developed to increase the use of seatbelts, firefighters find a way to defeat the technology. Despite the installation of seatbelt alarms in newer fire apparatus, firefighters still find ways to try and defeat or silence the alarm. I can almost guarantee that the same member who defeated the seatbelt sensor will be the first person to sue the driver for not ensuring the passengers were wearing their seatbelts just before the vehicle crashed.

While seatbelt sensors and occupant protection systems are useful tools, nothing will compare to old fashioned peer pressure. Instead of relying on a display screen in the front of the cab, the driver should always look to ensure that everyone is seated properly and wearing a restraint. The NFPA requirement for high visibility seatbelts makes it easy for a driver to look back and be sure that everyone is in compliance before pulling out the door. New apparatus should include a mirror in the cab to assist the driver with this task.

Keep in mind that this responsibility should not fall solely on the driver. The officer and everyone else in the rig must look around and make sure that every firefighter on the apparatus is properly restrained. Even if a member is wearing a seatbelt during a serious crash, they could still be seriously injured by a member sitting next to them who was too lazy to put theirs on. The culture of the fire service must shift from the days of turning a blind eye, to the days of telling someone to put a seatbelt on or get off the rig.

NFPA specifically requires all fire apparatus to be equipped with working seat belts. If there is a malfunction with the seat belt, that riding position is to be taken out of service. If the driver's seat belt is malfunctioning, the entire rig must be placed out of service. To those that scoff at this idea, I ask how many firefighters would be willing to fight a fire if they weren't provided with an SCBA and turnout gear by their employer or volunteer company. Few firefighters are willing to do this. Just as many firefighters are killed as a result of a vehicle crash as smoke inhalation, yet no one complains when a fire department is given a truck that does not have functioning seat belts. If a company is given a truck with malfunctioning seatbelts, the firefighters should refuse to ride it.

It should also be noted that none of these standards provide exemptions for special activities. In fact, the standards specifically state that seat belts must still be worn during parades, funerals, and community relations events. Should someone be hurt or killed during such an event, there is no exemption which will provide protection from the law. While firefighters will take exception to this fact and demand that the fire service continue to honor what some consider to be tradition, the fire service must abandon the long-standing tradition of dying in the line of duty from preventable accidents.

Notes

1. Public Citizen. "Rolling over on Safety: The Hidden Failures of Belts in Rollover Crashes." April, 2004
2. Public Citizen. "Rolling over on Safety: The Hidden Failures of Belts in Rollover Crashes." April, 2004
3. Public Citizen. "Rolling over on Safety: The Hidden Failures of Belts in Rollover Crashes." April, 2004
4. National Center for Statistics and Analysis, National Highway Traffic Safety Administration, Analysis of Ejection in Fatal Crashes Research Note, Washington, DC: NHTSA, Nov. 1997. 1.
5. NFPA 1500, Section 6.1.5

Railroad Crossings

Introduction

In recent years, there have been several serious crashes involving fire apparatus and trains. While it should seem like common sense to not cross in front of a train, there may be times when a fire apparatus operator is tempted to challenge a train in an attempt to reduce response time. This is an extremely dangerous practice that must be discouraged.

Railroad Crossing Design

There are two types of warning systems used at rail crossings: passive warning systems and active warning systems.

Passive warning systems

Passive warning systems are used at unguarded rail crossings, which are rail crossings not equipped with safety gates. These crossings are common on private driveways and in rural areas where there is little traffic. Passive warning systems usually consist of advance warning signs, posted some distance before the crossing, as well as signs and road markings directly at the crossing (fig. 9–1).

FIGURE 9–1. This is an example of a passive warning system on a private roadway. The only warning for the railroad track is a sign.

Active warning systems

Active warning systems are used at busier rail crossings and include gates that close and lights that flash when a train is approaching (fig. 9–2). The decision to install these systems is usually left to state highway authorities. Considerations for using these systems include the following:

1. The numbers of vehicles that will cross the tracks
2. The types of vehicles that will cross the tracks
3. The number of trains that use the tracks each day
4. A history of collisions at the crossing

When faced with a rail crossing, fire apparatus operators must consider if the crossing is equipped with an active or passive warning system. In the event of an active system, common sense should prevail: if the gates are down, don't cross. Unfortunately, fire apparatus operators may succumb to tunnel vision and drive around the stop gates thinking they can improve their response time to an emergency. Under no circumstances should this occur. Fire apparatus operators must exercise patience and wait for the train to pass.

According to data provided by "Operation Lifesaver," it will take a train traveling 55 mph over one mile to come to an emergency stop. Regardless of the use of lights and sirens, a train **always** has the right-of-way. Even if a train engineer could see the fire apparatus some distance away, a high-speed train would not be able to stop before it reaches the railroad crossing.

It is important to note that some train crossings have low ground clearance. This is an important consideration for large vehicles with long wheelbases such as a fire truck. Fire departments must be aware of these situations and consider alternate routes of travel if there is the possibility that a fire apparatus with a long wheelbase may get hung up while traversing the low ground clearance at a railroad crossing.

Best Practice at Railroad Crossings

Every fire department driver training program should include a module that addresses the safe crossing of railroad tracks. Even if there are no tracks in a fire department's primary district, it is quite possible that drivers may face them on a mutual aid run. Fortunately, fire departments will be aware of rail crossing locations ahead of time and can preplan accordingly.

The obvious choice for railroad crossings is to avoid them all together. Whenever possible, fire departments should do their best to plan routes of travel that do not

FIGURE 9–2. This is an example of an active warning system. The system includes gates which will close when a train approaches.

involve crossing a railroad track. Not only does this planning remove any chance of a fire apparatus vs. train collision, it may also improve response times to an emergency. If a fire apparatus comes to a rail crossing while a train is driving through, the apparatus will be left waiting for the train to go by.

When approaching a railroad crossing, fire apparatus operators must take note of any stop lines on the near side of the tracks. These stop lines indicate where a vehicle should stop in order to prevent the vehicle from getting too close to the tracks. Getting too close to the tracks may result in part of the apparatus coming in contact with an approaching train. This is especially important for vehicles such as rear-mount ladder trucks which may have a ladder or bucket projecting in front of the crew cab (fig. 9–3).

If there are no obstacles on the far side of the tracks, such as a traffic signal, stop sign, or anything that may cause a vehicle to have to stop, the fire apparatus can cross the tracks and proceed on its way. However, if there is something on the far side of the tracks that may cause the apparatus to have to stop, such as traffic stopped at a red light, apparatus operators must ensure that there is enough room to cross and clear the entire apparatus from the tracks. When there is a limited amount of room on the far side of the tracks, the area is referred to as the containment or storage area.

Keep in mind that a train will be at least 3 ft wider than the tracks.[1] This is especially important because many fire apparatus have a long overhang in the rear. Drivers must ensure that the rear overhang of the apparatus is completely clear of the entire area around the tracks once the apparatus crosses.

Apparatus operators must control the vehicle's speed when approaching a railroad crossing. Rushing up to a railroad crossing at a high speed may not leave enough time and distance to bring the apparatus to a stop. Instead, the apparatus may crash into the gate, a train, or a line of stopped traffic waiting for a train to pass. Approaching a railroad crossing should be treated the same as approaching an intersection: with a great deal of caution. This is especially true when crossing multiple tracks, as one train may have passed by while another is still approaching.

The issue of railroad crossings and fire apparatus was highlighted during a major crash in Virginia. This crash, which resulted in the death of two firefighters and the derailment of an Amtrak train, was the subject of both a National Transportation Safety Board (NTSB) report and a United States Fire Administration (USFA) publication. These publications provide guidance for emergency responders regarding safe vehicle operations at railroad crossings.

FIGURE 9–3. Take note of the stop bars at this railroad crossing. Under no circumstances should any part of the apparatus encroach across these stop lines. This includes such things as aerial baskets and extended bumpers.

The following procedures should be used when arriving at the rail crossing:

1. Turn off sirens, air horns, and other sound-producing devices. Keep in mind that even if sirens and air horns are deactivated, other noise in the cab may make it difficult to hear an approaching train.
2. Slow down, open the vehicle's windows, and look both ways to see if a train is approaching. Keep in mind that if a train is approaching, it is difficult to judge the speed and distance of the train. When in doubt, don't go.
3. At crossings with sight obstructions or severe curves, stop the vehicle and ask a crew member to get out and check down the track.

Be especially aware of railroad crossings that have faulty signals or gates that activate for no reason. If a member should notice one of these issues, the crossing should be immediately reported to the railroad agency. A faulty gate that activates when no train is approaching will create a false sense of security for drivers who use the crossing. Assuming that the gate is activated due to a fault with the track sensors, instead of the actual approach of a train, can lead to disaster.

It is also important to keep a list of contact numbers for the rail agencies that maintain the track. Should the apparatus break down or get hung up on the tracks, the fire department must be able to quickly notify the rail agency. The rail agency can stop train traffic on the track and prevent a collision with a stalled apparatus. In the event that an agency does not have a contact number, there should be a placard at the crossing with a contact phone number for the appropriate agency. Also remember to never park on the tracks unless the rail agency has been notified and rail traffic has been suspended.

Should the apparatus stall or hang-up at a crossing and there is no way to contact the rail agency, use fire personnel to warn the train before it arrives at the crossing. Members from the apparatus, or members from another emergency agency that can be reached by radio (such as the police), should find a safe area approximately 2 miles from the crossing. From this safe location, a road flare or flashlight should be waved in an exaggerated fashion from side-to-side at waist height. This signal should catch the engineer's attention and signal him to stop.

If the fire apparatus operator makes a poor decision and decides to cross the tracks in front of an approaching train, it is important to understand that low ground clearance, stacked traffic on the other side of the tracks, or mechanical failure may cause the fire apparatus to get stuck across the track. If the apparatus becomes stuck across the track, sometimes the best answer is to abandon the truck and run away. Understand that while this solution may save the crew, it will lead to the complete destruction of the apparatus and the possible derailment of the train, injuring all those on board.

Fire department policies and procedures should include a requirement for fire apparatus to stop at all railroad crossings. The few seconds it will take to ensure that a train is not approaching will be minimal compared to the increased amount of safety and discipline that will be instilled in the crew.

Understand that an emergency vehicle cannot win a collision with a moving train. Any contact with a train will cause significant damage and injury to those riding the apparatus, as well as to those people riding the train. Whenever possible, response routes should avoid train crossings altogether. However, if crossing railroad tracks cannot be avoided, a fire department should have extensive driver training and standard operating procedures in place to mitigate the possibility of a serious incident. The website "Operation Lifesaver" (www.oli.org) is a useful tool for any fire apparatus operator or driver trainer.

Case Study—Virginia

On September 28, 1989, a Virginia fire department was dispatched to a car fire at 19:20 hours. A fire engine responded to the car fire with a crew of five firefighters. While en route to the car fire, the engine missed the driveway that accessed the fire location and drove 1.6 miles in the wrong direction.[2]

The fire engine responded back to the fire location and missed the driveway again. The engine finally located the driveway which led to the car fire and drove towards the fire. As the engine was driving towards the fire, the driver failed to notice a train approaching on a set of railroad tracks. Because the railroad tracks crossed a private driveway, the only warning signal was a standard buck sign. The train and the fire engine collided, causing the gasoline tank on the fire engine to ignite and an Amtrak commuter train to derail. As a result of the collision, two firefighters were killed and three firefighters were seriously injured. The train was carrying 399 people, 57 of which were injured as a result of the collision.

Subsequent investigation revealed that the train was traveling at 77 mph before the impact. The train engineer stated that he saw the fire apparatus approaching the rail crossing but believed that it was going to stop. Instead, the engine entered the crossing at which time the engineer sounded the horn and applied his emergency brakes. The engineer reported that the firefighter riding in the officer seat never even looked at the train as the engine crossed the tracks.

As a result of this extensive investigation, the National Transportation Safety Board (NTSB) made the following recommendations:

- If it is not practical to plan an emergency response route that avoids grade crossings, selection of crossings that are equipped with automatic warning devices is preferable to selection of those that are not. All planning should include identification of the location at the crossing from which a driver or other observer assigned to the apparatus can see the maximum available distance down the track(s) on both sides.

- Train horns may not be audible when a vehicle siren is operating. Engine and cab noise may be sufficient to obscure the sound of train horns. It may be necessary for drivers of large vehicles to stop the vehicle; idle the engine; turn off all radios, fans, wipers, and other noise-producing devices in the cab; lower the window; and listen for a train's horn before entering the grade crossing.

- A challenge-response protocol should be used when negotiating an unprotected rail crossing. (See the box on page page 302 for an explanation of "challenge and response.") After stopping the vehicle, the driver and passenger should ascertain that there is no train coming from either direction before proceeding. The challenge and response method would positively assure that the driver was taking proper precautions during the response.

- Because trains are not required to sound their horns at private grade crossings, it is necessary to visually determine that no train is approaching before entering a crossing. Reliance cannot be placed on listening for the train horn.

- In cases where an approaching train is obscured from direct observation by the driver of a safely positioned vehicle, such as crossings located on curves in the railroad track or multiple track crossings where a stopped train blocks vision, it may be necessary to have a member of the crew proceed ahead of the apparatus on foot to assure that there is no train approaching before signaling the vehicle to cross the track.

The "Operation Lifesaver" truck driver guide states the following:

"When a train approaches, you are required to stop the truck before the crossbuck sign or signal at the crossing. On gravel roads there are no pavement markings or stop Lines. The stop lines on each side of a single track grade crossing are at least 35 feet apart. Do not stop within this area. Remember to apply the emergency or parking brakes while waiting at the stop line, so your truck won't move or be shoved into the path of the train."

Notes

1. Operation Lifesaver. "Highway-Rail Grade Crossing Training for Professional Truck Drivers."
2. U.S. Fire Administration, "Fire Apparatus/Train Collision—Cartlett, Virginia," USFA-TR-048, Technical Report Series, (September 1989).

Air Brake Operations

Introduction

Federal guidelines require the driver of a commercial motor vehicle to possess a commercial driver's license (CDL). However, many states exempt fire apparatus operators from the CDL requirement. As few fire departments provide training on the proper operation of air brakes, this CDL exemption creates a serious training gap for firefighters. In order to help address this training gap, this chapter will provide a basic overview of air brake systems. This chapter is not designed as a comprehensive training program for air brakes. Instead, fire departments are encouraged to send their drivers to any one of the professionally taught classes that have been developed by air brake manufacturers.

Why Use Air Brakes?

There are several reasons why air brakes are used on large trucks. Air brakes have an infinite supply of air that will never run out. If there is a leak in a hydraulic brake system, the brake fluid may leak out and cause a loss of brakes. In a vehicle equipped with air brakes, the compressor will be able to generate a constant supply of air to power the brake system, even if there is an air leak.

Air brakes are also able to generate more stopping power than hydraulic brakes. In a hydraulic-braked vehicle, the stopping power of the braking system is generated by the driver's muscle power. In an air brake system, stopping power is generated by the energy of compressed air. In a vehicle equipped with air brakes, a much smaller application of the brake pedal generates a larger amount of stopping power.

The downside to air brakes is the time it takes to pressurize the system with enough power to engage the brakes. This is known as *lag time*. In a hydraulic-braked vehicle, pressure on the brake pedal engages the brakes almost instantaneously. In a vehicle with air brakes, it takes time for the air to pressurize the brake chambers and fully engage the system.[1]

Most vehicles equipped with air brakes utilize a dual system for redundancy and safety. In a dual-brake system, there is a single set of brake controls, but two separate air brake systems. One system will operate the rear brakes, while the other system will operate the front brakes. The two systems are referred to as the *primary* and *secondary* systems. The reason for this concept is simple: if one brake system fails, the entire vehicle does not lose all of the brakes.

Spring Brakes—Used for Parking and Emergency Braking

Air brake chambers are divided into two sides: the service brake side and the spring brake side (service brakes may also be called "foundation brakes"). During normal operations, the spring brake side of the chamber is pressurized with air. This air pressure compresses a powerful spring, which holds the brake in a released position and allows the wheel to spin freely.

When the driver wishes to engage the parking brake, a yellow knob in the cab is pulled out from the dashboard. Pulling on the knob allows the air to escape from the spring brake side of the chamber. As the air escapes, the spring expands and engages the parking brake. This explains the sound of whooshing air heard when the driver pulls the yellow knob to engage the parking brake.

The spring brake is also used as the emergency braking system. Air pressure on the spring brake side of the chamber keeps the spring brake compressed, allowing the wheel to spin freely. If there is a problem with the air brake system which results in air escaping and the pressure dropping below a predetermined level, there won't be enough air pressure to keep the spring brake in the released position. As a result, the spring will expand and engage the parking brake. The spring brake acts as a failsafe in the event of an air leak. Instead of losing the brake system, the spring brake will engage and stop the truck.

Service Brakes— Used for Routine Braking

The spring brake side of the brake chamber controls the parking brake and emergency brake. The other side of the brake chamber controls the service brakes, also called the foundation brakes. Service brakes are used to stop the truck during routine driving. Commercial motor vehicles can be equipped with drum or disc brake systems.

Operation of a Drum Brake System

1. To apply the brakes, the driver presses down on the foot valve, otherwise known as the brake pedal.
2. Air travels from the air tank reservoir, through the air lines, and into a brake chamber.
3. Inside the brake chamber, the compressed air presses against a rubber diaphragm. The air causes the diaphragm to expand and push on a pushrod.
4. The pushrod pushes on a slack adjuster, which turns a camshaft.
5. When the camshaft turns, it will twist an S-cam which will expand the brake shoes and press the brake pads against the drum.[2]
6. When the brake pads contact the drum, they create friction. The friction creates heat which burns off the vehicle's kinetic energy to bring it to a stop (refer to Chapter 1 to review a discussion of kinetic energy).
7. The size of the brake chamber depends on the size and weight of the truck, as well as on which axle it is located. Brake chambers can range in size from 9, 12, 16, 20, 30 and 36 sq. in.
8. The amount of force the brake chamber can create depends upon the size of the chamber and the air pressure being applied. Applying 100 pounds of air pressure to a size 20 brake chamber results in 2,000 pounds of force on the pushrod.
9. The distance the pushrod has to travel to properly apply the brakes is known as "stroke." Properly adjusted brakes have enough "stroke" so that when the brakes are applied, the brake shoes spread apart and press the brake pads in full contact with the drum.

Operation of a Disc Brake System

1. To apply the brakes, the driver presses down on the foot valve, a.k.a. brake pedal.
2. Air travels from the air tank reservoir, through the air lines, and into a brake chamber.
3. Inside the brake chamber, the compressed air presses against a rubber diaphragm. The air causes the diaphragm to expand and push on a pushrod.

4.   The pushrod is pushed out, which pushes a slack adjuster, which turns a power screw (instead of an S-cam as in drum brakes).

5.   The power screw engages the brake calipers which clamp the brake pads onto the brake disc.[3]

6.   When the brakes pads contact the disc, they create friction. The friction creates heat which burns off the vehicle's kinetic energy to bring it to a stop (refer to Chapter 1 to review a discussion of kinetic energy).

7.   The size of the brake chamber depends on the size and weight of the truck, as well as which axle it is located on. Brake chambers can range in size from 9, 12, 16, 20, 30 and 36 sq. in.

8.   The amount of force the brake chamber can create depends upon the size of the chamber and the air pressure being applied. Applying 100 pounds of air pressure to a size 20 brake chamber results in 2,000 pounds of force on the pushrod.

Inspecting the Air Brake System

While most fire apparatus operators are not professional diesel mechanics, there are several procedures they should be familiar with to ensure that the brake system is working properly. These procedures should be conducted on a routine basis, as part of a vehicle maintenance check.

The *Pennsylvania Department of Transportation—Commercial Driver's License Manual* recommends the following procedures for checking the air brake system:[4]

1.   Locate the air compressor. If the compressor is belt driven, check the belt and ensure that it does not appear to be worn or loose. If the belt appears frayed, damaged or excessively loose, have the truck serviced immediately.

2.   Check the slack adjusters. Park on level ground and **chock the wheels.** Release the parking brake so the inspector can move the slack adjusters. If the slack adjuster moves an inch or more, it needs to be adjusted. Slack adjusters are designed to ensure that the pushrod maintains the proper stroke to ensure full braking efficiency. If the slack adjuster is loose, there is an underlying problem that is causing excessive wearing of the brakes. Have the truck serviced immediately. Constantly adjusting an automatic slack adjuster is indicative of a larger problem. Don't cover up symptoms…find the cause.

3.   Check the drums, brake assemblies, and air hoses for cracks, signs of excess wear, or fluid leakage. If these indications are found, have the truck serviced immediately.

4.   Test the air leakage rate. Charge the air system to 125 psi, turn off the engine, release the parking brake, and time the air pressure drop. For a straight truck (not a tractor-trailer combination) there should be less than 2 psi of air leakage per minute.

5.   Next, apply at least 90 psi with the brake pedal. After the initial pressure drop, air pressure should not fall at more than 3 psi per minute. If there is excess air loss within the system, have the truck serviced immediately.

6.   Test the low pressure warning signal. Turn off the engine when the vehicle has enough air to ensure that the low pressure warning light does not activate (usually more than 60 psi). Turn on the electrical power and step on and off the brakes to reduce the air pressure in the brake system. The low air pressure warning signal should activate when the air pressure drops below 60 psi. If the warning signal does not activate, the driver will have no warning that the brakes are losing air and the spring brake is about to activate the emergency brake system. Have the truck serviced immediately.

7.   Check for spring brake activation. Park on level ground and **chock the wheels.** Release the parking brake and shut off the engine. Step on and off the brake pedal to reduce the air pressure in the brake system. The parking brake knob should pop out when the air pressure reaches 20–40 psi (depending on the manufacturer's specifications). This will cause the spring brakes to engage and activate the emergency brake system.

8.   Check the rate of air pressure buildup. When the engine is on and operating at normal RPMs, air pressure should build from 85 psi to 100 psi within 45 sec for a dual air system. For vehicles built before 1975, which still have single air systems, the rate of buildup is 50 psi to 95 psi in 3 min. If the air compressor is unable to generate this much air in the needed timeframe, it may not be able to keep up with the use of the air brakes. Have the vehicle serviced immediately.

9.   Check the air compressor governor. The air compressor should turn on when air pressure in the braking system drops to around 100 psi. The air compressor should shut off when pressure in the system builds to around 125 psi. Fast idle the engine and make sure that the air compressor turns off around 125 psi (depending on manufacturer specifications). Next, step on and off the brakes to reduce the air pressure. The compressor should turn on when the air pressure in the brake system reaches around

100 psi (depending on manufacturer specifications). If this does not happen, have the truck serviced immediately.

10. Test the parking brake. Apply the parking brake and gently apply the throttle in a low gear to ensure that the parking brake will hold.

11. Test the service brakes. Release the parking brake and slowly drive forward. Apply the brakes firmly and ensure that they work properly and the vehicle doesn't pull to the side or have any delayed stopping actions. If there is a problem, put the truck out of service immediately.

A similar methodology for braking inspection is provided in *NFPA 1911 Standard for Service Tests of Fire Pump Systems and Fire Apparatus Section* (7.12) "Braking Systems."

In addition to this inspection procedure, it is important to follow manufacturer guidelines related to the air tank reservoir. Moisture and oil may build up in the air tank which may freeze in cold temperatures or fill the tank with fluid which reduces the space for air. This is why it is important to make sure that the tank is drained properly. Some tanks have manual drains, while others are automatic. Even if the tank has an automatic drain, make sure that the system is working properly.

Brake Fade

The braking efficiency of a commercial motor vehicle is less than a passenger vehicle. This is due to the rubber composition of the truck tires and the lag time of the air brake system. Reduced braking efficiency will result in a longer stopping distance.

The braking efficiency of a commercial motor vehicle can also be affected by *brake fade*. Brake fade is a result of the braking system's inability to burn off excess kinetic energy. Excess kinetic energy is the result of an overweight vehicle or excess speed.

Fire departments that operate in districts with hills, long distance responses, or stop-and-go urban traffic should be especially wary of brake fade. This is especially true if a fire apparatus is overweight or driving too fast. While brake fade is more common in a vehicle equipped with air brakes, it can also occur in a vehicle equipped with hydraulic brakes.

Mechanical Fade—Drum Brakes

An overweight or speeding vehicle may create more kinetic energy than the brakes were designed to dissipate. As a result of the excess kinetic energy, the brakes may overheat and start to fade. To understand how brake fade occurs, a driver must understand how fire apparatus manufacturers design an air brake system.

When members of a fire department design a fire truck, they will usually meet with a sales person. During this meeting, they will explain to the salesperson what type of fire truck they are looking to build and how much water, equipment, and personnel the vehicle is going to carry.

The salesperson will bring this information back to the factory and present it to a group of engineers. The engineers will determine how much the truck is going to weigh and how much kinetic energy the vehicle will have as it is driven on the highway. The engineers will then specify the proper size brake chambers which can safely dissipate that amount of kinetic energy.

The problem occurs when the truck shows up and starts backing into the station. The firefighters start to look at the rig and say "we could fit an extra 100 feet of 5 inch up there," or "we can carry extra equipment in this compartment." Soon after, the truck is overloaded.

A few weeks later, the apparatus is en route to a call. The apparatus is overweight and the driver is driving faster than is appropriate. As a result of the overweight vehicle and excess speed, there is more kinetic energy present than the brakes were designed to dissipate. Due to the excess kinetic energy, the metal brake drums begin to overheat when the brakes are applied. When the metal brake drums overheat, they expand.

Now the driver wants to bring the apparatus to a stop. The driver places a foot on the brake pedal and presses down, but nothing happens. The driver presses the pedal all the way to the floor, but still nothing happens. Even though the driver's foot is all the way down to the floor and the brake chamber has pushed the pushrod as far as it can go, the pushrod only has so much stroke. It can only expand and spread the brake shoes so far. However, due to the overheated and swollen brake drum, the brake shoes are no longer able to effectively press the brake pads against the lining of the brake drum. As a result, the vehicle loses its braking efficiency. This is a mechanical brake fade situation.

Friction Fade

With friction fade, the friction of the brake pad is reduced. There are two theories behind friction fade, both of which revolve around the binding agents that hold the composite brake pad together. Some believe that when the brake pad overheats, the binding agent in the brake pad starts to give off gas. This gas forms a layer between the brake pad and the brake drum (or rotor) resulting in reduced braking efficiency. This theory is disputed among experts, as some experts don't believe that enough gas could be generated to cause this type of fade.

Other experts believe that the binding agents in the brake pad will begin to melt at higher temperatures and create a slick surface on the brake pad. The slick brake pad will result in reduced braking efficiency as it is no longer able to effectively grip the drum or rotor. This is known as *glazing*.

Fluid Fade

In a hydraulic-braked vehicle, brake fluid is used to activate the brakes. When the driver presses on the brake pedal, the pedal pushes on the brake fluid and activates the brakes. The braking system works quickly because brake fluid is not compressible. When the driver presses on the brake pedal, it pushes on the fluid, which causes the brakes to activate.

While brake fluid is not compressible, vaporized brake fluid **is** compressible. If the brake system starts to overheat, the brake fluid will start to vaporize. When the driver presses on the brake pedal, the vaporized brake fluid must first be compressed before any pressure is able to reach the brake fluid and activate the braking system.

A similar situation can result if water is allowed to enter the brake system and contaminate the brake fluid. If water mixes with the brake fluid, it may turn to steam, causing a fluid fade situation. This is why mechanics recommend that the brake system be flushed every few years.

Cascade Failure

If one brake begins to fade, it is no longer doing its share of work to burn off the vehicle's kinetic energy. As a result, the other brakes will have to work harder to burn off the energy. The brakes that are still working will be more prone to overheat as they try to burn off all of the vehicle's kinetic energy. Eventually, all of the brakes may enter a fade situation and the vehicle will lose its braking ability. This is known as a cascade failure.

Preventing Brake Fade

There are several ways to prevent brake fade. Some methods involve vehicle maintenance, while others involve vehicle operations.

Annual Weight Certifications

Brake fade is a result of too much kinetic energy. Excess kinetic energy is often the result of an overweight vehicle. As indicated in NFPA 1911 (16.2.1), fire departments must have each fire apparatus weighed on an annual basis to ensure that the vehicle is not overweight. The standard provides detailed instructions and worksheets on how to properly weigh the apparatus. This includes adding the weight of the firefighters who may be riding the apparatus.

There are many ways a fire department can find scales to weigh the fire apparatus. Many industrial facilities, such as quarries, junk yards, and landfills, have fixed scales that are used to weigh vehicles as they enter and leave a facility. Many police departments are also equipped with portable scales and would be more than willing to come to the firehouse to weigh the vehicles.

The key is to make sure the vehicle is weighed with a **certified** scale by a **certified** weigh master. It is also important to make sure that the fire department keeps detailed records of every weight certification. Hopefully, these records will never be needed, but if the fire department finds itself involved in a lawsuit, the fire department will want to make sure that they have properly documented each annual weight certification as required by NFPA.

Proper Brake Adjustment

In addition to an annual weight certification, it is important to make sure that the vehicle's air brakes are properly adjusted. If one brake is out of adjustment and not working properly, the other brakes will have to work harder to burn off the vehicle's kinetic energy. The other brakes on the vehicle will be more prone to overheat and enter a brake fade situation. An out-of-adjustment brake may lead to a cascade failure of the braking system.

In September, 2016, a major fire apparatus skid testing exercise took place in suburban Philadelphia. During this exercise, ten fire apparatus were examined by certified truck inspectors, weighed, and then skid to a stop to determine the braking efficiency of each apparatus. During this test procedure, two of the ten apparatus were placed out of service by the DOT truck inspectors due to out of adjustment brakes.

It is imperative that each fire department have the brakes examined on a regular basis by a certified mechanic or truck inspector. In both of the aforementioned cases, the apparatus which were placed out of service had been recently examined by mechanics. Yet, the brakes were still found to be out of adjustment. Proper inspection of the brake system by trained personnel must occur on a regular basis to ensure that the brake system is working to its full potential.

Brake Imbalance

Improper brake adjustment or improper repairs can lead to a brake imbalance. A vehicle can experience brake imbalance for many reasons, including mismatched parts and failing relay valves. There are two common types of brake imbalance: air imbalance and torque imbalance.

Air Imbalance

The same amount of air pressure must arrive at each brake chamber at the exact same time. When this doesn't happen, one brake may activate before the others, causing a brake imbalance and possible loss of control.

Each valve in the air brake system has a *crack pressure*, which is the amount of air pressure needed to open the valve. Contaminants and other failures could cause the crack pressure of a valve to change, leading to an air imbalance in the braking system.

In combination vehicles (such as a tractor trailer or tractor-drawn aerial), the trailer crack pressure is often set lower than the tractor crack pressure. In situations where the driver is using light braking, the trailer brakes may be activated, while the tractor brakes are not. This may cause the trailer brakes to overheat, as the tractor is not doing its fair share of work.

Torque Imbalance

It is important that the same amount of torque is applied to each service brake. If one brake is receiving more torque, it will generate more braking power. A difference in braking power on one wheel will lead to an improperly balanced braking system which may cause the brakes to fade. If a vehicle is repaired improperly and a different sized slack adjuster or brake chamber is installed on one brake chamber and not the others, it will lead to a torque imbalance.

Check the Pushrod Stroke

Properly adjusted brakes will have a minimal amount of gap between the brake pad and the drum. As the vehicle is used over time, the brake pad will wear down, causing this gap to widen. Extended application of the brakes (such as descending a long hill) will also increase the gap between the pad and the drum as the drum heats up and expands. As the gap grows larger, the pushrod will have to travel farther before the brake pad contacts the drum. If the pushrod stroke is too long, it will result in delayed application of the brakes, as well as reduced braking force. Less braking force is applied to the wheel as the pushrod stroke grows longer.

If the gap between the pad and the drum is large enough, the pushrod will bottom out. When this happens, the pushrod has traveled as far as it can go, and the brake shoes have expanded as far as possible, but the brake shoes are no longer able to effectively press the pads against the drum. The vehicle will suffer reduced braking efficiency and a longer stopping distance.

An ideal brake adjustment will result in no more than a 90 degree angle between the pushrod and the slack adjuster. As the brake pads wear down, or the brake drum expands from heat, the distance the pushrod has to travel to press the brake pads against the drum grows longer. As the pushrod stroke grows longer, the angle between the pushrod and the slack adjuster will exceed 90 degrees (figs. 10–1 and 10–2).

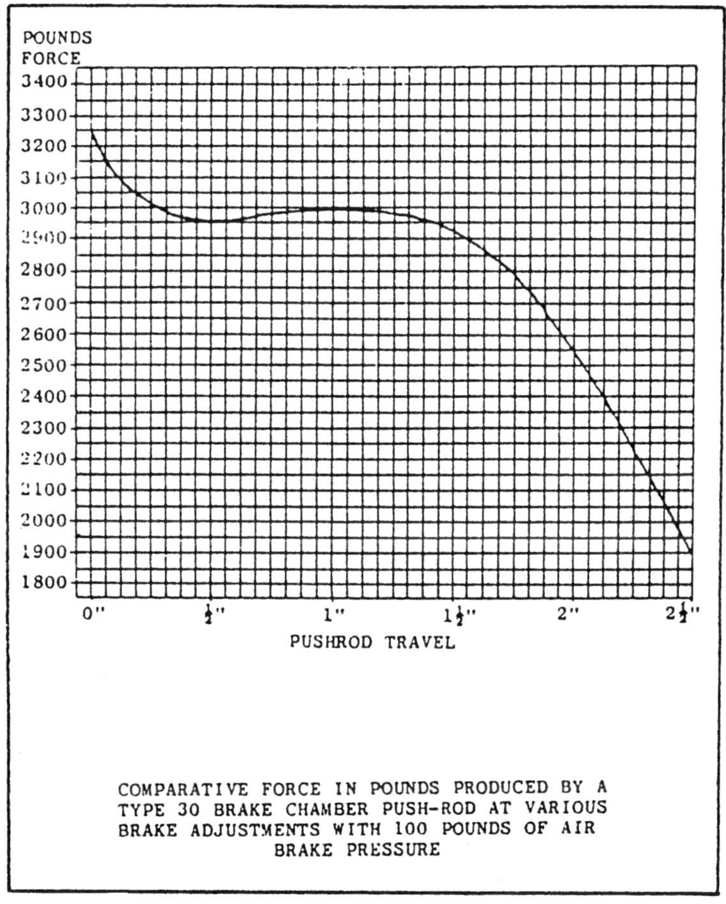

FIGURE 10–1. Take note of how the braking force on the X axis goes down as the pushrod travel distance increases. If the brakes are out of adjustment, there will be reduced braking force at the wheel. (WSR 13-21-102: Proposed Rules Washington State Patrol, Filed October 21, 2013. http://lawfilesext.leg.wa.gov/law/wsr/2013/21/13-21-102.htm)

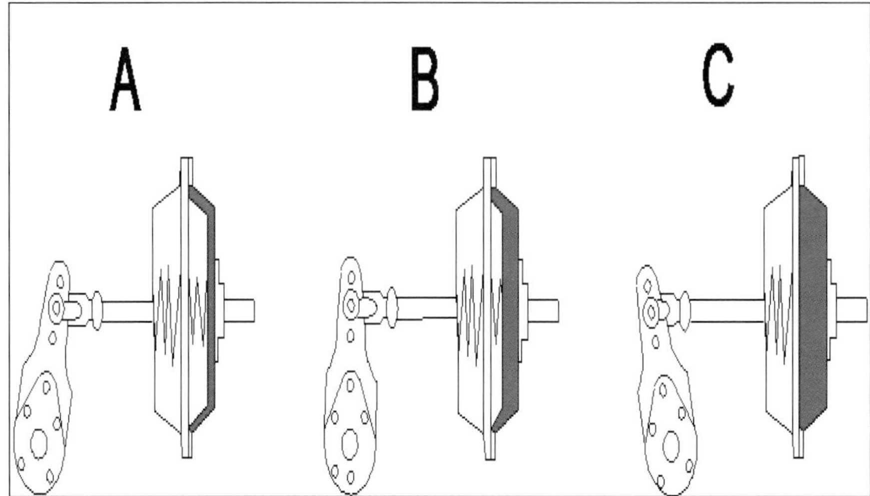

FIGURE 10–2. The angle between the pushrod and the slack adjuster should not exceed 90 degrees. Picture A shows the air brake when it is not applied. The wheel is spinning freely. In Picture B, the air brake is applied and the angle between the pushrod and the slack adjuster is 90 degrees. This will provide the maximum braking force. In Picture C, the angle between the pushrod and the slack adjuster has exceeded 90 degrees. At this point, the braking force will go down and the vehicle will take a longer distance to come to a stop. When the angle exceeds 90 degrees, the brakes will need to be readjusted.

Checking the pushrod stroke should be a routine part of every vehicle maintenance program. This is not to say that drivers should adjust the brakes, but they should be trained to recognize when there are problems. If a brake is found out of adjustment, fire apparatus operators should place the vehicle out of service and have a qualified technician make the proper repairs.

Even if a vehicle is equipped with automatic slack adjusters, the pushrod stroke should be checked on a regular basis. There may be times when a mechanical failure or broken component causes an issue that the automatic slack adjusters cannot compensate for. If an automatic slack adjuster is constantly out of adjustment, there is an underlying mechanical issue that must be identified.

Fire departments are encouraged to have their fire apparatus operators professionally trained on how to check the adjustment of the brakes. The following section will give a brief overview of how this can be done using the *applied stroke method*, which is the same method used by roadside inspectors:

1. Chock the tires, turn off the engine, and release the brakes. Make sure to stay clear of the wheels and tires and wear safety glasses, as dirt and debris may fall in your eyes. You will need to be under the truck where the brake chambers are, while a second person sits in the driver's seat.

2. Have your assistant apply the brakes a few times until the reservoir pressure is between 90 and 100 p.s.i.
3. Mark the pushrod where it meets the face of the brake chamber. A piece of chalk or soapstone works best. Make sure that the pushrod is fully retracted by tapping it with your hand. Sometimes the pushrod does not retract all the way, resulting in a faulty measurement.
4. Have your assistant make a full brake application.
5. Measure the distance between the mark you made on the pushrod and the face of the brake chamber. This distance is the chamber applied stroke. The allowable stroke distance depends on the size of the brake chamber. The allowable stroke distances are provided in table 10–1.

It is important to be sure that the brakes are properly inspected on a regular basis by a certified emergency-vehicle technician. If the department trains its fire apparatus operators to check the brake system in-house, the check should be done as part of a daily, weekly, or monthly vehicle inspection. If the department does not have certified technicians, the vehicle will have to be sent to a specialized facility, or a qualified technician will have to come to the station to make sure the inspection is completed. I can assure you that if your vehicle is involved in a crash, one of the first things a crash investigator is going to do is to check the condition and adjustment of the brake system.

Case Study—Missouri

On September 5, 2004, an engine company from the Kansas City (MO) Fire Department was en route to an apartment fire. As the engine was responding to the scene, it encountered a civilian vehicle which had come to a complete stop in the left travel lane. The fire apparatus operator assumed the civilian was going to stop and yield the right-of-way, so the engine passed the vehicle in the left lane. As the engine began to pass, the civilian vehicle turned left into a driveway. The two vehicles collided and the engine veered across the oncoming lane, where it struck a stopped vehicle head-on. The engine then drove through a utility pole and struck a tree. During the crash, the driver of the fire engine was seriously injured and the officer of the rig was killed.

During the crash investigation, the police department determined that the fire engine was traveling 51 mph when it struck the first vehicle, 34 mph when it struck the second vehicle and 24 mph when it struck the tree. A post-crash inspection of the apparatus revealed that the brakes had not been serviced for 16 months and were out of adjustment. The investigating officer determined that if the brakes had been working properly, the engine would have been able to come to a stop before it struck the tree.

TABLE 10–1.   Allowable pushrod stroke distances based on chamber size and type.

Clamp Type Brake Chambers*		
Brake Chamber Size	Marking	Adjustment Limit
6	None	1-1/4″
9	None	1-3/8″
12	None	1-3/8″
16	None	1-3/4″
16-L	L stamped in cover, stroke tag	2″
20	None	1-3/4″
20-L	L stamped in cover, stroke tag	2″
24	None	1-3/4″
24-L	L stamped in cover, stroke tag	2″
24-LS	Square ports, Tag & cover marking	2-1/2″
30	None	2″
30	DD3 (Bus/Coach)	2-1/4″
30-LS	Square Ports, Tag & Cover Marking	2-1/2″
36	None	2-1/4″

L denotes the long-stroke pushrod design.

LS denotes the long-stroke pushrod design with square ports.

*Ontario Ministry of Transportation, The Official Air Brake Handbook, retrieved from http://www.mto.gov.on.ca/english/handbook/airbrake/section12-9-0.shtml

Bolt-Type Brake Chambers*	
Type	Brake Readjustment Limit
A	1-3/8″
B	1-3/4″
C	1-3/4″
D	1-1/4″
E	1-3/8″
F	2-1/4″
G	2″

*https://www.govinfo.gov/content/pkg/CFR-2015-title49-vol5/pdf/CFR-2015-title49-vol5-sec393-47.pdf

Rotochamber-Type Brake Chambers*	
Type	Brake Readjustment Limit
9	1-1/2″
12	1-1/2″
16	2″
20	2″
24	2″
30	2-1/4″
36	2-3/4″
50	3″

*https://www.govinfo.gov/content/pkg/CFR-2015-title49-vol5/pdf/CFR-2015-title49-vol5-sec393-47.pdf

The engine was equipped with size 30 brake chambers and the post-crash inspection revealed that the pushrod stroke was 2-1/8" for the driver's side and 2-1/4" for the passenger side. As noted in Figure 10–3, the allowable pushrod stroke for size 30 brake chambers is 2". This case study is an example of how poor brake maintenance led to two out-of-adjustment brakes. The out-of-adjustment brakes were a causative factor in the severity of this crash.

Downshifting and Auxiliary Braking

Brake fade is caused by overheated brakes. Reducing the amount of time the brakes are used will help reduce the chance of brake fade, while also increasing the lifespan of the brake components. There are typically two methods of slowing a vehicle without using the brakes: downshifting and the use of an auxiliary braking device.

Downshifting

Downshifting uses the vehicle's transmission to slow the vehicle. Anyone who has descended a long hill with a manual transmission knows that shifting to a lower gear will help to hold the vehicle at a low speed.

Auxiliary Braking Device

An auxiliary braking device will slow the vehicle and reduce the need for the service brakes. Because the service brakes are not used as often, the brake components remain at a cooler temperature, which reduces the chance of brake fade.

NFPA 1901 requires an auxiliary braking device on vehicles with a gross vehicle weight rating over 36,000 lbs. It is important for fire apparatus operators to identify the type of auxiliary braking device their vehicle is equipped with, as well as the manufacturer's specifications and limitations of each device.

Compression Release Engine Retarder

In very simple terms, an engine is made of pistons which move up and down inside a piston chamber. These pistons are attached to the vehicle's driveline, which are attached to the wheels. The pistons move up and down as a result of the engine firing or combusting fuel inside each piston chamber. When the fuel combusts, it creates energy which forces the piston down. As the piston moves down, it turns the driveline, which turns the wheels.

Fire apparatus diesel engines have multiple pistons and at any given time each piston is performing one of four functions. These four functions are as follows:

1. *Induction*: The piston is sucking air into its piston chamber.
2. *Compression*: The piston is compressing the air inside its piston chamber.
3. *Power*: The piston is in the midst of firing the air/fuel combination and creating power.
4. *Exhaust*: The piston is pushing the exhaust out of the piston chamber.

When the driver's foot is on the accelerator pedal, fuel is flowing into the piston chambers and combusting. As combustion occurs, the pistons are working to drive the crankshaft and turn the wheels.

If the vehicle is equipped with an engine brake, it will activate when the driver takes his foot off the accelerator pedal. When the engine brake activates, a valve will open and allow the air to escape the piston chamber during the compression cycle. When the air escapes, there is no energy left to push the piston back down. However, the piston is still moving up and down because wheels are still rolling and rotating the driveline. Because the driveline is pulling the piston back down inside the chamber, the pulling action absorbs the vehicle's forward momentum. The engine has turned from an energy producing combustion engine into an energy absorbing air compressor.

Exhaust Retarder

In an exhaust retarder, there is a butterfly valve attached to the exhaust manifold of the motor. This butterfly valve will open and close accordingly to block exhaust gases from leaving the manifold. This creates a back pressure which restricts the movement of the piston, slowing the driveline and wheels.

Electro-magnetic Retarder

In an electro-magnetic retarder, there are a series of electrical coils placed around a flywheel which is attached to the driveline on the apparatus. When the device is activated, the electrical coils energize and slow the flywheel with magnetism. Slowing the flywheel will slow the driveline and wheels.

Transmission Retarder

In a transmission retarder, a flywheel is mounted inside a chamber and attached to the driveline of the apparatus. When the device activates, transmission fluid will enter the chamber and slow the flywheel. Slowing the flywheel will slow the driveline and wheels.

Some retarders are set to activate when the driver's foot comes off the throttle. Others are set to activate when the brake pedal is depressed. And finally, some retarders have a separate hand control so that the device can be manually activated. Each retarder also has different power settings, such as "high" vs "low," or percentage points. A driver training program must address when to change the power settings and when to shut the device completely off.

It is also important to remember that engine or driveline retarders should not be used in inclement weather. Using the retarder in wet weather may lead to wheel lock-up, causing the vehicle to skid. If the apparatus is equipped with an auxiliary braking device that should not be used in inclement weather, make sure the driver training program includes time behind the wheel with the auxiliary braking device deactivated. This is because it is not uncommon for new fire apparatus operators to complete an entire driver training program on sunny days with dry roads. As luck would have it, the driver's first real call is in the rain and the officer reminds them to shut off the auxiliary brake. The new driver is shocked at the sudden increase in stopping distance, as they never had the opportunity to drive the vehicle without the protection of an auxiliary brake. For this reason, make sure the driver training program includes several hours behind the wheel with the auxiliary braking device turned off. Also make sure to check the retarder's operating manual for information on adverse weather issues and include these issues in the written driver training program.

Snub Braking

There was a time when driver training courses advocated the use of light, sustained brake applications when traversing a long, downhill grade. However, due to differences in the crack pressures of the air relay valves, a light application of the brake system may not charge the brake system with enough air pressure to activate all of the brakes. If all of the brakes are not activated, some brakes may be doing more work than the other brakes, causing them to overheat and fade.

Snub braking is now the recommended method of braking when descending a long grade. Snub braking will produce the same amount of heat as a long, soft application of the brakes. The advantage to snub braking is that it will prevent an air imbalance caused by differences in the crack pressures of the air relay valves. Snub braking with a hard press on the brake pedal will ensure that enough air pressure reaches all of the brake components and overcomes the crack pressure of each relay valve.

Consider a situation in which the driver is gently riding the brakes down a long hill and only applying 10 psi of brake pressure. However, one of the brake valves won't open until the air pressure reaches 20 psi. Because one of the brakes has not activated, the other brakes must work harder to burn off the vehicle's kinetic energy. If the driver had charged the system with 30 psi of air with a hard press of the brake pedal, the stuck valve would have opened. Snub braking will ensure that all of the air relay valves open appropriately, even if one of the brake valves has an improper crack pressure[5] (figs. 10–3 and 10–4).

The recommended practice for snub braking is as follows:[6]

1. Apply the brakes hard enough to feel a slowdown in the vehicle.
2. When vehicle speed has been reduced to around 5 mph below a safe speed, release the brakes (about 3 seconds).
3. When speed has increased to a safe speed, repeat steps 1 and 2.

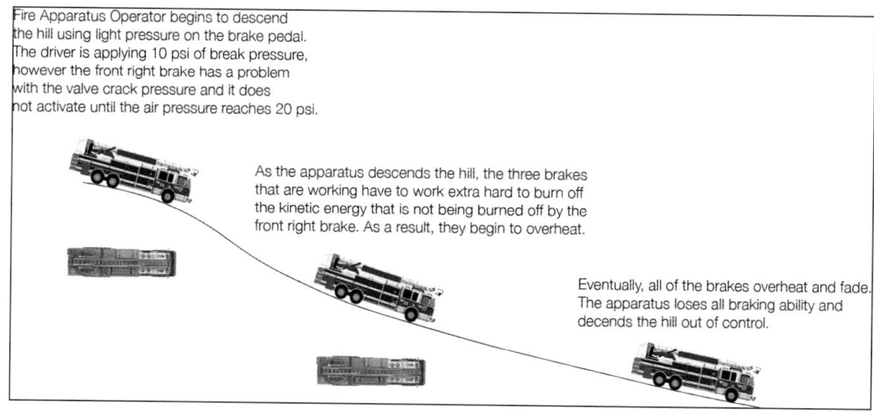

FIGURE 10–3.   Snub braking not applied.

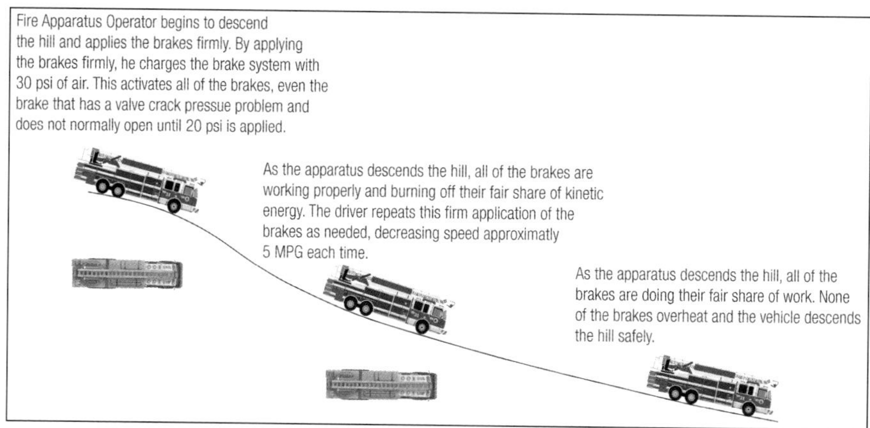

FIGURE 10–4.   Snub braking applied.

Anti-lock brakes (ABS)

Chapter 5 discussed the concept of "The Friction Circle." In summary, there is only so much friction available at the tire contact patch. This friction can be used in the longitudinal direction (front to back), lateral direction (side-to-side), or a combination of the two. If the driver forcefully applies the brakes and the tires lock, all of the friction will be used in the longitudinal direction. There will be no lateral friction leftover to steer the vehicle from side-to-side. As a result, a vehicle that is skidding will lose all steering control.

Those who attended an EVOC class before the wide-spread use of anti-lock brakes may recall the concept of threshold braking. This technique was designed to prevent tire lock-up and the subsequent loss of steering control. Drivers were instructed to apply the brakes to the point where they felt the tires start to lock. As the tires started to lock, the driver was instructed to release pressure on the brake pedal and manually pump the brakes. This technique kept the tires from locking, allowing the driver to maintain steering control.

The problem with threshold braking is that it is not easy to instruct and very difficult for a driver to remember. As a result of the difficulties associated with threshold braking, engineers developed anti-lock braking systems. In an ABS equipped vehicle, the ABS system pumps the brakes automatically, even when the driver forcibly applies the brakes. This allows the driver to maintain steering control, even during a panic stop (figs. 10–5 and 10–6).

Fire apparatus operators must know which trucks are equipped with ABS and which are not. Fire trucks are typically replaced every 15, 20, or 25 years. As a result, the fire service is in a period of mixed fleets, where older apparatus do not have ABS while the newer apparatus do.

Each fire apparatus must be clearly identified as to whether or not it is equipped with anti-lock brakes. This can be accomplished by applying a label to the dashboard or taping a laminated sign within the driver's view. Identifying each apparatus will allow the fire apparatus operator to know which method should be used to stop the apparatus during panic braking.

If the vehicle is **not** equipped with ABS, the driver will need to remember threshold braking. If the vehicle **is** equipped with ABS, the driver will need to remember a different stopping method known as **stomp, stay, and steer.** The driver should **stomp** down on the brake pedal, **stay** on the brake pedal (keep the brake pedal depressed), and **steer** around the hazard. Just keep in mind that a fire truck has a high center of gravity. Any sudden steering input to the left or to the right may cause the vehicle to upset itself and rollover.

It is important for apparatus operators to practice proper emergency braking. Driver training instructors should find a large, empty parking lot and obtain permission from the owner to use the lot for training. After a suitable location has been found, each driver should be given the opportunity to take the vehicle up to 25 or 30 mph and stomp on the brakes. Why? If a driver has never experienced the ABS engaging, the activation of the ABS system may come as a startling surprise.

When the ABS system engages in a hydraulic braked vehicle, it can cause the brake pedal to pulse and make strange noises. For someone who has never experienced this phenomenon, they may become startled and react inappropriately. This is why it is imperative for fire apparatus operators to feel the ABS engaging under controlled circumstances.

The driver of a newer vehicle with anti-lock **air** brakes may not feel anything pulsing in the brake pedal. However, they may hear a hissing noise coming from the wheels as the ABS engages. The driver of a smaller vehicle equipped with hydraulic brakes will still feel the ABS cycling in the form of a vibrating or pulsing brake pedal.

It's much cheaper to send the rig out for maintenance after this type of training exercise than to have the ABS kick-in at 2:00 in the morning, startle the operator, and have him crash the truck. However, it is extremely important for the driver to remember that they are driving a vehicle with a high center of gravity. A sudden steering input during an evasive maneuver may be enough to upset the stability of the truck. Smooth, controlled steering inputs should always be used during an evasive maneuver in a high-center-of-gravity vehicle. This is especially true when the brakes are engaged. From a real-life perspective, skid avoidance is the key. By looking far down the road and anticipating hazards, a driver will be able to avoid a skid altogether. This will prevent the fire apparatus operator from having to decide whether to slam on the brakes and steer, thus risking a rollover, or simply slamming on the brakes and waiting for a crash.

Evasive maneuver training in a high center of gravity vehicle can be dangerous. If the driver reacts inappropriately, it may cause the vehicle to trip itself and rollover. For this reason, it is best to practice these maneuvers with smaller vehicles that have a low center of gravity. If the department has the finances to purchase a specialized system that can be mounted to a large apparatus to prevent a rollover during this type of training, it would be ideal. However, most departments do not have these resources. Instead, creativity is the key for driver trainers.

Understanding the ABS warning light

If the ABS system has a fault or a problem, the ABS light will illuminate on the dashboard. The purpose of this light is to inform the operator that the ABS is not working and needs to be repaired. Report the problem immediately and have the system checked. Keep in mind that while the ABS may be disabled, the brakes should still work.

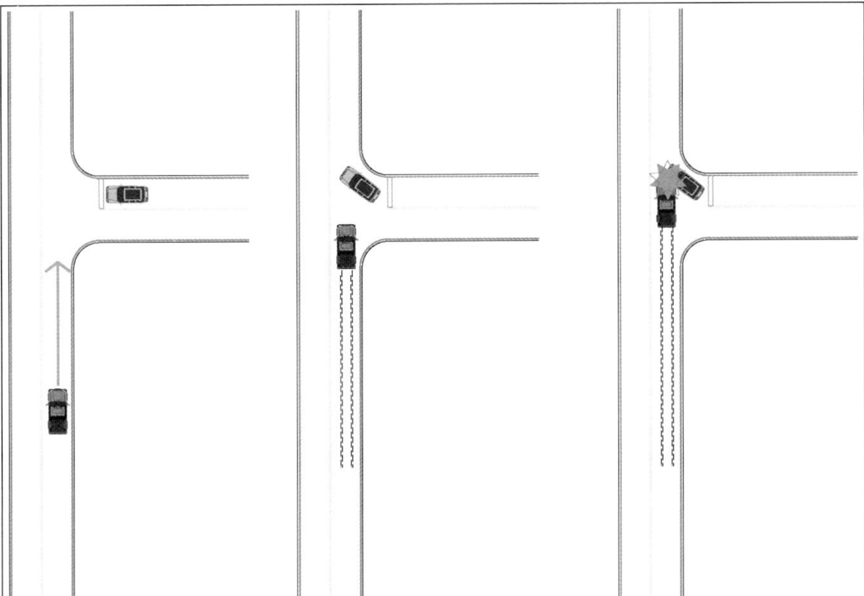

FIGURE 10–5. In a vehicle that is not equipped with anti-lock brakes, the driver will lose steering control when the wheels lock. The vehicle will skid in a straight line until it runs out of energy and stops, or slams into something.

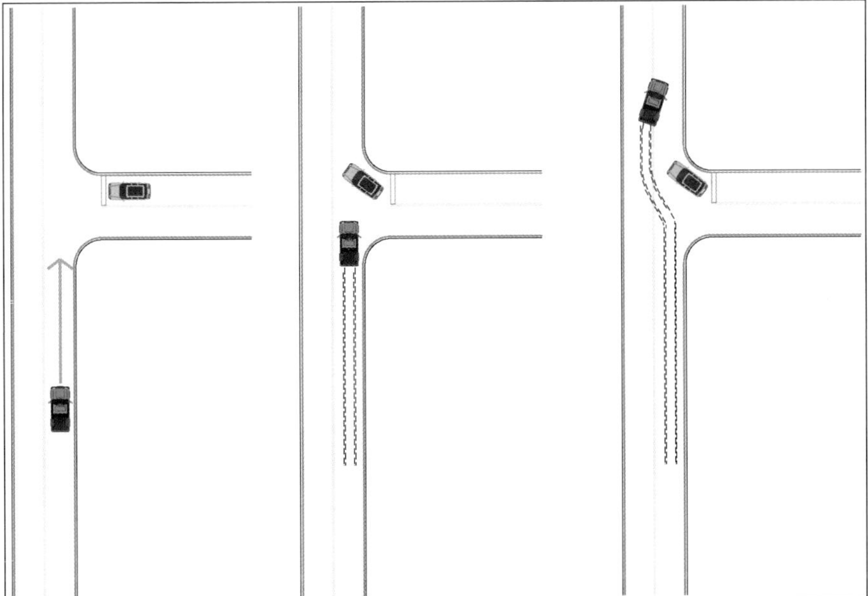

FIGURE 10–6. In a vehicle equipped with anti-lock brakes, the wheels will not lock and the driver will **not** lose steering control. The driver will be able to steer around the hazard and avoid a collision.

More than likely there is a bad wheel sensor, a wire may be corroded or broken, or the system is sensing an electrical fault. These issues are more common with older vehicles, especially on front axles, as the wires move back and forth thousands of times over the life of the vehicle as the steer axle moves from left to right.

> When I first became a police officer it was not unusual to go to a crash scene involving an older driver. When I asked them what had happened, they would say "I was driving down the street and a car pulled right out in front of me. I slammed on the brakes, but the brakes stopped working." I would then ask "What do you mean the brakes stopped working?" They would then explain, "I slammed down on the brake pedal and the brake pedal started bouncing up and down. So I let go and I hit them."
>
> What was actually happening wasn't that the brakes had failed, they were working just fine. What was happening was the anti-lock braking system was kicking in, but the driver had been driving vehicles for 40 or 50 years without ABS. As a result, when the ABS kicked in, the pulsing brake pedal startled the driver and they didn't know what was happening.

Notes

1. This issue was addressed in Chapter 5 during our discussion on braking efficiency and total stopping distance.
2. Pennsylvania Department of Transportation, *Commercial Driver's License Manual*, pg. 5-1–5-1.1
3. Pennsylvania Department of Transportation, *Commercial Driver's License Manual*.
4. Pennsylvania Department of Transportation, *Commercial Driver's License Manual*.
5. Keep in mind that this assumes the crack pressure issue is solvable. There may be times where even charging the system to a higher pressure may not open the valve due to other issues.
6. PENNDOT, *CDL Manual*, Chapter 5, pg. 5–10.

Tire Safety

Hydroplaning

Wet weather is a common cause of vehicle crashes. Wet weather reduces the coefficient of friction of the roadway, which results in longer stopping distances and reduced critical curve speeds. In addition to friction issues, rain-soaked roads can induce a hydroplane. When a vehicle hydroplanes, it will lose all steering and braking control. Drivers must understand how and why hydroplaning occurs.

Contact patch

The part of a tire in contact with the road is called the *contact patch*. The contact patch creates the friction between the vehicle and the roadway, allowing it to stop, steer, accelerate, and make evasive maneuvers. The contact patch is one of the most important safety components of a vehicle. It is literally where "the rubber meets the road."

A typical vehicle has a contact patch that is approximately the size of a man's hand. Considering the size and weight of a fire apparatus, there is a limited amount of contact between the apparatus and the highway. Because of this limited amount of contact between the rig and the highway, it is important to ensure that the contact patch is as healthy as possible.

In a properly inflated tire, the contact patch is flat. The vehicle will get as much traction as possible from the tire because there is as much rubber as possible touching the road. In an underinflated tire, the contact patch will start to cup. As a result of this cupping, there is not as much rubber touching the road. As a result of the smaller contact patch, the vehicle will not have as much traction and may lose control at a lower speed during inclement weather. Underinflated tires may also lead to a blowout.

In an overinflated tire, the tire may develop a "U" shaped contact patch. There may be a wear ring around the center of the tire and the vehicle may give a stiff ride. While most people believe that tires blow out because there is too much air in them, this is not the case. Most tires blow out because they are underinflated, causing them to overheat and explode (fig. 11–1).

Tire Tread

A bald tire will have better traction on a **dry** road than a treaded tire. This is because there is more rubber touching the road surface. However, fire apparatus operate in areas with rain, snow, and inclement weather, so tire manufacturers put tread on tires. Tire tread is designed to channel and push rain, slush, and snow out of the way and allow the rubber tire face to come in contact with the road surface.

Problems arise when the vehicle is driven too fast during bad weather. If the vehicle is traveling too fast, the tire is spinning too quickly, and the tread does not have enough time to push all of the water out of the way. Instead, a wedge of water builds up in front of the tire and eventually the tire climbs on top of the wedge of water. When this happens, the tire is no longer in contact with the road; it is hydroplaning (fig. 11–2).

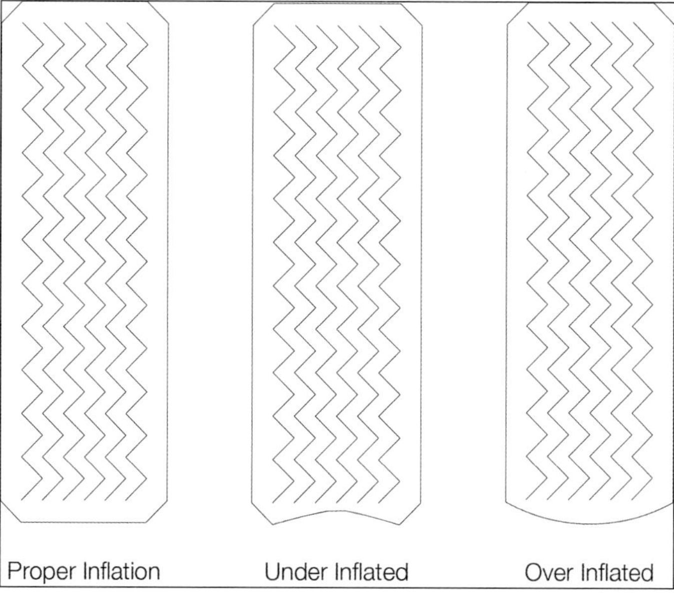

FIGURE 11–1. Notice how the size and shape of the contact patch changes with inflation pressure. It is important to maintain proper tire inflation pressure to maximize the size of the contact patch.

Hydroplane Speed

The speed at which a vehicle will hydroplane depends on three major factors: the depth of the tire tread, the air pressure inside the tire, and the depth of the water the vehicle is driving through.

Tire Tread Depth

If a tire has shallow tread, the tread will not be deep enough to properly move water out of the way. NFPA 1911 requires the tires on the front steering axle of a fire apparatus to have no less than 4/32" tread depth at any two adjacent major tread grooves anywhere on the tire. The standard requires any non-steering axles to have no less than 2/32" tread depth at any two adjacent major tread grooves anywhere on the tire.[1] However, studies have shown that tires with less than 5/32" of tread depth will have a significantly longer stopping distance in wet weather.[2] It is important that any weekly or monthly apparatus check involve a thorough examination of the tire tread.

Air Pressure

An underinflated tire will not have enough air pressure to properly push the water out of the way. As a result, an underinflated tire will hydroplane at a lower speed. Crash reconstruction experts use a formula to calculate the speed at which a vehicle will hydroplane based on the air pressure inside the tire.

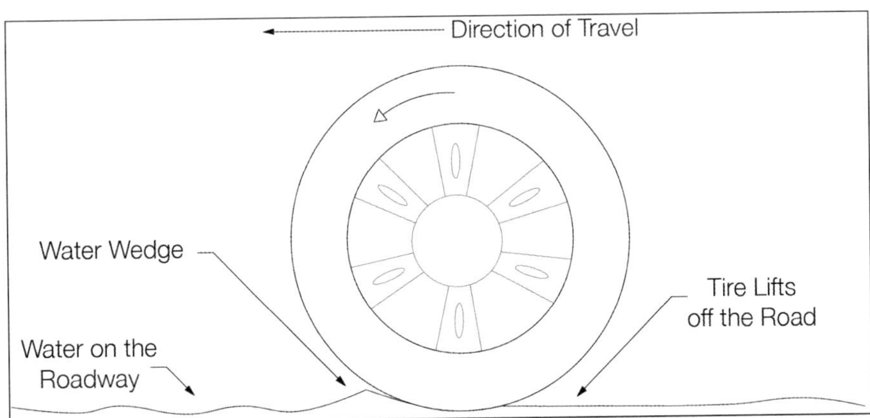

FIGURE 11–2.   When a tire is spinning too quickly because the vehicle is traveling too fast, a wedge of water will build up in front of the tire. Eventually, the tire may lift off the road and hydroplane.

TABLE 11–1.   Calculating Hydroplaning Speed[3]

$\text{Hydroplaning Speed} = (10.35)\sqrt{\text{InflationPressure}}$	
Tire Pressure	**Hydroplaning Speed**
30 psi	56.7 mph
25 psi	51.8 mph
20 psi	46.3 mph

Note: This formula assumes that the water depth exceeds the ability of the tire tread to remove water from the path of the tire. These speeds are approximations.

The formula in table 11–1 demonstrates that a tire with a lower tire pressure will hydroplane at a lower speed. It should be noted that this formula assumes that the water is not deeper than the tire tread. If the vehicle drives through water that is deeper than the tire tread, it may hydroplane at an even lower speed. Some vehicles may hydroplane at speeds as low as 30 to 40 mph.

Studies have shown that despite the high pressures associated with truck tires, commercial motor vehicles may still hydroplane in certain situations. It is important that any weekly or monthly apparatus check involve a thorough examination of the tire pressures.

Depth of Water

Deep water can overwhelm the tread of the tires, causing the tire to float up and rise on top of the water surface. As a result, fire apparatus operators should take care to avoid standing puddles or deep pockets of water. Fire apparatus operators should also try to stay in the tracks of the vehicles in front of them, as the other vehicles will have already helped to sweep the water from the path of the apparatus.

Handling a Hydroplane

Should the vehicle start to hydroplane, it is important to avoid slamming on the brakes, as this could lead to a loss of control. Professional drivers must remember to remain calm, cool, and collected behind the wheel. When things start to go south, that is when the term "professional" kicks in. Knowing what to do ahead of time can be the difference between a hydroplane crash and a successful evasive maneuver.

If the vehicle starts to hydroplane, take a firm grip of the wheel and remove your foot from the accelerator pedal. Use as little steering as possible and allow the vehicle to slow on its own. Once the vehicle's speed drops enough, the tires should regain contact with the road surface. Should you find yourself in a situation where you absolutely have to apply the brakes, brake normally if your vehicle has anti-lock brakes, but don't slam on them. If your vehicle doesn't have anti-lock brakes, use light and rapid presses on the pedal and threshold brake to a stop.

Unfortunately, there is little a driver can do once the vehicle enters a hydroplane. The key to avoiding a hydroplane crash is to avoid putting the vehicle in a hydroplane situation. Remember to keep a close eye on the condition of the tires, especially the tread depth and air pressure. Most importantly, slow down when it's wet out.

Tire Blow Outs

Lifting air bags are used in the fire service as rescue tools. Those familiar with lifting bags understand that the rubber bag does not lift the load. The air inside the bag lifts the load. The amount of weight the bag can lift is determined by the size of the bag and the amount of air pressure inside the bag.

The same principle holds true for vehicle tires. The rubber tire is not meant to support the weight of the vehicle. The air inside the tire is what supports the weight of the vehicle. When there is not enough air inside the tire, the rubber sidewalls of the tire will bear the weight of the vehicle. When this happens, the sidewalls of the tire bulge out, while the rim of the tire may sink down and contact the road. This is referred to as a "flat tire."

A normally inflated tire will be round at the top and slightly compressed at the bottom, where the tire meets the ground. As the vehicle drives forward, the tire will rotate around the axle, constantly changing shape from round at the top to compressed at the bottom. The tire changes shape by flexing the sidewalls of the tire. This flexing action creates heat, and the tire is designed to tolerate a certain amount of heat.

Problems arise when the tire is underinflated and there is not enough air pressure in the tire. When this occurs, the tire sidewalls will over-flex as the tire rotates around the axle. The over-flexing sidewalls will eventually generate more heat than the tire was designed to handle. As a result of this excess heat, the tire could fail and blow out.

In addition to low air pressure, a tire blowout can also result from a vehicle that is overweight. If the vehicle is overloaded, there is not enough air pressure in the tire to properly support the excess weight. As a result of the excess weight, the tire sidewalls may over-flex, and thus overheat, resulting in a tire blowout. This is another reason why the NFPA 1911 standard requires fire apparatus to be weighed every year.

Proper Tire Pressure

Proper examination of the tire pressure is more than a simple glance to see if the tire looks flat. Every routine apparatus check should include a close examination of the tire pressures using a calibrated pressure gauge on all of the tires, especially the inside dual tires. Even if there are no outward signs of damage or air loss, a tire can lose air in other ways. In tubeless tires, an improper seal between the tire and the rim can cause air to escape. Punctures caused by nails, screws, or other roadway hazards may lead to small leaks that can slowly rob a tire of its air. Few people realize that air can actually leak through the rubber membrane of a tire at a rate of 1 to 2 psi per month.[4] Over the course of a year, a tire may be 12 to 24 psi underinflated, just from sitting in the firehouse.

In order to properly check the vehicle's tire pressure, a fire apparatus operator must first know where to find the recommended tire pressure for the vehicle. Most people believe that the recommended tire pressure is printed on the tire sidewall. This is not correct. The tire pressure that is printed on the side of the tire is the **maximum** tire pressure for the tire. The recommended pressure for the vehicle is not always the maximum pressure that is printed on the sidewall (fig. 11–3).

To understand this concept, a driver must consider how tires are used on different vehicles. A vehicle owner can go to any local tire store and select from a number of tire brands, regardless of the type of vehicle that they own. One person may own a vehicle that weighs X pounds, while another person may own a vehicle that weighs Y pounds. Even though each person owns a different vehicle, they could both purchase the same tires for their respective vehicles. However, even though the same tires were purchased for two different vehicles, they will probably have different recommended air pressures. These recommended air pressures are determined by the tire manufacturer and will vary based on how much weight each tire is supporting.

The purest method for determining the proper air pressure for the tire is to have the vehicle weighed and determine how much weight each tire is supporting. Once it is determined how much weight each tire is supporting, refer to the tire manufacturer's tire load and inflation pressure chart. These charts are provided by tire manufacturers and can be obtained when the tires are purchased.

A shortcut to using the tire manufacturer's tire load and inflation chart is to use the vehicle manufacturer's information placard, which is usually found inside the cab of the apparatus, near the operator's knee (fig. 11–4). For a civilian vehicle, this information is usually found on a sticker somewhere on the door jamb, on the trunk lid, or perhaps inside the gas fill cover. This plate or sticker will include information related to proper tire size, wheel size, gross vehicle and axle weight ratings, and recommended tire pressures (figs. 11–5 and 11–6).

FIGURE 11–3. Take note in top image that the sidewall of this tire indicates that the maximum inflation pressure is *35 psi*. However, in the bottom image the tire loading information plate on the door sill indicates that the recommended inflation pressure is *30 psi*. Many drivers believe that the maximum tire pressure printed on the side of the tire is the recommended pressure for everyday driving. This is not the case. Always check the tire loading information to determine the proper tire inflation pressure.

Pierce

Manufactured by: Pierce Manufacturing, Inc.
- Custom Designed and Manufactured Exclusively For -
GOSHEN FIRE COMPANY

Mo./Yr of Mfgr	Oct - 2005	Job No.	17076		WO No.	5793996
GVWR 32,115 KG (70,800 LB)		Tire-Limited Max Speed	68	mph	Chassis	Dash

GAWR		TIRES	RIMS	COLD TIRE INFLATION
Front 10,342 KG (22,800 LB)		425/65R22.5 (L)	22.50x12.25	827 kPa (120 PSI) SINGLE
Rear 21,773 KG (48,000 LB)		11R22.5 (H)	22.50x8.25	827 kPa (120 PSI) DUAL

THIS VEHICLE CONFORMS TO ALL APPLICABLE U.S. FEDERAL MOTOR VEHICLE SAFETY STANDARDS IN EFFECT ON THE DATE OF MANUFACTURE SHOWN ABOVE.

VIN	4P1CD01HX5A005592	TYPE	Emergency Vehicle

CUSTOM HIGH GRADE PAINT FINISH

Red	Pierce No.	80	Sikkens Autocoat LV	FLNA3021
				FLNA4041

FIGURE 11–4. Fire apparatus will have a tire load and weight plate which is usually mounted near the driver's seat. Make sure to check this plate for tire inflation information and weight ratings. If the department changes the tires during the life of the vehicle, check with the tire manufacturer to ensure that the information has not changed.

FIGURE 11–5. Make sure to check the sidewall of the tire to ensure that the tire is properly rated for the amount of weight supported by the tire.

FIGURE 11–6. Many fire departments also overlook the weight rating on the wheel rims. The maximum weight rating should be engraved in the metal.

The vehicle manufacturer should have done all the work and provided the information on the plate inside the cab. However, fire departments must remember that if new tires are purchased and installed on the vehicle, it is important to check the tire load and inflation charts for the new tires, as the inflation pressures may be different from the tires that originally came with the vehicle. If necessary, the engineering staff should update the placard in the cab to reflect the new recommended inflation pressures.

It is also important to note that the information contained on the plate or sticker is the *cold air pressure* of the tire. Do not check the air pressure if the vehicle has been driven within the past three hours. Driving the vehicle for a length of time will cause the air in the tires to heat up and expand, providing a different air pressure reading than if the vehicle was sitting cold for several hours.

Fortunately, most firehouses have the ability to put air in a tire with on-site compressors, so there is no need to drive the apparatus to a different location to put air in the tires. Should the fire apparatus operator find themselves in a situation where they must drive somewhere to put air in the tire, it is recommended that the vehicle not be driven more than one mile so as to prevent a false air pressure reading.

Should the vehicle have to drive more than one mile for air, make sure to measure the air pressure in the tire and record the information before you leave the firehouse. Once you arrive at your destination, check the air pressure again. This will provide you with an accurate depiction of how much air you really need to put in the tire. For instance, let's say you check the air pressure at the firehouse and it is 110 psi. You know that you need to fill the tire to 120 psi, based on the tire manufacturer's recommended air pressure. After driving a few miles to your air fill station, you recheck the tire pressure and find that it is now 114 psi. This means that you will have to fill the tire to 124 psi to ensure that is properly matched to the recommended cold air pressure reading. At the firehouse, you were 10 psi underinflated.

Dual Tires

When checking tire pressures, drivers must also remember to take the time to check the inside tires on the dual tire assembly. According to a Bridgestone/Firestone study which was conducted on EMS vehicles, 39% of the inside dual tires couldn't be checked because there was no access to the valve stems. Of the tires that could be checked, two-thirds were found to be underinflated by at least 20 psi, or 25 % capacity.[5] This is extremely concerning because tire manufacturers consider any tire that is more than 20% underinflated to have been run flat. Running on a flat tire may cause damage to the tire and lead to a catastrophic blowout.

TABLE 11–2.   Bridgestone/Firestone EMS Vehicle Tire Suvey[6]

Condition	Percent
Inside Dual Over 20 psi Underinflated	33%
Outside Dual over 20 psi Underinflated	15%
Couldn't Check	39%
Checked OK	13%

When checking the air pressure on the inside dual tires, it is also important to make sure that dual tire assemblies are matched appropriately. Dual tires should differ by no more than one-quarter of an inch in diameter and three-quarters of an inch in circumference. Both tires should be of the same make and size and should be kept at the same inflation pressure. If the tires on a dual assembly are different sizes, the larger tire will drag the smaller tire. The Bridgestone/Firestone studies found that a dual tire assembly that differed by just a 5-psi inflation pressure resulted in a 5/16 of an inch difference in circumference. This minor difference in size would cause the smaller tire on the dual axle to be dragged 13 feet over just one mile.[7]

A proper check of the tire pressures should also include an inspection of the valve stem caps. Most manufacturers recommend metal valve stem caps with a strong rubber gasket to ensure that air cannot leak out through the valve stem. Keeping a few extra valve stem caps in the engineering room is a good idea. Also make sure that the valve stems on rear dual tire assemblies are easy to access and give a proper seal for the air pressure gauge.

NFPA 1901—4.13.4 requires each tire to have a visual indicator or monitoring system that indicates tire pressure. In addition to old-fashioned visual methods for checking tire air pressure, there are some excellent automatic monitoring systems now coming onto the market. These systems include a tire pressure monitor for each wheel, which can wirelessly transmit the information to a computer. I once taught at a fire department in West Virginia which had such a system. Anytime a tire pressure dropped to a predetermined PSI, the chief engineer received a text on his smart phone. These systems are excellent tools to help keep a constant eye on proper tire pressures.

Handling a Tire Blowout

Tire blowouts are common in the fire service and sometimes lead to fatalities. According to statistics provided by Michelin, tire blowouts result in 23,000 collisions and 535 fatalities each year. Most of the time, it is not the blowout that causes the crash, but the driver's improper reaction to the blowout. While NFPA 1451 states that driver's must be trained in the proper handling of a tire blowout, few fire departments provide this type of driver training.

The University of Michigan Transportation Research Institute conducted a study of fatal truck crashes caused by tire blowouts.[8] This analysis of blowout-related crashes revealed some interesting crash outcomes based on which tire blew out. The results of this study are summarized below:

> *In general, the fatal truck crashes precipitated by a blowout can be divided into the following three scenarios:*
>
> 1. *If the front left tire blows, the truck loses control to the left, veers into oncoming or adjacent traffic, and rolls immediately or after a collision with another vehicle. These crashes are primarily multiple-vehicle crashes; 15 of the 22 involved two or more vehicles.*

> 2. *If the right front tire blows, the truck loses control to the right, veers off the road and rolls or collides with roadside structures, or both. These are typically single-vehicle crashes. Ten of 13 right front tire blowouts were single-vehicle crashes.*
>
> 3. *If a drive or trailer axle blows, the truck typically, though not always, remains under control. [Note: In some cases, the crash is entirely unrelated to the flat tire. For example, in one case, the truck was rear-ended by an alcohol-impaired driver. The truck driver said he was driving normally and was on his way to get the flat repaired, remarking that "one tire being flat did not slow the truck down that much.]*

This study addressed only crashes that resulted in fatalities. It is unknown how many other crashes occurred that did not result in a fatality.

The proper handling of a tire blowout is almost counterintuitive to what might be expected. Most drivers hear a loud bang and the vehicle starts to shake violently and pull to one side. The natural response is to immediately engage the brakes and try to get to the side of the road. This response could mean disaster.

According to studies and demonstrations performed by tire manufacturers, the proper response to a tire blowout is to *accelerate* and keep the vehicle in a straight line. If the driver accelerates, they will regain forward momentum which will help bring the vehicle back under control. Once the vehicle is back under control, the driver can then gently decrease speed and come to a controlled stop.

Recovering from a tire blowout requires the driver to give the vehicle forward acceleration. Because of this, it is important to remember to not drive the vehicle at the governed speed. If the vehicle is governed at 65 mph and the driver is traveling at the governed speed, there is no room to accelerate if a tire blowout should occur. Always make sure to travel at a speed that is reasonable and prudent for the existing conditions and never run the vehicle against the governed speed. Running against the governed speed will leave no room for error should a tire blowout occur.

While hands-on training for tire blowouts may be difficult, there are ways to simulate these events. Have the driver find a safe, open space to bring the vehicle up to a typical road speed. Once the vehicle reaches a realistic speed, the driver training instructor can do something to simulate a tire blowout: yell, blow the air horns, or pop a balloon. When the simulated tire blowout occurs, ensure that the driver reacts appropriately by accelerating, regaining control and then gently bringing the vehicle to a safe stop.

Inspecting and Reading a Tire

NFPA 1911 states that a fire service tire should be replaced when the tread wear exceeds state or federal standards.[9] Due to a limited number of calls, many fire apparatus do not wear down the tire tread to a point that requires replacement. As there is no outward indication that the tires are being worn or broken down, many fire departments never replace the tires over the entire life of the vehicle. This is a dangerous practice, as advanced age and poor condition can lead to a tire blowout.

Tires degrade just by sitting in the firehouse. Each tire sits stationary for weeks at a time, waiting for the next run. Over time, the rubber tire begins to age and crack, just like a rubber band sitting in a desk. The rubber slowly ages and dries out, all while supporting the weight of a fire truck. Often, the largest vehicles that are used the least are the vehicles that have the oldest tires.

For these reasons, the NFPA 1911 standard further states that the maximum life of a tire shall be no more than 7 years. While there is controversy over the necessity and cost effectiveness of replacing tires every 7 years, most tire manufacturers recommend tire replacement every 6–10 years. The replacement schedule of a tire will depend on the climate, use, and wear on the tires. Because the condition of the tires is difficult to evaluate without destroying them, fire departments are placed in a tough situation when it comes time to decide whether or not to replace vehicle tires. Regardless, fire apparatus operators must make it a point to check the manufacture date of the tire during routine inspections. The manufacture date will be printed on the side of the tire as a four digit code (fig. 11–7). The first two digits are the week of the year and the second two digits are the year the tire was manufactured.

In addition to the age of a tire, the maintenance of a tire will have a significant impact on its lifespan. Factors such as tire pressure, tread wear, sidewall weathering, wheel alignment, tire balance, tire rotation, and tire load rating will dramatically affect the condition of a tire over its lifetime. The way in which a tire is maintained and cared for has a dramatic effect on vehicle safety, as well as on the tire replacement plan for the entire fire department.[10]

Daily or weekly truck checks must include a comprehensive evaluation of each tire, including the inside duals. The tires must be examined for cracks, bulges, uneven tread wear, or other indications of subtle problems that are not readily apparent as the vehicle is driving on the road. This type of inspection is especially important for fire departments that don't replace tires every 7 years as required by NFPA. Countless fire apparatus crashes have been caused by old or brittle tires.

Fire apparatus operators must be able to read or decipher the information printed on the sidewall of a tire. The following information is provided in the

FIGURE 11–7.   The date of manufacture is printed on the side of the tire as a four digit code. The first two digits are the week of the year and the next two digits are the year the tire was manufactured. In this case, the tire was made during the 46th week of 2015.

National Highway Traffic Safety Association publication "Tire Safety: Everything Rides on It."[11] It should be noted that this information pertains to car and light truck tires. Check with your tire dealer for information on heavy truck tires.

P

The "P" indicates the tire is for passenger vehicles. "LT" would indicate a light truck tire.

Next Number

This three-digit number gives the width in millimeters of the tire from sidewall edge to sidewall edge. In general, the larger the number, the wider the tire.

Next Number

This two-digit number, known as the aspect ratio, gives the tire's ratio of height to width. Numbers of 70 or lower indicate a short sidewall for improved steering response and better overall handling on dry pavement.

R

The "R" stands for radial. Radial ply construction of tires has been the industry standard for the past 20 years.

Next Number

This two-digit number is the wheel or rim diameter in inches. If you change your wheel size, you will have to purchase new tires to match the new wheel diameter.

Next Number

This two- or three-digit number is the tire's load index. It is a measurement of how much weight each tire can support. You may find this information in your owner's manual. If not, contact a local tire dealer. Note: you may not find this information on all tires because it is not required by law.

M+S

The "M+S" or "M/S" indicates that the tire has some mud and snow capability. Most radial tires have these markings; hence, they have some mud and snow capability.

Speed Rating

The speed rating denotes the speed at which a tire is designed to be driven for extended periods of time. The ratings range from 99 miles per hour (mph) to 186 mph. These ratings are listed below. Note: You may not find this information on all tires because it is not required by law.

Q	99 mph	H	130 mph
R	106 mph	V	149 mph
S	112 mph	W	168 mph*
T	118 mph	Y	186 mph*
U	124 mph		

*For tires with a maximum speed capability over 149 mph, tire manufacturers sometimes use the letters ZR. For those with a maximum speed capability over 186 mph, tire manufacturers always use the letters ZR.

U.S. DOT Tire Identification Number

This begins with the letters "DOT" and indicates that the tire meets all federal standards. The next two numbers or letters are the plant code where it was manufactured, and the last four numbers represent the week and year the tire was built. For example, the numbers 3197 means the 31st week of 1997. The other numbers are marketing codes used at the manufacturer's discretion. This information is used to contact consumers if a tire defect requires a recall.

Tire Ply Composition and Materials Used

The number of plies indicates the number of layers of rubber-coated fabric in the tire. In general, the greater the number of plies, the more weight a tire can support. Tire manufacturers also must indicate the materials in the tire, which include steel, nylon, polyester, and others.

Maximum Load Rating

This number indicates the maximum load in kilograms and pounds that can be carried by the tire.

Maximum Permissible Inflation Pressure

This number is the greatest amount of air pressure that should ever be put in the tire under normal driving conditions.

UTQGS Information—Tread Wear Number

This number indicates the tire's wear rate. The higher the tread wear number is, the longer it should take for the tread to wear down. For example, a tire graded 400 should last twice as long as a tire graded 200.

Traction Letter

This letter indicates a tire's ability to stop on wet pavement. A higher graded tire should allow you to stop your car on wet roads in a shorter distance than a tire with a lower grade. Traction is graded from highest to lowest as "AA","A", "B", and "C."

Temperature Letter

This letter indicates a tire's resistance to heat. The temperature grade is for a tire that is inflated properly and not overloaded. Excessive speed, underinflation or excessive loading, either separately or in combination, can cause heat build-up and possible tire failure. From highest to lowest, a tire's resistance to heat is graded as "A", "B", or "C."

In the case of a commercial motor vehicle tire, the information found on the tire sidewall is different from that found on the sidewall of a passenger car or light truck tire.

Snow Tires

For most drivers, switching to snow tires seems like a hassle, but it is an excellent way to increase safety during winter driving. While all-season tires can handle an occasional light snow, they are not designed to handle the freezing

FIGURE 11–8. A true snow tire will have this symbol imprinted on the sidewall. This symbol is in addition to the "M+S" imprint.

temperatures and deep snow seen in most of the country. Even if all-season tires are marked on the sidewall with an "M+S," indicating that the tire can be used in snow, it does not mean that the tire will handle very well in deep snow.[12] Instead you must look for the special symbol created by the rubber manufacturer's association which is depicted in Figure 11–8.

Snow tires are made of a different rubber compound that is more pliable in cold weather. While the rubber in an all-season tire may harden in cold weather, snow tires will remain soft to better grip the road. Many experts recommend owning a set of snow tires in regions where the winter temperature regularly drops below 45 degrees.[13] Snow tires also have a more aggressive tread pattern which is deeper and wider than an all-season tire. This is because snow and slush may clog the tread on an all-season tire, turning the tires into skis. The larger tread of a snow tire will allow the tire to expel the snow and slush to give the tire a better grip as it rotates around the axle and back down onto the snow covered road.

It is important to equip the vehicle with four snow tires instead of just two. Equipping the vehicle with only two snow tires could affect the handling of the vehicle as one part of the car will grip the road better than the other. If one part of the vehicle maintains better traction than another, it may induce a spinout, or loss of control.

Tire Chains

For large fire apparatus, changing tires during a snow storm is not feasible. Instead, most fire departments equip each vehicle with tire chains. There are different chain systems, including cable chains, traditional chains, and automatic chains. Regardless of what chain system the fire department uses, it is important to understand the proper operation of the vehicle when chains are in use. Most chain manufacturers recommend keeping the vehicle under 25 mph.

Fire apparatus operators must be familiar with proper chain installation procedures or how to deploy the automatic tire chains as the vehicle leaves the firehouse. This is the type of skill that should be honed and practiced before a snow storm strikes. Trying to figure out how to use a traction control device as a storm is hitting will be too late. While some may roll their eyes at a summer-time drill on installing or deploying traction control devices, it's much better to learn in the warm weather than to fight with the traction system during a fierce snow storm.

Notes

1. NFPA 1911—Section 6.3.1(4)
2. "How Safe Are Worn Tires?" *Consumer Reports*, retrieved from http://www.consumerreports.org/cro/cars/tires/dangers-of-worn-tires-204/overview/index.htm. Date accessed: Oct. 17, 2006.
3. "NPRM on Tire Pressure Monitoring System FMVSS No. 138," U.S. Department of Transportation, Office of Regulatory Analysis and Evaluation Planning, Evaluation and Budget, Sept. 2004.
4. "Ready to Roll: The Shocking Truth!" Bridgestone/Firestone, publication BF50919, July 2001.
5. "Ready to Roll: The Shocking Truth!" Bridgestone/Firestone, publication BF50919, July 2001.
6. "Ready to Roll: The Shocking Truth!" Bridgestone/Firestone.
7. "Preventing on the Road Tire Failures," Tire Retread Information Bureau, retrieved from http://www.retread.org/Inflation/index.cfm/ID/288.htm. Date accessed: Nov. 29, 2006.
8. Z. Bareket, D.F. Blower, and C. MacAdam, "Blowout Resistant Tire Study for Commercial Highway Vehicles," The University of Michigan Transportation Research Institute, Aug. 31, 2000.
9. NFPA 1911 Section 7.3.4
10. Sreenivasan Ranganathan and Minchao Yin. "Automotive Fire Apparatus Tire Replacement." Fire Protection Research Foundation, March 2015.
11. "Everything Rides On It Tire Safety." NHTSA. Web http://www.nhtsa.gov/cars/rules/tiresafety/ridesonit/images/brochure.pdf.
12. TireBuyer. "Everything you need to know about snow tires". Web. http://www.tirebuyer.com/education/all-season-tires-vs-winter-tires#.VYu8REb3jrY.
13. TireBuyer. "Everything you need to know about snow tires". Web. http://www.tirebuyer.com/education/all-season-tires-vs-winter-tires#.VYu8REb3jrY.

Drunk and Impaired Driving

Introduction

Many will deny that impaired driving is an issue in the fire service. However, alcohol, illegal drugs, and prescription medication have resulted in numerous fire apparatus crashes. Some of these crashes have resulted in serious injuries and fatalities to both firefighters and civilians.

Drugs and alcohol will have a detrimental effect on a person's coordination, balance, and ability to follow directions. Drugs and alcohol can also affect a person's ability to multitask, operate machinery, and think clearly. Ingesting large amounts of drugs or alcohol can result in double vision, drowsiness, unconsciousness, and even death. When a person ingests drugs or alcohol and then gets behind the wheel of a vehicle, these effects can be compounded, resulting in a serious crash.

Impaired Driving Defined

Different states refer to impaired driving using different terminology, such as the following:

1.   Driving while impaired (DWI)
2.   Driving while intoxicated (DWI)
3.   Driving under the influence (DUI)

Regardless of the terminology, the legal limit for operating a vehicle under the influence of alcohol is the same in every state. Laws prohibit drivers from operating a non-commercial motor vehicle if their blood alcohol concentration (BAC) is greater than 0.08%. A driver who is operating a commercial motor vehicle, defined as having a gross vehicle weight rating in excess of 26,001 pounds, may not operate a vehicle with a BAC greater than 0.04%. Keep in mind that most front-line fire apparatus weigh more than 26,001 pounds. Therefore, the legal limit for operating a larger fire apparatus is 0.04%.

A blood alcohol concentration greater than 0.04% (commercial motor vehicle) or 0.08% (non-commercial vehicle) is considered a "per se" rule, which means that if a driver is over this limit, they can be criminally charged regardless of the circumstances. It should also be noted that many states have laws that address lower BACs. These laws typically address the issue of general impairment. In these situations, if it can be demonstrated that a crash or unsafe driving was caused by the ingestion of alcohol, illegal narcotics, or prescription medication, charges can still be brought against a motor vehicle operator, even with a BAC below the per se rule of 0.04% or 0.08%.

Drivers under 21 years of age are prohibited from driving with any alcohol in their system. Most states set the per se rule for underage drunk driving at 0.02% BAC. However, even if the operator's BAC falls under 0.02%, they can still be charged with other offenses if it can be proven that the driver consumed alcohol. These laws are commonly referred to as "minors operating with alcohol in their system."

Impaired driving is defined as "driving a vehicle while under the influence of alcohol or a controlled substance to a point which renders a person unfit to safely operate a motor vehicle." While many people immediately associate DUI with alcohol, it also includes illegal narcotics and prescription medication.

Most people are aware that they shouldn't ingest alcohol or drugs before getting behind the wheel of a vehicle. What many people do not realize is the impact prescription medication can have on the safe operation of a motor vehicle. If you are taking a medication that was prescribed by a doctor, and the side of the bottles says that you shouldn't operate a vehicle or heavy machinery—don't! If you are involved in a crash and blood tests reveal a prescription medication or illegal narcotic that can affect your ability to drive, you will face legal sanctions.

Understanding Blood Alcohol Concentration

A person's blood alcohol concentration (BAC) will rise based on the quantity and type of alcohol that they ingest. The rise in BAC also depends on other factors, such as gender, body type, and body weight. Due to differences in human physiology and body composition, a 200-lb male will see a different rise in BAC than a 100-lb female. Figure 12–1 demonstrates the approximate rise in BAC based on weight and gender. It should be noted that this figure does not include differences in the strength of the alcoholic drink being ingested.

APPROXIMATE BLOOD ALCOHOL CONCENTRATION - *FEMALE*									
BODY WEIGHT IN POUNDS									
DRINKS	90	100	120	140	160	180	200	220	240
0	0.00	0.00	0.00	0.00	0.00	0.00	0.00	0.00	0.00	Only Safe Driving Limit
1	0.05	0.05	0.04	0.03	0.03	0.03	0.02	0.02	0.02	Impairment Begins
2	0.10	0.09	0.08	0.07	0.06	0.05	0.05	0.04	0.04	Driving Skills Affected
3	0.15	0.14	0.11	0.10	0.09	0.08	0.07	0.06	0.06	Possible Criminal Penalties
4	0.20	0.18	0.15	0.13	0.11	0.10	0.09	0.08	0.08	
5	0.25	0.23	0.19	0.16	0.14	0.13	0.11	0.10	0.09	
6	0.30	0.27	0.23	0.19	0.17	0.15	0.14	0.12	0.11	
7	0.35	0.32	0.27	0.23	0.20	0.18	0.16	0.14	0.13	LEGALLY INTOXICATED
8	0.40	0.36	0.30	0.26	0.23	0.20	0.18	0.17	0.15	
9	0.45	0.41	0.34	0.29	0.26	0.23	0.20	0.19	0.17	
10	0.51	0.45	0.38	0.32	0.28	0.25	0.23	0.21	0.19	

APPROXIMATE BLOOD ALCOHOL CONCENTRATION - *MALE*								
BODY WEIGHT IN POUNDS								
DRINKS	100	120	140	160	180	200	220	240
0	0.00	0.00	0.00	0.00	0.00	0.00	0.00	0.00	Only Safe Driving Limit
1	0.04	0.03	0.03	0.02	0.02	0.02	0.02	0.02	Impairment Begins
2	0.08	0.06	0.05	0.05	0.04	0.04	0.03	0.03	
3	0.11	0.09	0.08	0.07	0.06	0.06	0.05	0.05	Driving Skills Affected
4	0.15	0.12	0.11	0.09	0.08	0.08	0.07	0.06	Possible Criminal Penalties
5	0.19	0.16	0.13	0.12	0.11	0.09	0.09	0.08	
6	0.23	0.19	0.16	0.14	0.13	0.11	0.10	0.09	
7	0.26	0.22	0.19	0.16	0.15	0.13	0.12	0.11	
8	0.30	0.25	0.21	0.19	0.17	0.15	0.14	0.13	LEGALLY INTOXICATED
9	0.34	0.28	0.24	0.21	0.19	0.17	0.15	0.14	
10	0.38	0.31	0.27	0.23	0.21	0.19	0.17	0.16	

FIGURE 12–1. This chart demonstrates the approximate rise in blood alcohol concentration based on gender, weight, and the amount of alcohol consumed. While there are many variables that may affect the values in this chart, it provides a good guideline for training and awareness.

As we consume alcohol and our BAC rises accordingly, our body is trying to burn off the alcohol by metabolizing it. Studies indicate that an average person will metabolize alcohol at a rate of approximately 0.015% per hour. Note that this rate is for the average person and may differ depending on the person and the situation.

Understanding how the body metabolizes alcohol, as well as how long it takes to metabolize this alcohol, is an important teaching point for a driver training program. The hazards associated with a late night out, followed by an early morning fire run, are demonstrated in the following scenario:

Let's assume that it is a Friday night and you and your colleagues head out for happy hour. After a long night of partying, you return home at 0200 hours with a BAC of 0.25%. You are supposed to head to work at 0700 hours, so you lay down to "sleep it off."

After waking up, you head to work and begin your shift. At 0800 hours, the tones drop and you jump behind the wheel. While en route to the call, you find yourself involved in a minor crash. During the course of the crash investigation, the police officer can't help but notice the obvious odor of an alcoholic beverage about your person. When he questions you about it, you explain that you were out late the night before, but had plenty of time to "sleep it off."

Despite this explanation, the police officer has no choice but to detain you and take you for a blood test. More than likely, you will have to submit to a blood test anyway because it is required by your fire department anytime there is a crash. A few days later, you are notified that you blood alcohol concentration at the time of the crash was 0.16% and drunk driving charges are pending. "But wait," you ask, "how can this be?"

Fire apparatus operators must understand that their body needs time to remove alcohol from their system. Sleep, coffee, cold showers, or eating greasy foods do not help to lower a person's BAC any quicker. These methods are myths and do nothing to speed-up the process. Time is the only thing that will allow your body to do its job and remove the alcohol from your blood, returning your BAC to a safe level.

This fact is demonstrated in table 12–1, which shows how long it will take an **average** person's body to completely remove alcohol. Considering most fire apparatus weigh more than 26,001 pounds, and the per se BAC limit is 0.04%, it would be late into the next afternoon before a driver is once again fit for duty. In these situations, it is best for a career firefighter to call out on a personal day, or for a volunteer firefighter to rollover and reset the pager.

TABLE 12–1.   Example of possible blood alcohol concentration after a late night of drinking. Notice how long it takes a person to burn off the alcohol to a point where it is safe to drive again.

Time	Blood Alcohol Concentration*
0200 hours	0.25%
0800 hours	0.16%
1500 hours	0.05%

*Assuming you "burn off" the alcohol at 0.015% per hour

Prescription and Over-the-Counter Medications

Impaired driving includes more than just alcohol and illegal drugs. Many prescription medications can also have a negative effect on a person's ability to safely operate a motor vehicle. Pain relievers, muscle relaxers, and other controlled substances can cause side effects such as fatigue, dizziness, and impaired cognitive function. Apparatus operators who are prescribed these medications must have an in-depth discussion with their medical doctor and station supervisor to ensure that the use of these medications will not put other people at risk.

Even if a medication is properly prescribed, legal sanctions may result if a person is found to be operating a motor vehicle with a substance in the blood stream that impairs the ability to drive. While no one wants to use sick time or be out of work for an extended period of time, it is safer to remove yourself from duty than risk a crash as a result of prescription medication.

NIOSH, IAFC, and NFPA Recommendations

A review of the National Institute of Occupational Safety and Health (NIOSH) Firefighter Fatality Reports reveals several fatal crashes that were the result of fire apparatus operators who were impaired by alcohol or drugs (F2003-20 and F2008-22). As a result of these incidents, NIOSH has made the following recommendations related to drug and alcohol use by fire department members:

Adopt the International Association of Fire Chief's (IAFC) Zero-Tolerance Policy for Alcohol and Drinking to prohibit the use of alcohol by members of any fire or emergency services agency/organization at any time when they may be called upon to act or respond as a member of those departments. Departments should develop written policies and have procedures in place to enforce this policy.

In addition to the zero tolerance policy established by the IAFC, the issue of alcohol and motor vehicle operation is specifically addressed in the "Guide to IAFC Model Policies and Procedures for Emergency Vehicle Safety." In this document, the IAFC states the following:

Fire department members are not permitted to be on duty, to respond to emergency incidents, to drive or operate fire department vehicles, or to perform any other duty-related functions while under the influence of alcohol or drugs.

Fire department members shall not perform any duty-related functions for a minimum of eight (8) hours following the consumption of any alcoholic beverages. A longer waiting period may be required to ensure that the individual is free of impairment. A blood alcohol concentration of 0.02% or higher, while on duty, shall create the presumption that the member is under the influence of alcohol.

> *The driver and the officer in charge of any fire department vehicle that is involved in an accident that causes measurable property damage, injury or death shall be tested for the presence of alcohol or drugs with the least possible delay. In addition, a chief officer may require a member to be tested for the presence of drugs or alcohol at any time, upon reasonable suspicion that the member could be under the influence of such substances.*

IAFC Zero-Tolerance for Alcohol & Drinking in the Fire & Emergency Service

A Leadership Policy Statement from the International Association of Fire Chiefs (Adopted by the Board of Directors, August 14, 2003)

This policy statement is most easily described as a "zero-tolerance" standard about the use of alcohol by members of any fire or emergency services agency/organization at any time when they may be called upon to act or respond as a member of those departments.

Basically, if someone has consumed alcohol within the previous eight (8) hours, or is still noticeably impaired by alcohol consumed previous to the eight (8) hours, they must voluntarily remove themselves from the activities and functions of the fire or emergency services agency/organization, including all emergency operations and training.

No member of a fire & emergency services agency/organization shall participate in any aspect of the organization and operation of the fire or emergency agency/organization under the influence of alcohol, including but not limited to any fire and emergency operations, fire-police, training, etc.

No alcohol shall be on the premises of any operational portion of the fire department including, but not limited to, the apparatus, the apparatus floor, the station living areas, etc.

Fire & emergency services agencies/organizations which raise funds by operating and/or renting social halls must provide a clear and distinct separation of facilities to help insure the zero-tolerance standard of alcohol consumption by their members who may be called upon to perform official duties.

Take note that while the IAFC recommendations make reference to "eight hours following the consumption of any alcoholic beverages," this chapter has discussed how long it can take a person's body to remove alcohol or controlled substances from the bloodstream. If a person consumes a large amount of alcohol or illegal or prescription drugs, it could take much longer than eight hours for these substances to completely clear a person's blood stream.

NFPA also addresses the use of alcohol and controlled substances in NFPA 1500—"Fire Department Occupational Safety and Health Program," which states the following:

> *NFPA 1500, Section 10.1.5—Members who are under the influence of alcohol or drugs shall not participate in any fire department operations or other duties.*

> *NFPA 1500, Section A10.1.5—If any member, either career or volunteer, reports for duty under the influence of alcohol or drugs, or any other substance that impairs the member's mental or physical capacity, this situation cannot be tolerated. Evidence of substance abuse could include a combination of various factors such as slurred speech, red eyes, dilated pupils, incoherence, unsteadiness on feet, smell of alcohol or marijuana emanating from the member's body, inability to carry on a rational conversation, increased carelessness, erratic behavior, inability to perform a job, or other unexplained behavioral changes. The possibility of liability exists if a member who is under the influence of alcohol or drugs is allowed to remain on duty, to operate or drive vehicles or equipment on duty, or to drive a private vehicle from the duty site. A member who is believed to be under the influence of alcohol or drugs cannot be allowed to operate equipment or drive a vehicle, including a private vehicle, until the condition of the member has been determined and verified.*

NFPA also addresses drug and alcohol use specific to apparatus operations in the NFPA 1451—"Standard for a Fire and Emergency Service Vehicle Operations Training Program." This standard addresses post-crash blood testing and states the following:

> *4.1.5—The FESO [fire and emergency service organization] shall institute a program of post-crash drug and alcohol testing for the drivers of vehicles involved in crashes.*

A4.1.5—The program should include, as a minimum, drug and alcohol testing in any crash involving an injury or fatality. Additional testing criteria should be considered for all incidents where property damage has occurred or when FESO vehicles sustain a set minimum of cost to repair. The department should also consult with risk management resources and consider state law requirements in setting program criteria.

While some would argue that these nationally recognized standards are only recommendations, these are the standards that a driver and a fire department will be judged against should they find themselves involved in a crash that involves drugs or alcohol. Detailed policies must be written and enforced throughout the entire fire department to prevent impaired driving. In the event that impaired driving does occur, contingency plans must be in place to properly deal with the offender, as well as provide them with a safe way home and long-term rehabilitation.

Supervisor Considerations

Few scenarios are more difficult for a supervisor than the need to discipline a crewmember. The nature of a firehouse creates a tight-knit environment where almost everyone looks out for each other. It is important to understand that looking out for each other does not always mean turning a blind eye to a problem. This is especially true when dealing with drug and alcohol abuse, as these issues can be career and life altering due to their severity.

Supervisors should be trained to properly recognize and identify the signs and symptoms of drug and alcohol abuse. These symptoms could be as obvious as detecting the odor of an alcoholic beverage or controlled substance in the proximity of a member. These symptoms could also include dilated or pinpoint pupils, staggered gait or walk, difficulty remembering or following directions, and fatigue. As most firefighters have some level of training in emergency medicine, it should be obvious when a member arrives at the firehouse in an impaired state.

Not only should supervisors be trained to recognize potential substance abuse problems, they must also be thoroughly trained on how to handle a situation should it arise. What is the required reasonable suspicion for a supervisor to initiate an investigation? At what point can a supervisor remove a member from duty and request a blood test? What is the policy and procedure for getting an impaired member home? These are all questions that must be specifically addressed in a policy and explained to department members **before** an issue arises. Shift change on a Saturday morning is not the time to try and figure out what to do, or who to call, because there is a problem.

These issues relate to volunteer fire departments as well as career departments. In fact, the situation may be much more complex in a volunteer department, which may or may not have a substance abuse policy or procedure in place. Volunteer departments may not have any disciplinary recourse should a member refuse to cooperate with an investigation. While a career firefighter usually shows up at shift change and is readily observed by others in the firehouse, what happens when the volunteer firefighter runs in the back door of the firehouse at 3:00 am? Does this firefighter grab their gear, jump in the driver's seat, and go? What happens when the odor of an alcoholic beverage suddenly wafts across the doghouse to be noticed by others in the apparatus? All of these questions must be answered before the situation occurs; not in the middle of a potential disaster.

A person who shows up at the firehouse under the influence of drugs or alcohol needs help. This is especially true in career departments where the member should have been aware that they were scheduled to work. To risk a career, benefits, and a pension by showing up to work while under the influence of alcohol or a controlled substance is a clear sign that something is amiss. Even in the volunteer world, members should be aware of the ramifications of responding to a call while under the influence of alcohol or a controlled substance. This type of behavior shows a clear lack of judgment.

Looking out for each other does not mean turning a blind eye. Looking out for each other means having the courage to stand up and address the problem. This type of courage will require a supervisor to immediately initiate disciplinary action and prevent an impaired member from hurting himself, other members of the department, or the public we are sworn to protect.

Looking out for each other also means providing the proper support and treatment for a member in need of assistance. In large departments, this may mean putting the member in contact with an employee assistance program. For smaller departments or volunteer departments, it may mean scheduling an intervention or providing the moral support to get the affected member into a treatment program. In the long-term, it is more beneficial to address the problem head-on, rather than look the other way.

Case Study—Wyoming

Turning a blind eye to a member who is under the influence of drugs or alcohol is in no way helping that member. In fact, it may cause death or serious injury to the member, other fire department members, or someone in the general public.

This fact was proven during an apparatus crash in Wyoming. The details of this crash are provided in NIOSH Firefighter Fatality Report F2003-20.

According to the NIOSH report, the local fire department was dispatched for a report of a brush fire at approximately 2219 hours. The driver of the water tanker involved in the crash had left a local drinking establishment approximately 15 minutes before the dispatch. The driver drove his personal vehicle to the firehouse, arrived at the firehouse, and asked another firefighter to start the truck for him. The second firefighter was aware that the tanker driver was forbidden to operate emergency vehicles for the department, so he ignored this request.

The tanker driver then started the truck himself. As the driver got ready to leave the station, another firefighter got in the truck but then changed his mind and got off the truck. A third firefighter got in the truck to respond to the call. At no point did anyone prevent the driver from operating the tanker, despite knowing that he was not allowed to drive.

As the tanker was preparing to leave the station, a 16-year-old firefighter asked if she could ride in the tanker. The firefighter in the tanker agreed to give up his seat, and the victim firefighter boarded the rig. The tanker left the station and responded to the call.

While driving along a graded gravel roadway, the tanker driver allegedly swerved to miss an antelope. When the driver swerved, the tanker left the right side of the roadway, causing the driver to overcorrect. The vehicle entered a skid and then rolled over nearly two times before coming to rest 50 ft from the roadway. The victim firefighter was partially ejected and trapped beneath the tanker.

While at the scene of the crash, a highway patrol officer administered a portable breath test to the tanker driver. The portable breath test revealed that the driver's blood alcohol concentration was nearly twice the legal limit. When blood was drawn 2½ hours later, the driver's BAC was 0.086%. The driver was arrested for DUI.

Sadly, no one prevented the driver from operating the tanker despite the fact that other firefighters knew he was suspended from operating emergency vehicles. Regardless of whether anyone noticed that the driver was intoxicated or smelled an alcoholic beverage about his person, the simple fact that he was suspended from driving should have resulted in someone telling him he was not allowed to drive. In this case, turning a blind eye had tragic results.

In another incident at a major urban fire department, the intoxicated driver of a ladder truck allegedly drove through a steady red light, striking and seriously injuring a motorcyclist. The other crew members on the apparatus brought the driver to a nearby bar and gave him water to drink. A subsequent investigation not only revealed that the driver had nearly four times the legal limit of alcohol in his system at the time of the crash, but also the crew members' attempts to conceal the drunk driving. As a result, not only did the driver of the apparatus find himself under investigation and face charges, but so did the other crew members on the rig.

Operating a vehicle while under the influence of alcohol or a controlled substance is a guaranteed way to face legal and disciplinary sanctions. The criminal and civil liabilities associated with this type of behavior cannot be underestimated. Fire departments must have clear policies in place that address how to recognize and deal with this type of behavior. There is no excuse for impaired driving within the fire service.

Overweight, Oversized, and Modified Vehicles

Introduction

In many areas of the United States, fire departments are often under-funded and poorly equipped. These fire departments are forced to use obsolete fire apparatus or modified vehicles that were not originally designed to fight fires. Modified vehicles are often refurbished in-house and do not meet current NFPA safety specifications. To further compound the problem, modified vehicles are often placed in service as water supply apparatus. When the inherent dangers of a fire apparatus are combined with modified or out of date equipment, tragedy often results.

The most common examples of modified fire apparatus include military surplus vehicles or liquid transport vehicles such as petroleum tankers or milk trucks. While these vehicles may appear to provide a cost effective means of delivering fire department services, this is often not the case.

Because modified vehicles were not originally designed for the fire service, they often lack safety components that are important for the dangerous job of fire suppression. Furthermore, these vehicles are often retired from other agencies and are at the end of their useful lives. Subsequent aftermarket modifications often lead to overweight and unstable vehicles that are prone to crash.

Overweight Issues

Fire departments will often take delivery of a retired civilian vehicle and simply fill it with water. If this is done without examining the manufacturer's specifications and making appropriate modifications to the suspension, tires, and related vehicle components, a dangerous overweight condition may arise. This is because many modified water tankers were first used as petroleum transportation vehicles, carrying products such as gasoline, fuel oil, and jet fuel, which are lighter than water. If the vehicle was designed to carry 3,000 gallons of gasoline and is later filled with 3,000 gallons of water, the vehicle may become dangerously overweight (table 13–1).

The difference between 4,000 gallons of gasoline and 4,000 gallons of water is approximately 8,800 lbs. If the vehicle was already near its maximum load carrying capacity as a gasoline tanker, it will certainly be overweight when filled with the same amount of water.

The "Deuce-and-a-half" truck also provides a common example of this scenario. Many fire departments use the "Deuce-and-a-half" because they are easily obtained through military surplus dealers. While these vehicles can provide an inexpensive alternative to a brand-new custom truck, it is important to understand that they are called a "Deuce-and-a-half" for a reason. According to military specifications, the "Deuce-and-a-half" is designed to carry 2½ tons (5,000 lbs) off-road and five tons (10,000 lbs) on a paved road.

Many fire departments purchase a surplus "Deuce-and-a-half" and equip it with a large water tank. The truck is then used as a tanker or off-road brush vehicle. If the vehicle is equipped with anything greater than a 1,200 gallon tank, it will be overloaded. Keep in mind that if the vehicle is equipped with a 1,200 gallon tank, it will be at the very limit of its weight-carrying ability. Adding a pump, hose or other firefighting equipment will cause the vehicle to exceed its load carrying abilities. If the vehicle is going to be used off-road, it should carry no more than 600 gallons of water.

TABLE 13–1.   This table shows the difference in weight for different types of liquid products. 4,000 gallons of water weighs considerably more than 4,000 gallons of gasoline. If a fire department purchases a used gasoline tanker and simply fills it with water, the vehicle may be dangerously overweight.

Product	One Gallon Weight	(1000) Gallons	(2000) Gallons	(3000) Gallons	(4000) Gallons
Gasoline	6.1 lbs	6,100 lbs	12,200 lbs	18,300 lbs	24,400 lbs
Fuel Oil	7.2 lbs	7,200 lbs	14,400 lbs	21,600 lbs	28,800 lbs
Water	8.3 lbs	8,300 lbs	16,600 lbs	24,900 lbs	33,200 lbs

Brake Fade

Overweight vehicles, whether modified or brand-new, are dangerous for many reasons. As discussed in previous chapters, an overweight vehicle will have more kinetic energy than the brake system was designed to dissipate. This excess kinetic energy will cause the brakes to overheat and may result in a brake fade situation. When brake fade occurs, the vehicle will lose much of its braking efficiency.

Skids and Handling

In addition to brake fade, an overweight vehicle may have significant handling issues. This is because the vehicle's weight distribution has been altered, resulting in a brake imbalance which may cause the vehicle to pull to one side. A brake imbalance is dangerous as it may cause the tires to lock up during panic stops or evasive maneuvers, especially during poor weather conditions. Depending on which tires lock-up, the driver may experience a loss of control and crash the vehicle.

Tank Baffles

As a fire apparatus rounds a curve or makes an evasive maneuver, the water in the tank will want to continue moving in a straight line. The energy and inertia of the sloshing water will push the vehicle along its original heading, fighting against the driver who is trying to steer the vehicle elsewhere. Depending on the severity of this liquid surge, the fire apparatus may lose control or rollover.

In order to dampen the energy of liquid surge, baffles are installed in the water tanks of fire apparatus. The National Fire Protection Association (NFPA) 1901—18.2.6, Standard for Automotive Fire Apparatus, 2016 edition states "All water tanks shall be provided with baffles or swash partitions to form containment cells or dynamic water movement control."

Despite this requirement, it is not unusual for a fire department to take delivery of a used or modified vehicle that is not equipped with a proper baffling system. This is especially true in vehicles that served as fuel trucks or milk trucks. Fuel trucks are equipped with simple compartments designed to carry different types of fuel. Milk trucks have no baffles at all, as it is too difficult to sanitize a baffled tank. Prior to placing any modified vehicle in-service, there must be a thorough inspection of the water tank to ensure that a proper baffling system is in place.

Tire Blowouts

Many modified vehicles are placed in service without replacing or inspecting the tires. As a result, the vehicle is operated with old and degraded tires that are prone to tire blowouts. Combine old and degraded tires with an overweight vehicle and a catastrophic tire failure is almost certain to occur.

When a vehicle is retired from its original profession and then placed in service as a fire truck, it is important to reassess the weight and load ratings of the tire. This process should include weighing the vehicle at each tire to determine the proper air pressure as indicated in the manufacturer's tire load and inflation chart. Properly weighing the vehicle in this manner will ensure that fire service modifications did not exceed the load carrying ability of the existing tires.

Case Study—Utah

On June 21, 2005, these issues were highlighted during a fatal tanker crash in Utah. A volunteer firefighter was driving a 1981 2 ½-ton military surplus tanker truck to a neighboring town to have the vehicle inspected. As the firefighter drove down a straight gravel road, the front driver's side tire blew out, causing the driver to lose control. The driver was able to keep the vehicle on the road for approximately 100 ft before striking a ditch on the east side of the roadway and rolling over three times. The driver was ejected from the cab and pronounced dead at the scene.[1]

Investigation of this fatality revealed that the GVWR for this surplus military truck was 16,530 lbs off-road. Considering that the empty weight of the vehicle was 13,530lbs, the vehicle was rated to carry only 3,000 lbs of payload. However, the vehicle was equipped with a 1,200-gallon tank. When filled with 1,200 gallons of water (which weighs 8.3 lbs per gallon), the vehicle exceeded the off-road GVWR by 6,960 lbs.

In addition to being nearly 7,000 lbs overweight, eight out of ten tires on the vehicle were 24 years old at the time of this crash. It is not surprising that the overweight condition of this vehicle, in conjunction with old and degraded tires, resulted in a serious vehicle crash.

Case Study—New Mexico

On June 26, 2003, a volunteer firefighter was responding to a wildfire on a remote, unpaved road in New Mexico. While attempting to negotiate a downhill curve, the vehicle left the roadway and rolled over four times while traveling 123 feet into a canyon. During the course of the crash, the unrestrained driver was ejected from the vehicle and sustained fatal injuries when the truck rolled over on top of him.[2]

The vehicle involved in this incident was a surplus 1954 2 ½-ton fuel truck, commonly known as a "Deuce-and-a-half." The vehicle was equipped with a 1,200-gallon tank that had one centralized tank baffle. According to the NIOSH report, the vehicle's loaded weight was 24,930 lbs, and the vehicle was not equipped with seat belts. A National Forest Service officer who arrived shortly after the crash stated, "The brake pedal could be depressed completely to the floor with no resistance," indicating a complete failure of the braking system.

Case Study—Texas

A surplus military tanker truck lost control and killed the firefighter who was driving it. The vehicle involved was a 1979 cab-over tractor pulling a trailer that contained 5,000 gallons of water. According to NIOSH Report F2006-06, the trailer was designed to carry a payload that weighed less than 35,250 lbs. Had the vehicle been carrying 5,000 lbs of gasoline, for which it was originally designed, the weight of the gasoline would have only been 28,000 lbs. However, when the fire department took delivery of the truck, it filled it with 5,000 gallons of water. The weight of this water was 41,500 lbs, which was well in excess of the maximum allowable payload weight. In the case of the modified 5,000-gallon fuel trailer, the only baffle found inside the tank was a partition that divided the tank into two separate compartments. Although it is not known if the lack of baffles was the primary cause in this crash, this fact is an important learning point for the fire service.[3]

Height, Weight, and Width of Vehicles

Even nonmodified fire apparatus can present challenges that result from their size and weight. A thorough knowledge of a vehicle's height, width, and weight should seem like common sense. However, it is not unusual to see a fire apparatus stuck inside a tunnel or sitting atop a collapsed bridge. The height and weight of the vehicle should be clearly marked inside the crew cab and every driver should be fully aware of the safe routes each fire apparatus can take to an emergency (fig. 13–1).

Vehicle height and width becomes an issue when a fire apparatus attempts to drive under a low bridge, enter a parking garage, or pull into a neighboring fire station while on a standby. Many fire apparatus operators don't have a complete understanding of how bridge heights are measured and find their apparatus severely damaged from driving into one. I have personally observed a brand new fire truck wedged under a railroad bridge in the vehicle's first-due territory.

FIGURE 13–1. It is important to mark the size and weight of each apparatus inside the vehicle so the driver can readily check if needed.

Bridge heights are especially important for arched bridges. Drivers must know whether the posted height of the bridge represents the lowest point along the entire width of the roadway, or if the bridge was measured in the center. If a fire apparatus with a high ladder rack enters an arched tunnel, they may be unpleasantly surprised when the ladder rack is suddenly ripped off. While the vehicle's measured height may fit in the center of the bridge, it does not fit along the fog line (fig.13–2).

Fire stations are also an issue. I once witnessed a fire engine back into its own station with a telescoping deck gun raised in the air. The deck gun struck the top of the bay door, severely damaging the deck gun, the pump, and the roof of the cab. Ensure that every driver conducts a thorough walk-around of the vehicle prior to driving and checks to make sure any telescoping equipment, such as pole lights or deck guns, are properly stowed.

Exceeding the weight restrictions of a bridge or road can also result in a fire apparatus crash. If the bridge says no vehicles are allowed over XXXXX lbs, it is posted this way for a reason. Driving an overweight vehicle over a restricted bridge can cause failure of the bridge and loss of the vehicle; both of which are extremely expensive when it comes time for repair.

While posted bridge weights are self-explanatory, a less obvious issue is a road surface that cannot support the weight of a fire apparatus. Fatal crashes have resulted when the road edge gave way causing the fire apparatus to roll off

FIGURE 13–2. Apparatus operators must know the height of their apparatus and the potential route restrictions within their jurisdictions and mutual aid areas.

the road. Crashes have also occurred when the driver attempted to avoid a collapsing road, overcorrected, and rolled into the oncoming lane.

Road failure is more common when there is a significant drop off at the road edge. Erosion, weather, and poor drainage may cause the road edge to become weak or non-existent if it is washed away. While the road appears normal from above, there is no support under the asphalt surface. As a large fire truck drives across this weakened portion of the road, the asphalt collapses and the vehicle loses control. The same holds true for unpaved roads, where loose or eroded soil may create a weakened portion of roadway. When the apparatus drives over the weak spot, the soil gives way causing the vehicle to pull or roll to one side.

A proper reconnaissance of the first-due area may help to identify unsupported road edges before the road is used. If there are roads in the district that may be unsafe to drive on, alternate routes or alternate vehicles should be considered for response.

Road Defects

Road defects can occur on any type of road. However, these defects are more common on rural or unpaved roads. Fire apparatus operators must understand each of these defects and how they may affect the handling characteristics of a vehicle.

Potholes

Potholes are large holes in the road which may cause severe damage to tires, rims and suspensions systems. In extreme cases, potholes may cause a tire blowout, resulting in a complete loss of control, or cause the vehicle to change direction and drive off the road.

Rutting

Rutting results in longitudinal depressions which follow the path of the roadway. Rutting is usually found on roads which have high moisture content in the sub-surface soil. Ruts may cause a vehicle's tires to become trapped, making it extremely difficult to steer (fig. 13–3). Drivers may also be prone to oversteer as they attempt to escape the ruts. If the vehicle oversteers, it may travel into the oncoming lane or entirely off the roadway. Fire apparatus operators may also encounter ruts caused by deep snow or slush.

FIGURE 13–3. Road defects can create substantial safety issues for fire apparatus operators. In this case, ruts in a dirt road pose significant issues related to steering and rollover stability. Fire apparatus operators must use caution while driving on roads that have defects (photo provided by TheDrivingCompany.com).

Corrugations

A corrugated roadway has depressions or ridges that travel across the roadway, resulting in a washboard effect. Unlike ruts, which run longitudinally along the roadway, a corrugated road has ruts that run across the road from left to right. In addition to causing an extremely uncomfortable ride, a corrugated road can lead to a loss of control as the vehicle starts to bounce up and down uncontrollably. This bouncing effect can have a significant impact on steering, braking, and use of antilock brakes.

Soft spots and depressions

Soft spots are often caused by poor drainage. When water can't run off the road, it tends to pool and create areas of mud or soft soil. These soft spots can have a detrimental effect on a vehicle, as it causes one or more of the tires to stick or get hung-up while driving. The stuck tire may cause the vehicle to pull to one side, resulting in a crash.

Different surface types

Apparatus operators must be cautious when turning from one road onto another road if the roads are made of different materials. The sudden change in friction may cause the vehicle to slide or lose control. As an example, drivers must slow down and use care when turning off of an asphalt road and onto a gravel or dirt road.

Drivers must know every bridge, tunnel, and poor roadway within their first-due and mutual aid districts. A regular inspection of roads, bridges, and tunnels should be completed by drivers and members of the training staff. Any potential changes, such as newly paved roads (which may affect bridge heights), signs of degradation, or any other problems, should be immediately reported to the local highway department. All fire apparatus drivers should be notified immediately if there are any issues. Keeping abreast of these changes and knowing the safest route to an emergency will help prevent road related crashes from occurring.

Notes

1. NIOSH Firefighter Fatality Investigation and Prevention Program Report F2005-27, "Volunteer Fire Chief Dies from Injuries Sustained During a Tanker Rollover—Utah," July 24, 2006.
2. NIOSH Firefighter Fatality Report #2003-23, "Volunteer Assistant Chief Dies in Tanker Rollover—New Mexico."
3. NIOSH Firefighter Fatality Investigation and Prevention Program Report #2006-06, "Volunteer Firefighter Dies in Tanker Rollover Crash—Texas."

Retiring Apparatus, Out-of-Service Apparatus, and Annex D

Introduction

NFPA 1451 states that the fire department shall consider the health and safety of vehicle occupants as the primary concern in the specification, design, construction, acquisition, operation, maintenance, inspection, and repair of all fire department vehicles. However, the high cost of modern fire apparatus makes it difficult to retire and replace older fire apparatus that lack critical safety systems. Fire departments are often forced to keep vehicles for decades or purchase used apparatus from other fire departments and military surplus stores. Often, there is little known about the history, maintenance, and past use of these vehicles. This lack of knowledge about the vehicle's history may hide significant mechanical and safety defects.

Purchasing used or outdated vehicles may also mean that the vehicle lacks up-to-date safety equipment such as seat belts, electronic stability control, anti-lock brakes, or occupant protection systems. When this lack of advanced safety equipment is combined with poorly maintained tires, brakes, or other vital components, the severity of a crash may be compounded. While safety equipment can be retrofitted onto the apparatus, at what point does a retrofit become cost-prohibitive for the department?

It is understood that the financial limitations placed on the leaders of a fire department are often out of their control. Fire chiefs in large urban departments must compete with other city agencies to secure the limited funding that is available. Fire departments in rural areas are faced with smaller populations which limit the tax base from which to draw. Ironically, it's these rural fire departments that require larger and more dangerous vehicles, such as water tenders, which should be equipped with advanced safety equipment.

While most fire chiefs are well aware of their aging fleets, politicians may not understand that the high costs associated with keeping an out-of-date vehicle on the road may surpass the long-term cost of replacing the apparatus. These facts must be carefully detailed and explained to those in control of the finances.

In order to assist fire chiefs and administrators who are trying to justify a new apparatus, NFPA provides a detailed process to determine if a fire apparatus is obsolete. This evaluation is provided in Annex D of the NFPA 1901 standard and is an excellent resource for chiefs or administrators trying to fund the replacement of an aging apparatus. The annex makes a general observation that fire departments should give serious thought to the risks versus benefits of keeping a fire apparatus in front-line service that is more than 15 years old. This is due to the substantial improvements in fire apparatus safety that have been made in more recent editions of NFPA 1901. Annex D also provides detailed considerations that will help a fire department to determine the efficacy of keeping, retrofitting, or replacing an older apparatus.

While reading the considerations listed in Annex D, take note that many of the listed items are specific to fire service vehicles. When considering the purchase of a retired military or civilian vehicle, fire departments must consider how much work is needed to bring the vehicle anywhere near NFPA compliance. Would it be cheaper to purchase a used fire apparatus, rather than purchase a retired civilian vehicle and attempt to bring it into compliance? Chiefs and administrators should complete a thorough cost-benefit analysis when researching the purchase of a used or retired vehicle.

Even if a new vehicle is purchased which is 100% NFPA compliant, it is important for fire apparatus operators to understand that a brand new vehicle is only as safe as the driver behind the wheel. Electronic stability control and an occupant protection system will do nothing to prevent a crash if the driver still speeds and drives recklessly to calls.

Newer apparatus may be **more** dangerous if the driver is not properly trained. Older apparatus were not as powerful and were more forgiving when driven by an inexperienced or reckless driver. Instead of double-clutching a split transmission, today's driver sits in a cockpit very similar to that of a civilian car. Automatic transmissions and 500 HP motors can easily deceive drivers into believing they are sitting behind the wheel of a small sports car, instead of a large fire apparatus that is prone to rollover.

Volunteer Firemen's Insurance Services (VFIS) studied 36 fatal accidents that occurred between 1999 and 2005. Of the vehicles involved in these fatal crashes, 47% were manufactured after 1990. Further investigation revealed that 83% of the vehicles involved were manufactured after 1980. These statistics demonstrate that the chance of a fatal crash is still quite real, even when operating a newer fire apparatus.[1]

Obtaining Funding for New Apparatus

The pride and ownership inherent in the fire service can sometimes be its own worst enemy. While I am in no way advocating that we abandon these long-standing traditions, we must understand how it may affect us in a negative way—namely, trying to justify the need for a new fire apparatus.

No matter how old or outdated a fire apparatus might be, firefighters take great pride in keeping their vehicles in tip-top appearance. It is not unusual to walk into a firehouse and see a 30-year-old fire truck which appears to have rolled off the assembly line within the past few weeks. The pride that keeps the apparatus waxed, shined, and parade-ready may deceive the politicians in charge of funding.

When the average person sees a fire truck, they are not looking for an enclosed cab, lateral acceleration warning device, or occupant protection systems. They are judging the book by its cover. A city council person who visited the fire station last week will only remember the excellent condition of the vehicle as it appeared from the outside. Due to this lack of knowledge, fire department administrators must take strides to properly educate the public on the importance of NFPA compliance (fig. 14–1).

FIGURE 14–1.   While a firefighter will recognize an older apparatus, those who control the purse strings may not recognize an out of date vehicle because it is so well cared for. It is important to educate those who oversee the finances and explain the important differences between an old and new vehicle.

Before proposing an apparatus purchase, I strongly recommend taking the time to prepare a presentation which will help a civilian observer to understand the age of the current apparatus. Start the presentation with a picture of the fire truck and explain what year it was built and what was happening in the world at the time. If the truck was built in 1981, explain that Reagan was in office and the Russians were our adversaries. Provide examples of movies or music that may have been released that year. Most importantly, show other vehicles that were new in 1981. This will help the audience put the age of the vehicle into a context that they can understand.

My favorite ending is to show a police car that was manufactured the same year as the fire apparatus. When citizens and council members see a 1981 Dodge Grand Fury, it drives the message home (fig. 14–2). Would they be embarrassed to see their police officers driving such a vehicle? How is the fire service any different?

FIGURE 14–2. While a civilian may not recognize an older fire apparatus, they will definitely recognize an older police car. Provide a photograph of a police car manufactured the same year as the fire apparatus. Would the city or town be proud to have their police officers driving around in such "up-to-date" equipment?

Once the groundwork has been laid, provide a comprehensive explanation as to what is missing from the current vehicle. A comparison of a modern day, NFPA compliant apparatus versus the existing apparatus can be a powerful tool. Does the current vehicle have power steering? What about antilock brakes or rollover protection. Showing video of a non-ABS fire truck spinning out of control can be a powerful tool when shown next to a video of an ABS-equipped fire truck sliding to a safe stop.

The final part of the presentation should include a detailed breakdown of maintenance costs and expenses for the vehicle. A detailed examination of these costs often reveals that the cost of owning an older fire truck is more than the cost of owning a new one. Trying to find outdated parts and someone who knows how to repair an older vehicle can be a costly venture. By presenting a detailed cost analysis, fire department leaders can show how a new fire apparatus will actually save money in the long-run.

I have heard countless firefighters, fire chiefs, and administrators complain about NFPA standards as they relate to fire apparatus. While I understand the consternation, especially for those on limited budgets, it is important to remember why these standards are in place. Most of these standards relate directly to firefighter safety and cannot be ignored. Instead, these requirements should be embraced by fire service leaders and used as powerful tools to help move a fire department and its equipment into the future.

Out of Service Criteria

While many states do not require a commercial driver's license to operate fire apparatus, the driver will still be held to the same legal requirements as a CDL driver. Because of this, fire apparatus drivers must be thoroughly versed in how to complete a pretrip inspection and place the vehicle out of service if out-of-service criteria are met.

NFPA 1451 requires vehicles to be inspected at least weekly and within 24 hours after being used in an emergency response. These types of inspections should include all of the vehicle's safety features, such as brakes, tires, warning lights, and seat belts. The truck should be started to make sure it is properly functioning and all fluid levels should be checked on a regular basis. For career departments with staffed stations, the vehicles should be checked daily. For volunteer departments, the trucks should be checked at least weekly. Members who complete these inspections should be well versed in the basic operation of the vehicle and be able to make minor repairs such as changing a light bulb or windshield wiper. The fire department must maintain instructions and manuals for each apparatus to ensure that the members are performing proper inspections or repairs.

In the event that a problem is found during an inspection or while operating the vehicle, the fire department must have procedures for correcting the unsafe condition or making a repair. If the repair affects the safe operation of the vehicle, it should be taken out of service immediately and the proper authority must be notified. Repair orders must be documented in writing and must include the date and time the problem was discovered, a description of the issue, any action taken at the time the issue was discovered, and the date and time that the repair was made. Copies of this documentation must be sent to the fire department safety officer.

Once a repair is complete, the fire department must ensure that the vehicle is re-inspected before placing it back in service. Repairs and preventative maintenance should only be completed by qualified personnel, preferably an emergency vehicle technician.

NFPA 1911—Standard for Inspection, Maintenance, Testing, and Retirement of In-Service Emergency Vehicles contains an entire chapter detailing when a fire apparatus must be taken out of service for mechanical problems. While most fire apparatus operators do not have extensive knowledge related to fire apparatus repair, they should be well versed in the NFPA requirements for taking a vehicle out of service.

The NFPA 1911 requirements are quite extensive. In addition to providing a detailed list of out-of-service criteria, it states that the apparatus must be taken out of service if there is an issue that violates state, provincial, and local regulations, specific manufacturer recommendations, or requirements established by the fire department.[2] Apparatus operators must have knowledge of the manufacturer's recommendations and of federal, state, and local laws. These issues should be addressed as a required module in a driver training program and modules should be developed for each piece of apparatus. In some situations, the NFPA standard may require that a single part of the vehicle to be taken out of service, but not the entire vehicle. For instance, if the aerial device on a ladder truck is out-of-service, the vehicle can still be driven. If there is an issue with the engine company pump system, the pump must be taken out of service, but the vehicle can still be used. In fact, NFPA goes so far to say that if a seatbelt is broken or damaged, only the damaged riding position must be taken out of service, not the entire vehicle. However, if the damaged seatbelt is the driver's seat the vehicle cannot be used.[3]

If a fire apparatus meets out-of-service criteria and is still placed on the street, the fire department, its officers, and the driver could face significant legal ramifications if the vehicle is involved in a crash. If an NFPA standard is violated, these legal ramifications may include increased civil penalties. If a state or federal law is violated, the legal ramifications may include criminal charges.

Out of Service Procedures

It is important for every fire department to have a well-established and well-recognized procedure for placing an apparatus out of service. Chapter 6 of NFPA 1911 provides detailed requirements on how to place a vehicle out of service so that the vehicle is not inadvertently driven by another member.

I can speak from personal experience regarding an incident in which a fire apparatus was placed out of service at 0300 hours due to a serious brake problem. While the county radio room and chief engineer were notified of the problem, the only notice given to other members was a hand written note on a white board in the radio room. At 0800 hours, before the day crew had a chance to take a walk around the station and see the note on the white board, a call was received and the apparatus hit the street. Fortunately, the driver recognized the brake issue before the vehicle entered a major highway. However, this type of Murphy's Law scenario is exactly why NFPA has outlined specific procedures for properly placing an apparatus out-of-service. These procedures must include one of the following:

1. A sign on the outside of the driver's door near the door handle
2. A special bag that covers the steering wheel
3. A large sign on the driver's window
4. A highly visible mechanism at the driver's position on the fire apparatus that all members of the fire department recognize as an out of service indicator

These procedures don't have to include an expensive sign or designer bag over the steering wheel; they just have to be effective. A handwritten sign on an empty cardboard box will meet the criteria as long as the message is sent. There are also notification requirements if a specific part of the apparatus is placed out of service, such as the water pump or aerial device. If the vehicle is being used while a component is out of service, one of the following requirements must be met:[4]

1. Distinctive color sign located on the inside of the driver's door identifying which component is out of service
2. Highly visible device provided at the component control(s) indicating that the device is out of service

Fire apparatus operators must have a thorough understanding of when a fire apparatus is not safe to drive. Effectively communicating to other fire department members when and why an apparatus is out-of-service is an important aspect of a vehicle operations program. Only by ensuring that unsafe apparatus are kept off the street can we ensure the safety of our members and the public.

Notes

1. William Jenaway, "Safety 101—Lesson 17: Vehicle Safety: The Big Picture," http://cms.firehouse.com/web/online/Vehicle-Operations-and-Maintenance, November 22, 2007.
2. NFPA 1911, Section 6.1.3
3. NFPA 1911, Section 6.2.2
4. NFPA 1911, Section 6.2.2

Tiller Ladders

Many fire departments operate tractor drawn aerials which require the use of a *tiller operator.* These vehicles pose unique and sometimes difficult issues when it comes to vehicle operations and driver training. Ensuring that the tractor driver and tiller operator are well-trained and proficient are crucial components of a tractor drawn aerial operations program.

Tractor Trailer Dynamics

A tractor trailer is defined as a combination trucking unit that has a tractor hooked to a trailer.[1] These vehicles are unique because they are *articulated vehicles.* Articulated vehicles are able to bend at a connection point, unlike most vehicles which do not bend anywhere along the chassis (fig. 15–1). Articulated vehicles, such as tiller ladders, have unique issues related to vehicle dynamics and crash causation.

One of the greatest concerns with tractor-trailer operations is a jackknife. Jackknifing results when some of the tires lose traction with the roadway, while the rest of the tires do not (fig. 15–2). To understand jackknifing, consider the following circumstances:[2]

1. *Front tires on the tractor lock:* In this case, the vehicle will lose steering control and will usually skid in a straight line.
2. *Rear tires of the tractor lock*: In this case, the rear of the tractor will want to spin around and pull the trailer with it. This scenario will result in a jackknife.[3]
3. *Rear tires of the tractor lose traction due to a sudden increase in power*: This is caused by aggressive acceleration by the driver, which causes

the drive axle to lose traction with the road. The tractor will want to spin around and pull the trailer with it, resulting in a jackknife .

4. ***Rear tires of the tractor lose traction because the engine brake is left in the "ON" position during inclement weather***: When the driver lets off the accelerator pedal and the engine brake engages, it may cause the drive axle of the tractor to lock-up and break traction with a slippery road. The rear of the tractor will want to spin around and pull the trailer, resulting in a jackknife.

5. ***Trailer tires lock***: In this case, the trailer will be at risk of swinging out from behind the tractor, causing a loss of control. Because the trailer has a tendency to swing slower than the tractor, the driver can often see the trailer start to swing and attempt to correct the problem.

These scenarios are made worse when rounding a curve or making an evasive maneuver. If the driver enters a curve too fast and lets off the accelerator, the engine brake may activate and cause the drive axle of the tractor to break traction with the road.[4] If the rear tires of the tractor lose traction, the trailer may push through the tractor causing a jackknife or loss of control. Once a vehicle begins to jackknife, it is very difficult to regain control.

FIGURE 15–1. Tiller ladders can articulate or bend in the middle. This creates unique issues related to vehicle dynamics and crash causation (photo by Steve Crothers).

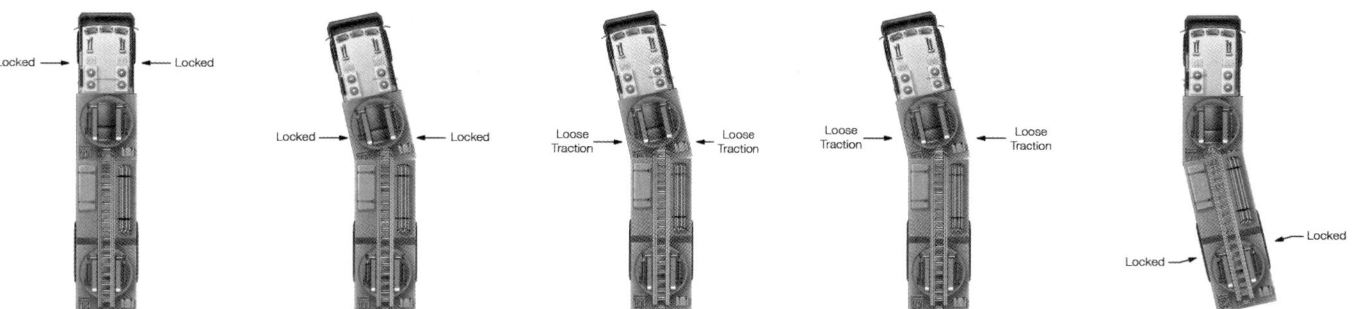

FIGURE 15–2. A tiller ladder will jackknife when any of the tires break traction with the road. How the tiller will lose control depends on which tires break traction.

Preventing Jackknifes

Newer tiller ladders will be equipped with antilock brakes which will prevent wheel lock-up and help keep the vehicle under control. However, many fire departments still operate older tiller ladders that are not equipped with antilock brakes. In addition, excess speed, aggressive driving, inclement weather, or slick surfaces may cause a tiller ladder to jackknife without any brake application at all. This is true even if the vehicle is equipped with antilock brakes.

Tractor drivers and tiller operators must understand situations that may cause a jackknife and know how to prevent them. These situations include:

1. *Over application of power to the tractor:* If the tractor driver is driving aggressively and applies too much power, the drive axle of the tractor may lose traction with the road and start to slip. In this situation, the trailer may start to push through the tractor, especially if the vehicle is rounding a curve. Aggressive use of the throttle must be avoided. Smooth acceleration appropriate for the existing road conditions is essential to preventing a loss of control.

2. *Failure to turn off the engine brake in inclement weather:* An engine brake will slow the drive axle of the tractor but will not slow the other axles on the apparatus. If the engine brake slows the drive axle too much on a slippery road, it may cause the rear of the tractor to break traction, spin out, and jackknife the trailer.

3. *Aggressive braking by the tractor operator:* This is especially true in vehicles that are not equipped with antilock brakes. If the tractor operator rushes to an intersection and slams on the brakes, one of the axles may lock-up, resulting in a loss of control. Drivers should remember to drive at an appropriate speed and avoid aggressive braking, especially during wet weather. Drivers must use smooth, progressive braking and anticipate situations where they may need to decelerate well ahead of getting there.

4. *Driving on two different surfaces:* Tiller ladders are very long. There may be times when the tractor wheels are on a surface that is different from the trailer wheels. When this occurs, one set of wheels may be more prone to lose traction. Tractor drivers and tiller operators must carefully watch the road conditions and respond accordingly if the conditions are more slippery for one axle than another. This is especially true when turning from one road onto another.

Tiller Ladder Skid Control

As with any skid control, the first step is to let off the offending pedal. If the vehicle is skidding because the brakes are applied, let off the brakes. If the vehicle is skidding because it is accelerating, let off the accelerator pedal. Letting off the offending pedal will give the vehicle a chance to regain traction and give the tractor driver and tiller operator time to take action.

Trailer tires skid

If the rear tires on the trailer have lost traction and the trailer is swinging out, the tiller operator may be able to steer the trailer back in line when the trailer tires regain traction. Keep in mind that this will be a very sudden and very stressful situation. Do not rely upon the trailer operator to regain control if the trailer starts to swing. Don't swing the trailer in the first place.

Tractor tires skid

If the tractor has lost traction, the tractor operator should let off the offending pedal, look where they want the vehicle to go, and steer in that direction. The tractor will tend to jackknife very quickly, and there will not be much time for the driver to regain control. Again, avoid driving aggressively and putting the vehicle in this situation in the first place.

Rollovers

The fact that the trailer portion of a tiller ladder is isolated from the tractor portion can cause problems for the driver. In a straight truck, the driver has a full sense of the vehicle. If the vehicle is breaking traction or beginning to roll, the driver will sense these forces, even if it's too late to do anything. In a straight truck, if the driver senses that the vehicle is breaking traction or rolling over, it is still possible to make corrections that can prevent the vehicle from losing control.

The same cannot be said for a tractor trailer. In a tractor trailer, the driver is sitting in a completely separate portion of the vehicle. If the rear trailer tires start to break traction or the body of the trailer starts to roll, the driver will not sense these warning signals. Instead, the first warning the driver may have of an impending rollover or skid will be watching the trailer in the rearview mirror as the rear axle attempts to swing around in front of the tractor. Or worse, the driver may feel the tractor start to rollover as the rolling trailer pulls the tractor with it.

The tiller operator must stay in constant contact with the tractor operator. Using vehicle-based intercom systems, or a manual system such as a horn, the tiller operator must keep the tractor operator constantly updated on the situation. If the driver is going too fast, the tiller operator must speak up and tell him. While no one wants a back-seat driver, there will be times when the tiller operator must speak up and ensure the safety of the vehicle. This is especially important when dealing with speed related safety issues.

The tractor driver must also be sure to communicate with the tiller operator and keep them constantly apprised of the planned route of travel and upcoming turns. Should the tiller operator make an incorrect assumption and swing the trailer too early for a turn, it may result in a crash when the tractor driver turns in a different direction than anticipated. The vehicle will not be positioned properly for the turn and the tiller operator will be fighting the steering wheel in an attempt to reposition the trailer in the correct direction.

Communication is the key to safe tiller operations. The tiller operator is at the mercy of the tractor driver. The tiller operator has no control over speed or braking or anything occurring at the front of the truck. Instead, the tiller operators must anticipate and handle whatever the tractor driver should throw their way. For this reason, it is helpful to maintain operating teams for tiller ladders. By working together on a regular basis, each driver can learn and anticipate the actions of the other.

Backing Operations

Backing operations involving a tiller ladder can be quite complex. However, the tiller operator is a set of eyes in the rear of the vehicle that will be able to monitor and control where the rear of the trailer is going. While this second set of eyes is very useful, tiller operators must be familiar with the blind spots associated with sitting high on top of the vehicle. Installation and use of a backup camera is still recommended as an additional tool for backing safety.

Tiller operators must also be aware of the difficulties that may be associated with backing an apparatus that steers with a tiller wheel. There will be times, especially with inexperienced drivers, when the brain has trouble comprehending what the arms want the steering wheel to do. Constant drilling on backing maneuvers is necessary to ensure that tiller operators have complete control over the rear of the vehicle.

Turning a Tiller Ladder

Tractor drivers must be mindful of their speed as they enter a turn. Excess speed may force the tiller operator to oversteer and lose control. Remember that the tractor is pulling the trailer: if the trailer is being pulled too fast, the tiller operator will not have enough time to correct the steering angle and bring the rear of the tiller ladder back in line with the tractor. If the tractor driver takes a turn at 25 mph, the vehicle will travel 73 ft in 2 seconds. If the driver takes the same turn at 10 mph, the vehicle will only travel 29 ft in 2 seconds. While 44 ft might not seem like a lot, it may be the difference between the tiller operator steering around a hazard instead of hitting it.

After making a turn, the tractor operator must ensure that the tiller operator has brought the trailer back in line with the tractor and the wheels have been straightened. Accelerating too quickly may not give the tiller operator enough time to bring the trailer axle back in line, causing the trailer to oversteer and strike an object. Tractor drivers must also remember to watch the rear wheels and not just the trailer. While the trailer may be back in line, the tiller operator may not have had enough time to straighten the wheels. A sudden acceleration may cause the trailer to swing out and strike an object.

As discussed in Chapter 4, the stability of a vehicle is directly related to the vehicle's rollover threshold. A vehicle with a higher rollover threshold will be more stable and less likely to rollover than a vehicle with a low rollover threshold. The rollover threshold of a passenger car is often above the maneuvering limits of the vehicle, meaning it is next to impossible for a civilian driver to maneuver the vehicle in such a way as to trigger an untripped rollover on a normal highway. However, fire apparatus have low rollover thresholds when compared to passenger cars.[5] A fire apparatus could easily exceed its rollover threshold when performing an evasive steering maneuver, rounding a curve, or turning too fast. This fact is especially true for tiller ladders.

Tractor drivers and tiller operators must remember to disengage the 5th wheel lockout prior to leaving a scene. The 5th wheel lockout is designed to provide better stability to the aerial device when the aerial is in service. However, if it is not disengaged before driving, it may put a bend in the vehicle which will cause poor handling as all of the tires will not be in full contact with the road.

Tiller operators must be aware of *overtillering* and ensure that they only steer the rear of the vehicle as much as necessary to safely complete the turn. Ideally, the tractor driver will operate the vehicle in a manner that does not require the tiller operator to steer much. By turning at an appropriate speed and properly positioning the vehicle prior to entering the turn, the tiller operator will have to make minimal steering corrections.

It is important for tiller operators to remember that their primary purpose is to keep the vehicle straight and in line when driving down the street. Tillering the vehicle should only be done when there is a need. Driving down the road sideways or overtillering to look cool will invite disaster. If the tiller operator fails to keep the vehicle in line, the vehicle is much wider and at a higher risk for a crash. In tight quarters with parked vehicles, heavy traffic, or other roadside obstructions, increasing the turning radius of a tiller ladder may result in the rear of the vehicle striking an object.

In addition to oversteering, the tiller operator must be constantly aware of the vehicle's overhang. Overhang is the portion of the vehicle that runs from the midline of the rear axle to the rear of the truck. To avoid collisions with other objects, tiller operators must ensure that there is enough space for the vehicle's overhang. A comprehensive driver training program must include drills and maneuvering exercises which address the overhang of the vehicle.

Tractor drawn aerials are versatile apparatus that serve an important role in the fire service. However, the unique issues associated with this articulated vehicle must be fully understood and respected by those who drive them. Aggressive driving and aggressive braking should be avoided when driving any vehicle, but this is especially true when driving a tiller ladder. Only through safe, calm, and nonaggressive driving can these large vehicles be kept under control.

Notes

1. http://www.dictionary.com/browse/tractor-trailer
2. Frank Navin, "Reconstructing Truck Accidents from Tire Marks," IPT, 2004, pg. 61–62.
3. Navin, "Reconstructing Truck Accidents from Tire Marks."
4. Navin, "Reconstructing Truck Accidents from Tire Marks."
5. NFPA 1901 only requires a rollover threshold which may be less than half that of a passenger car

Fifteen Passenger Vans

Introduction

Many fire departments own and operate 15-passenger vans as transport vehicles. Some of these transport vehicles are limited to nonemergency use, such as training classes and parades; other departments use them in actual emergencies (fig. 16–1).

FIGURE 16–1. It is not uncommon for a fire department to use a 15-passenger van as a transport vehicle. It is important for fire departments to recognize the significant safety issues related to a 15-passenger van and train drivers accordingly (photo provided by the Kingston Volunteer Fire Department).

While a 15-passenger van can be a useful tool in the fire service, it is important for fire departments and apparatus operators to understand the safety issues related to these vehicles. There was a point in time when 15-passenger vans were crashing so frequently that the National Highway Safety Traffic Administration (NHTSA) conducted several studies to investigate their use.

Federal Studies

As a result of a relatively high number of fatal crashes involving 15-passenger vans, NHTSA began to examine crash statistics related to these vehicles. NHTSA statistics indicate that between 1990 and 2002 there were 1,576 fatal van crashes, resulting in 1,111 fatalities of van occupants.[1] Of these fatalities, 86% of those killed were not wearing seat belts. Of the passengers who were restrained during the crash, 92% survived.

NHTSA studies indicate that the chance of a rollover crash involving a 15-passenger van is 2.2 times higher when the van is loaded to more than half of its capacity. The odds of a rollover crash are 5 times more likely when the van is fully loaded, compared to when the driver is the only occupant. The odds of a rollover are also 5 times more likely on a high-speed road (speed limit greater than 50 mph) and twice as likely on a curved roadway.[2] These statistics provide clear direction for any department looking to implement or modify standard operating procedures related to the use of a 15-passenger van.

Rollovers

As more passengers are loaded into a 15-passenger van, the center of gravity shifts higher and towards the rear of the vehicle. An increase in the height of the vehicle's center of gravity will result in a less stable vehicle that is more likely to rollover during evasive maneuvers or while rounding a curve (fig. 16–2).

While most fire apparatus operators are cognizant of the dangers of driving a vehicle with a high center of gravity, many do not realize that a 15-passenger van falls into this category. Any fire department operating 15-passenger vans must ensure that there is an extensive training program to highlight the dangers of these vehicles.

NHTSA studies indicate that 90% of van-related rollover crashes occurred when the vehicle operator lost control and left the roadway. NHTSA lists the following three scenarios as common causes of van-related rollover crashes:[3]

1. The van goes off a rural road. When this occurs, the van is likely to overturn if it strikes a ditch or embankment, trips on an object, or runs into soft soil.
2. The driver is fatigued or driving too fast for conditions. A tired driver can doze off and lose control. The driver can also lose control when traveling at high speeds, causing the van to slide sideways off the road. When the van strikes the road edge, a grass or dirt median can cause the van to overturn when the tires dig into the dirt.
3. A driver induces a steering overcorrection as a panic reaction to a roadway emergency or a wheel dropping off the pavement. At highway speeds, this situation can cause the driver to lose control, resulting in the van sliding sideways and rolling over.

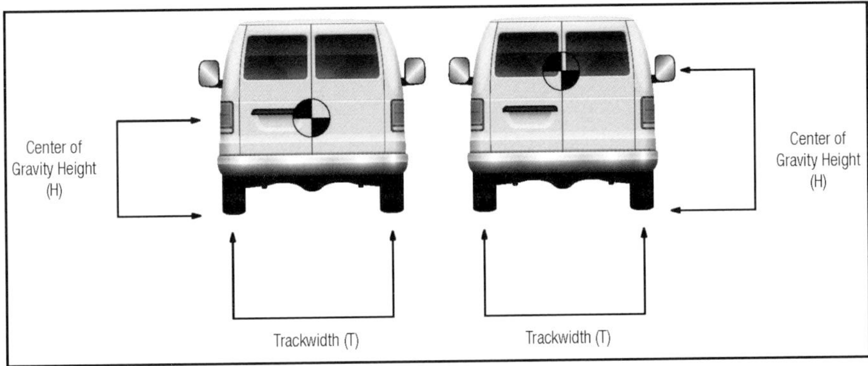

FIGURE 16–2. When passengers are loaded into a 15-passenger van, the center of gravity will shift upward. If the center of gravity shifts upward while the track width remains the same, the rollover threshold of the vehicle will be reduced. The vehicle is now more prone to a rollover.

Implementing Safety Procedures

Departments that operate a 15-passenger van must implement policies and procedures specific to these high-risk vehicles. Policies and procedures should consider the following:

Age of the Operator

While many fire departments have age requirements for driving large apparatus, the same cannot be said for smaller vehicles such as sport utility vehicles, squad vehicles, and 15-passenger vans. Most insurance companies recommend that only members over 21 years of age be allowed to operate fire apparatus. This age requirement should also include smaller vehicles and 15-passenger vans.

Younger drivers do not have the experience or maturity to operate specialized vehicles under stressful conditions. Throw in the distraction of operating a 15-passenger van filled with fellow firefighters who may be returning from a parade or other celebration, and a young driver may be faced with disaster.

Number of Passenger

Limit the number of people allowed in a 15-passenger van. As the number of passengers increases, so does the risk of rollover. Transporting more than six or seven passengers greatly increases the rollover risk.[4] Many studies suggest restricting the number of passengers in a 15-passenger van to nine people, including the driver.

To prevent more than nine people from riding in the van, remove the rear seat. This will prevent excess loading behind the rear axle, which contributes to reduced stability. Keep in mind that when loading passengers into a 15-passenger van, great care should be taken to ensure that everyone is sitting in front of the rear axle. This seating arrangement will help maintain vehicle control, especially during evasive maneuvers.

It is also important to make sure that no equipment or luggage is strapped to the roof of the vehicle, as this will further increase the height of the center of gravity and lead to a reduced rollover threshold. Do not use vans to tow trailers or equipment.

Speed

Drivers must be aware of safety issues related to high-speed roadways (speed limit greater than 50 mph), curves, and evasive maneuvers. As a vehicle's speed increases, the margin for error decreases. This is especially true in an unforgiving vehicle such as a 15-passenger van. A sudden steering input or jerk on the steering wheel may upset the vehicle's center of gravity and cause the vehicle to lose control.

As a vehicle rounds a curve or makes a sudden steering maneuver, g-force will act on the vehicle. As discussed in Chapter 3, the g-force will depend on the speed of the vehicle and the radius of the curve. Therefore, a higher speed will increase the g-force acting on the vehicle. If the g-force exceeds the rollover threshold of the vehicle, it will rollover.

Inertia compounds this problem. As the vehicle enters the curve, inertia will cause the vehicle's center of gravity to continue forward in a straight line. As the vehicle rounds the curve, inertia will cause the vehicle to dip down onto its suspension, and the center of gravity will shift towards the outside of the curve. As the center of gravity shifts, the rollover threshold of the vehicle will be reduced, making it easier for the g-force to cause the vehicle to roll on its side.

Seatbelts

Every fire department must have a seatbelt policy that is strictly enforced. This is especially important for passengers in a 15-passenger van. NHTSA studies show that 86% of the people killed in 15-passenger van crashes were not wearing their seatbelts.

Failure to use a seatbelt is dangerous for many reasons. First, the chance of being ejected from the vehicle is greatly increased. Once a person is ejected out of the passenger compartment, the odds of sustaining a fatal injury are 4 in 5. Ensuring that a person remains inside the vehicle will allow the occupant protection systems to do their job.

Also keep in mind that an unrestrained passenger can seriously injure or kill others who may be riding in the vehicle. During a violent crash, objects inside the passenger compartment move around violently. Imagine being struck in the head by a 200-lb firefighter.

Fatigue

15-passenger vans are often used for long trips to special events or training classes. It is not uncommon for these vans to be involved in trips that log hundreds, if not thousands of miles. Because these vans are used on such long journeys, it is not uncommon for drivers to drive for lengthy periods of time. It is important to ensure that the driver gets plenty of sleep before driving, makes plenty of rest stops during long trips, and shares the driving responsibilities with other members.

Policies should include limitations on how long a driver should be behind the wheel and require that more than one driver is along for the ride. Having multiple drivers will allow members to share driving responsibilities. Multiple drivers will also ensure that an unqualified driver is not used in the event of illness or fatigue. Many safety organizations recommend that driving responsibilities be shared every 2 hours and that overnight travel be avoided.

Distractions

Distracted driving is a serious safety issue, especially in 15-passenger vans. If a driver takes their eyes off the road for just a few seconds, they will have traveled hundreds of feet without looking to see what is happening in front of the vehicle. Should the vehicle drift off the road, it will be nearly impossible for the driver to bring the vehicle back onto the road in a safe manner. This is especially true for an unstable vehicle such as a 15-passenger van traveling at high speeds.

While there are many types of distractions, such as radios, siren use, and eating, the presence of multiple people inside a 15-passenger van is a distraction that must be avoided. Drivers must understand that their primary goal is to get firefighters to a destination safely. Animated conversations, arguments, or horseplay can lead to a dangerous driving environment. Special care must be taken to ensure that other members do not distract the driver and prevent the job from being done safely.

Tire Pressure

According to NHTSA studies, 74% of the 15-passenger vans involved in fatal crashes had tires that were not properly inflated.[5] Van operators must check tire pressures regularly to ensure that they are properly inflated. Underinflated tires can lead to blowouts and crashes. If a tire pressure warning light should activate while the vehicle is in use, drivers must stop immediately and evaluate the situation. If the tire is underinflated and unable to be fixed, the vehicle should be taken out of service until a new tire can be installed.

Case Study—Oregon

On June 21, 2002, 11 firefighters were traveling from Oregon to Colorado to fight the Hayman Wildfire. After stopping in Parachute, Colorado, for food and fuel, the extended van in which they were riding veered into a median strip and rolled over four times. Four firefighters were killed immediately, and another firefighter succumbed to his injuries three days later. Of the five firefighters killed, only one was wearing a seat belt. The investigation revealed that the driver had become distracted while reaching for a cup and subsequently lost control of the vehicle. The driver of the van was charged with careless driving and was sentenced to 50 hours of community service.

Notes

1. NHTSA, "Analysis of Crashes Involving 15-Passenger Vans," May 2004, 1.
2. NHTSA, "Analysis of Crashes Involving 15-Passenger Vans," May 2004, 1.
3. NHTSA Informational Flyer, "Reducing the Risk of Rollover Crashes in 15-Passenger Vans."
4. NHTSA, "Analysis of Crashes Involving 15-Passenger Vans," May 2004, 1.
5. NHTSA: 12&15 "Passenger Vans Tire Pressure Study: Prelimary Results." May, 2005—DOT HS 809-846

Trailer Operations

Introduction

Many fire departments use trailers as a cost-effective way to store and transport rarely used equipment. Trailers are often used for technical rescue, boat operations, traffic control, and mass-casualty incidents. While trailers provide a practical means of storing and transporting equipment, they pose significant safety issues.

Inadequate driver training is one of the safety issues posed by trailers. While some firefighters have a great deal of experience towing trailers, boats, and off-road vehicles in their personal lives, many other firefighters have never towed a trailer in their life. Regular practice and training on trailer operations must be mandatory for any fire department that uses a trailer.

Selecting a Tow Vehicle

The tow vehicle is the vehicle towing the load. If a pickup truck is towing the trailer, the pickup truck is the tow, or coach, vehicle. Knowing how much weight the tow vehicle can pull is the first step in trailer safety. Finding the vehicle's tow capacity is a simple process. The tow capacity of the vehicle will be clearly defined in the owner's manual. The tow capacity should also be printed on a plate which can be found on the vehicle's door sill. Under no circumstances should a vehicle tow more weight than it is rated for.

Towing more weight than the vehicle is rated to handle can create an unsafe and dangerous situation. Remember that the vehicle's braking system is only designed to burn off a predetermined amount of kinetic energy. If the driver chooses to overload the vehicle or hook to a trailer that exceeds the towing capacity of the vehicle, there will be more kinetic energy present than the brakes were designed to dissipate. The excess kinetic energy will result in a braking system that is no longer able to effectively stop the vehicle. The reduced braking efficiency will result in increased stopping distances and the possibility of complete brake failure due to brake fade.

If a vehicle attempts to pull a trailer that weighs more than the tow vehicle is rated for, it can also result in significant damage to the vehicle's engine, transmission, suspension, and braking system. This damage is caused by excess heat which is created as the tow vehicle works harder to move or stop the oversized load. Towing an oversized load may result in the following types of damage:

1. Brake components will be prone to damage and failure as they attempt to burn off excess kinetic energy created by the oversized load.
2. Transmissions may be subject to damage as the excess heat breaks down the transmission fluid, creating a lack of lubrication inside the transmission. This lack of lubrication may cause a breakdown of the transmission's mechanical components or even total failure if the transmission seizes up.
3. Overloaded suspension components may crack or fail completely under stress, such as driving off-road or striking a pothole.

Vehicles are usually sold with tow packages, which provide added safety and mechanical protection when towing a trailer. Tow packages typically include a heavy-duty engine with a transmission cooler, oversized mirrors, heavy-duty rear suspension, and specialized wiring and circuitry for the trailer. This beefed-up chassis will better handle the stress of towing a trailer, reducing the chance of damage. While purchasing a vehicle with a tow package may add to the purchase price, it could result in long-term savings as the stronger chassis will reduce maintenance costs over the life of the vehicle.

Before purchasing a tow vehicle, calculate the weight of the trailer and all of the loaded equipment. Once the fire department determines how much the loaded trailer will weigh, it can begin to research an appropriate-sized tow vehicle. The fire department must be sure to purchase a vehicle that can safely tow the fully loaded trailer.

Selecting a Hitch

After ensuring that the tow vehicle can handle the weight of the trailer, it is important to ensure that the vehicle is equipped with a proper hitch. There are many different types of hitch systems. Before discussing these different hitch systems, it is important to understand the following terms:

Tongue The V-shaped portion of the trailer that extends forward from the trailer frame. The tongue of the trailer also includes the coupler.

Tongue Weight The tongue weight is the amount of weight being carried by the trailer's tongue. Tongue weight is transferred to the tow vehicle at the hitch point. Most experts agree that the vehicle's tongue weight should fall between 9% and 15% of the gross weight of the trailer.

An improper tongue weight can result in an unstable vehicle. If the tongue weight is too light, the trailer may be more susceptible to sway, as there is not enough weight at the hitch point to maintain vehicle stability. When the tongue weight is too heavy, there will be an excess downward force placed on the rear axle of the towing vehicle. This excess downward force will push down on the tow vehicle's rear axle, causing the front of the tow vehicle to rise up. When the front of the tow vehicle raises up it will have a negative effect on the vehicle's ability to brake, steer, and accelerate. This situation is especially dangerous during panic braking, as the trailer will tend to push forward and cause the rear of the tow vehicle to dip down.

Gross Trailer Weight Rating The total weight of the trailer, includes equipment, fluids, and the weight of the trailer itself. When weighing the trailer, make sure that all equipment is accounted for. Once the trailer has been weighed, ensure that additional equipment is not added. Trailers often become storage places for equipment that is used infrequently. Every time equipment is added to the trailer, the trailer should be re-weighed.

Gross Combined Vehicle Weight Rating The total weight of the tow vehicle and trailer combined with all passengers, fuel, liquids, and equipment.

Gross Vehicle Weight Rating The recommended maximum weight of a vehicle when fully loaded, as specified by the manufacturer.

Hitch Ball   The ball-shaped piece of steel that the coupler latches to.

Draw Bar   The bar that the hitch bar attaches to, also called a ball mount. Draw bars come in two sizes—1 ¼ in. and 2 in. square.

Coupler   The forward-most part of the trailer tongue that drops over the hitch ball. The coupler latches and creates a tight fit with the hitch ball. There are four types of hitches:

Weight Carrying Hitch   A weight carrying hitch carries all of the trailer's tongue weight on the ball and receiver. The weight is then transferred to the tow vehicle through the hitch point. The vehicle's tongue weight should fall between 9%-15% of the gross weight of the trailer.

Weight Distributing Hitch   In a weight distributing hitch, the tongue weight is distributed across the tow vehicle's entire frame, not just the rear axle. By using long rods called "spring bars," some of the tongue weight is transferred to the front axle of the tow vehicle. Distributing the tongue weight will help prevent the rear of the tow vehicle from sagging and lifting the front end. Lifting the front end will result in reduced steering, braking, and acceleration control.

Fifth Wheel Hitch   This type of hitch is mounted in the bed of a pickup truck or commercial motor vehicle. The hitch is mounted forward of the rear axle, distributing the load across the entire truck body (instead of the rear axle). Fifth-wheel hitches improve stability and maneuverability. Because of the fifth-wheel design, trailer sway is nearly eliminated.

Gooseneck Hitch   This hitch is similar to a fifth wheel hitch, however it uses a ball hitch instead of a locking pin. Goose neck hitches are mounted forward of a pickup truck's rear axle, which distributes the trailer weight across the entire vehicle.

Trailer hitches are divided into classes depending upon their weight carrying capacities (table 17–1).

Keep in mind that hitch recommendations are ballpark estimates. Be sure to check with the manufacturer's recommendations for the type of hitch or trailer

TABLE 17–1.   Each trailer hitch has a different weight carrying capacity. Make sure the trailer hitch can handle the weight of the trailer.

	Class I	Class II	Class III	Class IV	Class V
Trailer Weight	2,000 lbs	3,500 lbs	5,000 lbs	10,000 lbs	18,000 lbs
Tongue Weight	200 lbs	350 lbs	500 lbs	1,000 lbs	1,800 lbs

the fire department plans on using, as some of the values may vary. Also keep in mind that values will differ between a weight **carrying** hitch and a weight **distributing** hitch. All of these issues must be considered when designing a fire vehicle that is going to tow a trailer.

Another consideration when selecting the hitch is the size of the hitch ball. Hitch balls come in three different sizes (table 17–2).

Always ensure that the hitch ball is the proper size for the coupler. Using a different size ball and coupler can cause the trailer to unhitch from the tow vehicle. When in doubt, consult your local dealer and ensure that the two components are matched properly.

TABLE 17–2.   Hitch balls come in different sizes. Make sure the hitch ball can handle the load of the trailer.

Hitch Ball Size	Weight Range
1 7/8 in.	Up to 2,000 lbs
2 in.	3,500 to 8,000 lbs (most common)
2 5/16 in.	6,000 to 30,000 lbs (heavy duty)

Selecting a Trailer

When selecting a trailer, the fire department should compile a detailed list of what will be carried, along with the exact weight of each piece of equipment. Knowing the weight of the equipment before buying the trailer will prevent problems in the future. The trailer must be able to safely carry all of the equipment without being overloaded.

Once the trailer is purchased, the equipment should be mounted in a way that ensures a safe tongue weight. If the trailer does not have the proper tongue weight, it can have a negative effect on handling and cause the trailer to sway. If equipment is added or removed, reweigh the trailer and make sure the tongue weight has not changed.

Understanding Trailer Brakes

Another important consideration when purchasing a trailer is the braking system. There are several types of braking systems on trailers, and each system has pros and cons. These systems include the following:

Electromagnetic Brakes

In an electromagnetic braking system, an electrical current is sent through an electromagnet found inside the brake drum. When the electromagnet is energized, it is attracted to the metal brake drum. As the magnet moves towards the brake drum, it causes an arm to move forward, spreading the brake shoes and applying the brakes.

Electromagnetic brakes use a brake controller which determines how much electricity is supplied to the electromagnet. The more electricity supplied by the brake controller, the more the magnet is attracted to the brake drum. The more the magnet is attracted to the brake drum, the greater the braking force.

An advantage to electromagnetic brakes is the ability to engage the trailer brakes separate from the tow vehicle's brakes. If the trailer begins to sway, the driver can apply the trailer brakes to help alleviate the trailer sway. The ability to apply the trailer brakes separately is also helpful when descending hills, as it will reduce the wear on the tow vehicle's brake components.

Disadvantages to electromagnetic brakes include problems when the trailer is used infrequently. If the system is not maintained, rust can accumulate on the brake components and reduce the effectiveness of the electromagnet. Another disadvantage is the fact that an electromagnetic braking system requires a brake controller to be installed inside the tow vehicle. Installing this brake controller may involve splicing into the tow vehicle's electrical system.

When using an electromagnetic braking system, the driver must know the two methods for activating the trailer brakes. During normal braking, the trailer brakes will activate when the brake pedal is depressed. The brake pedal will activate both the tow vehicle's brakes and the trailer brakes. However, the driver can choose to only activate the trailer brakes by using the brake controller. There are two types of brake controllers: time delay and proportional (fig. 17–1).

In a time delay controller, voltage is applied gradually to the trailer brakes using a time delay circuit. The amount of brake force applied by a time delay controller is preset by the driver. Regardless of how hard the driver applies the brakes in the tow vehicle, a time-delay controller will always apply a preset amount of brake force to the trailer.

In a proportional system, the controller will sense how fast the tow vehicle is decelerating and apply the trailer brakes at the same intensity. While this system

FIGURE 17–1. This is an example of an electromagnetic brake controller. This controller was installed by the manufacturer as part of a tow package. Other controllers may be installed after market. It is important that the fire apparatus operator understands how to properly adjust the trailer brakes using the brake controller.

is more expensive, there are clear advantages to a proportional controller, as the driver can regulate how forcefully the trailer brakes are applied.

Surge Braking Systems

In a surge brake system, the momentum of the decelerating trailer will cause the brake system to activate. When the tow vehicle applies its brakes, the trailer will push against the hitch and engage a hydraulic master cylinder. The master cylinder will engage a hydraulic brake system on the trailer, causing the trailer to slow down (fig. 17–2).

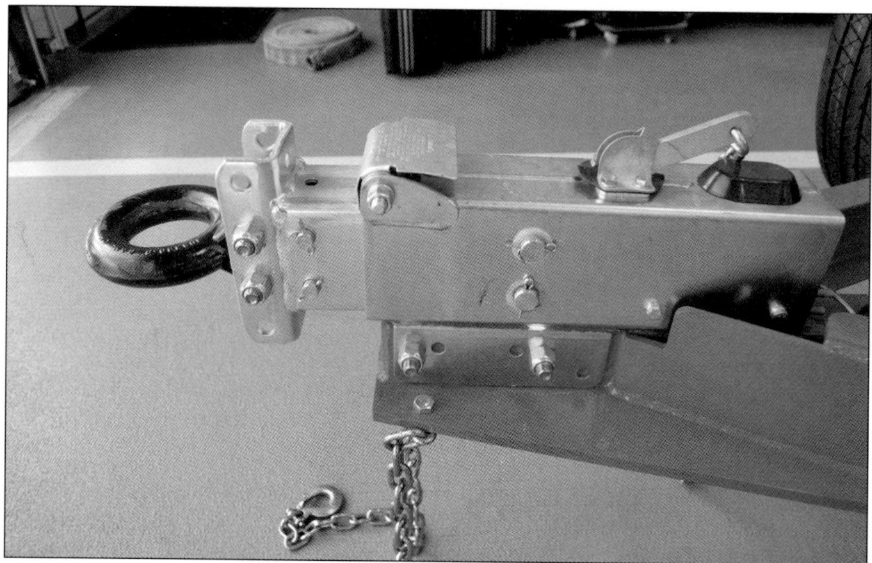

FIGURE 17–2.   This trailer is equipped with a surge brake system. It does not require a separate controller installed in the driver's compartment of the tow vehicle.

The advantage to a surge brake system is that there is no need for a separate controller inside the tow vehicle. This is why many rental car companies use surge brakes on their trailers. A person who rents a trailer has no need to install an electronic controller in the cab of the tow vehicle. Surge brakes are also used on boat trailers because there are concerns that dipping the trailer into salt water to launch a boat could have a negative effect on an electromagnetic brake system.

The disadvantage to surge brakes is the same as the advantage: there is no controller in the cab of the tow vehicle. Therefore, the driver of the tow vehicle has no means to engage the trailer brakes separately. If there is a situation where the driver needs to engage only the trailer brakes, such as a downhill grade or extreme trailer sway, it can't be done. Another issue with older surge brake systems was the fact that they may engage suddenly when going down steep hills or would not engage when going backwards.

Adjusting Trailer Brakes

Keep in mind that before towing a trailer, it is important to properly adjust the brakes. This will prevent the trailer brakes from locking up prematurely or not braking hard enough when needed. Brake adjustment procedures will depend on the type of brake system in use, as well as the manufacturer's recommendations.

Ensure that this type of adjustment is done before an emergency so that the tow vehicle and trailer are ready to go when needed.

Breakaway Systems

Trailers should also be equipped with a functioning breakaway system. In a breakaway system, a cord or cable is attached between the tow vehicle and the trailer. If the trailer breaks away from the tow vehicle, the brakes on the trailer will automatically engage, stopping the trailer. These systems are prone to malfunction, as they rely on a battery that is mounted to the trailer and the batteries often go bad. Before towing a trailer, the breakaway system should be tested by pulling the cable and ensuring that the trailer brakes properly engage.

Sway Control Systems

When purchasing a trailer, the fire department should also consider the purchase of an aftermarket sway control system. These systems use friction or a camlike device to reduce trailer sway. An aftermarket sway control system will pay for itself if it prevents just one sway-induced collision.

To understand how a sway control system works, consider how a hitch works. The trailer is connected to the tow vehicle at the ball hitch. When the tow vehicle steers from side to side, the trailer pivots around the ball hitch and follows the tow vehicle. However, if the trailer starts to pivot around the hitch ball due to wind or excess speed, it may result in an uncontrolled oscillation which causes the tow vehicle to lose control.

Sway control systems are constructed to provide lateral stability so the trailer no longer pivots around the hitch ball. There are too many systems to discuss individually. Nevertheless, it is recommended that these devices be considered for any tow/trailer combination that will be used for emergency operations. Once a system is purchased, tow vehicle operators should receive extensive training on how the individual system works.

Trailer Operations

Towing a trailer is one of the most difficult tasks a driver could face. The dynamics of a moving trailer can befuddle even the most experienced fire apparatus operator. This is especially true in fire departments that use trailers for specialized activities that are rarely needed.

The first step in safe trailer operations is to properly secure the trailer to the tow vehicle. Drivers must ensure that the connection between the two vehicles is secure and there is a proper amount of tongue weight on the hitch system. Drivers must also ensure that the trailer and tow vehicle are equally matched in height and that the tow vehicle and trailer are level and parallel to the ground. If the rear of the tow vehicle dips towards the ground, reevaluate the weight distribution of the trailer and make sure that the tow vehicle can properly handle the weight of the loaded trailer.

Make sure the vehicle is equipped with a hitch that is properly rated for the weight being towed. Drivers must also ensure that the vehicle's ball hitch is the proper size for the coupler and rated for the load. Using a different size ball and coupler could lead to a trailer separation. The latch on the coupler must be properly secured with a padlock or other locking device.

Ensure that chains are used at all times and that the chains are crossed. The reason for crossing the tow chains is to create a "nest" so if the hitch fails, the tongue of the trailer will drop into the nest instead of completely separating and slamming into the ground.

Ensure that all electrical connections are secure and working properly. If there is no electrical connection to the electromagnetic brakes, the brakes on the trailer won't work. If there is no electricity to the trailer, the brake lights and turn signals also will not work. Because the brake lights and turn signals on the tow vehicle are blocked by the trailer, there will be no warning lights visible to vehicles following the trailer.

Ensure that the breakaway system is intact and functioning properly. The breakaway cable needs to be attached to the rear of the tow vehicle and not to the chains. If the breakaway cable is attached to the chains, it will not pull away and trigger the brakes during a trailer separation. Instead, it will just drag along with the chains and never activate the emergency braking system.

Make sure that all hubs and bearings are well lubricated. This is especially true for trailers that are not used very often and for trailers that are used to launch boats into the water. If the trailer is used infrequently, or if the trailer is dipped in water, the bearings and hubs may corrode or lose lubrication. A bearing failure will result in an overheated axle which can result in mechanical failure or fire. It is a good idea to check the hubs by feeling them with your hand or by using a thermal imaging camera. If one hub is significantly hotter than the other, it could indicate a problem. Finally, make sure that tires are properly inflated and check the brakes before leaving the station.

Keep in mind that these procedures must be completed every time a trailer is towed. There is a strong temptation in the fire service to say "we're only driving a few miles and we need to get there quickly." As the saying goes, if you don't get there, you are no good to anyone. Taking shortcuts while hooking up a trailer may result in the trailer breaking free and crashing during an emergency response. If the trailer or tow vehicle crashes, the equipment is no good to anyone.

Once the trailer has been properly connected to the tow vehicle, the vehicle is ready for use. Fire apparatus operators who tow trailers must be well-versed in the handling characteristics of both vehicles. Understanding the difference in stopping distance, acceleration distance, and turn radius of the tow and trailer combination are important training points that must be addressed before driving the vehicle.

Acceleration

Because the tow vehicle is pulling more weight, the acceleration rate of the tow vehicle may be affected. A tow vehicle that has powerful acceleration when driven without a trailer may struggle to gain speed with a trailer attached. Vehicle operators must avoid pulling in front of an oncoming vehicle by allowing more time and space when entering highways or making turns. The tow vehicle may also struggle when climbing hills or driving on an unpaved surface. In these situations, it may be useful to downshift to a lower gear to increase power.

Towing a trailer requires an apparatus operator to slow down. Excess speed can cause trailer sway, longer stopping distances, and safety issues related to curves and lane changes. This is especially true when driving off-road or crossing uneven terrain such as construction zones or rail crossings. Ensure that the vehicle is kept at a safe speed at all times.

Drivers must remember that it is going to take much longer to pass another vehicle because of the added trailer weight. Make sure to leave enough room and allow extra distance behind the vehicle when getting back into the travel lane. Departments should consider policies that prohibit lights and sirens when towing a trailer.

Braking

Towing a trailer will affect the braking efficiency of the tow vehicle. The added weight may result in longer stopping distances, especially if the trailer brakes are poorly adjusted or are not working properly. In addition, the tow vehicle and trailer may be more susceptible to skids or jackknifes if the trailer brakes are adjusted too powerfully or if the trailer is lightly loaded. This is especially problematic with time-delay brake controllers and on slick roadways. When towing a trailer, the driver must look as far down the road as possible and anticipate problems. Looking ahead will allow the operator to slow down in advance of the problem, rather than waiting for the last minute and standing on the brakes.

When pulling a trailer, fire apparatus operators must slow down much sooner than they are used to. Braking should also be done in a smooth and controlled fashion so that the momentum of the trailer does not push on the tow vehicle and cause a loss of control. Smooth braking will also prevent the equipment in the trailer from shifting, which may result in damaged equipment or a loss of control.

When possible, downshift the tow vehicle and use the retarding force of the engine to slow the vehicle. Using the transmission for speed control will reduce wear on the brakes and ensure that the brakes are cool and ready to use should they be needed in an emergency. This is especially important when descending long grades or steep hills, and beware of brake fade.

Turning and Curves

A trailer will affect the vehicle's ability to turn, take corners, and change lanes. The position of the trailer axles will cause the trailer to *off-track*. When a vehicle off-tracks, the trailer wheels will travel in a different line from the tow vehicle. As a result, the trailer may be prone to hit curbs or roadside obstacles if the trailer is towed across a curb or corner. The tow vehicle must make a wide enough turn to account for the off-track of the trailer. There may be times where the driver will have to wait for a light to turn green so that traffic on a cross street clears, which will allow the tow vehicle to swing the trailer wide enough for the turn.

Many trailers are taller than the tow vehicle. As a result, the rollover threshold of the trailer may be less than that of the tow vehicle. It is not uncommon for a driver to forget about the trailer and drive the tow vehicle in a manner they normally deem safe. When this occurs, the driver may exceed the capabilities of the trailer and cause it to rollover while turning or rounding a curve too fast. Excess speed during curves, lane changes, or while turning corners, can lead to a rollover. Also, remember that the height and angle of departure of the trailer could cause issues when driving under low-hanging obstacles or when trying to maneuver up or down steep inclines or driveways.

Backing

Backing a trailer is tricky. When backing a trailer, drivers should place their hand on the bottom of the steering wheel. The direction drivers move their hand is the direction the trailer will go. If drivers move their hand to the left, the trailer will steer to the left. If drivers move their hand to the right, the trailer will move to the right.

When turning the trailer during a backing maneuver, the driver should use small movements of the steering hand, as the movement of the trailer will be exaggerated. The driver must not forget that the front of the tow vehicle will swing. Concentrating on the rear of the trailer and forgetting the swing of the front end can result in a crash.

Remember that the shorter the distance between the trailer axle and the hitch, the harder it will be to keep the trailer straight. Ironically, a larger trailer with a longer span between hitch and axles will back up more smoothly. This is because smaller trailers only require small inputs to make them change direction. As a result, backing a small trailer in a straight line can be more difficult. Use at least one spotter to ensure that the maneuver can be done with safety.

Trailer Sway

Another issue that can affect the safety of a trailer combination is *trailer sway*. Sway is excess side-to-side movement of the trailer as it drives down the road. Consider the fact that a typical hitch is located several feet behind the rear axle of the tow vehicle. Because the trailer hitch is located behind the rear axle, the trailer may have enough leverage to move the rear end of the tow vehicle back and forth, creating an oscillating effect. Severe oscillations can cause the tow vehicle's rear end to fishtail.

Trailer sway is more common in vehicles with a soft suspension and a longer distance between the rear axle and the hitch. If a vehicle has a stiffer suspension and a shorter distance between the rear axle and the hitch, the tow vehicle will be less susceptible to trailer sway.

There are many causes of trailer sway, including:

Weight Distribution

A common cause of trailer sway is improper weight distribution inside the trailer. If there is not enough weight on the trailer tongue (tongue weight), the trailer becomes unstable. More weight behind the axle of the trailer than at the hitch connection will cause the trailer to act like a pendulum. The trailer will sway from side-to-side. To increase the stability of the trailer, ensure that the trailer is loaded properly with a safe tongue weight.

Wind

If the trailer is larger than the tow vehicle, a breeze or wind will produce a much greater side force on the trailer. The trailer will act like a large sail as the wind force pushes it from side-to-side.

Tractor Trailers

A bow wind is a force of air created by a tractor trailer as it drives down the road. If the tractor trailer passes a vehicle towing a trailer, the bow wind created by the tractor trailer may cause the trailer to sway. These situations can prove quite startling if the driver of the tow vehicle is not prepared for it.

Roads

Uneven or bumpy roads can also induce trailer sway. This is especially true when driving off-road or where there may be two different road surfaces due to construction. The trailer may drop into a dip or off the road edge, causing it to sway.

If the trailer starts to sway, the driver should let off the accelerator, take a firm grip of the wheel, and try to keep the tow and trailer combination stable and in a straight line. Engaging the electromagnetic brake controller may cause the trailer sway to dampen. Applying the brakes on the tow vehicle during a sway may actually cause the trailer sway to worsen. This is another advantage of electromagnetic brakes over surge brakes, as the driver can control the trailer brakes separate from the tow vehicle. Trailer sway can be prevented by ensuring that the trailer tongue weight is properly distributed, a sway control system is in use, and the driver maintains a safe speed.

Maintenance and Inspections

It is not uncommon for a fire department to purchase a trailer and then park it behind the firehouse and forget about it. Keep in mind that a trailer is no different from any other fire apparatus. It must be inspected and maintained on a regular basis. This is especially true if the trailer is parked outside where it is under constant attack from the elements. Tire pressures, tire conditions, brake components, and other mechanical parts must be inspected to ensure that they are in good condition and ready to go to work. Failure to inspect and maintain the trailer may result in a component failure that causes a crash. When purchasing a trailer for fire department operations, fire departments should consult with NFPA 1901—Chapter 26. This chapter deals specifically with trailers.

Off-Road Driving

Introduction

Fires do not always occur in single family dwellings adjacent to a paved roadway. Fire apparatus operators are often faced with the difficult task of driving a large and unwieldy fire apparatus off the road and into dangerous terrain. Fire apparatus operators must have a keen understanding of off-road driving, especially operators faced with large amounts of rough terrain.

A thorough knowledge of the vehicle is the first step to safe off-road driving. Drivers must understand the vehicle's capabilities and, more importantly, its limitations. Driver training programs should include a thorough explanation of the vehicle and how it will handle when faced with adverse off-road conditions. Driving mistakes are magnified when driving off-road, leaving little room for error.

Understanding Four-Wheel Drive

While there is different terminology among manufacturers, drivers must understand how four-wheel drive works. Consider the following types of vehicles:

Front-Wheel Drive

In a front-wheel drive vehicle, the power generated by the engine is transferred to the two front wheels. The two front wheels drive the vehicle. Front-wheel drive vehicles tend to provide better traction on slick roads because the weight of the engine, transmission, and mechanical components rest on the front axle. Most front wheel drive vehicles provide better fuel economy because they do not weigh as much as other vehicles. The disadvantage of a front-wheel drive vehicle is that it tends to be nose heavy. A nose-heavy vehicle may not handle as well during high-speed or performance driving. Front-wheel drive vehicles are also susceptible to *torque steering*, which means the vehicle may pull to the right or to the left during heavy acceleration.

Rear-Wheel Drive

In a rear-wheel drive vehicle, the power generated by the engine is transferred to the rear wheels via a drive shaft. The two rear wheels drive the vehicle. Rear-wheel drive vehicles tend to have better acceleration. This is because weight shift puts more weight on the rear tires when the vehicle is accelerating, improving the rear tire's ability to grip the road and drive the vehicle. Rear wheel drive vehicles also tend to have better weight distribution, which improves stability and handling. The disadvantage of a rear-wheel drive vehicle is that it may have poor handling on a slick road. The rear end will tend to break traction and slide-out, inducing an oversteer (see Chapter 5).

All-Wheel Drive

Many manufacturers are now producing all-wheel drive vehicles. Many of these vehicles do not look like a typical four-wheel drive vehicle, as they are not equipped with a raised chassis and large, knobby tires. There are part-time and full-time all-wheel drive vehicles.

In a part-time all-wheel drive vehicle, the vehicle operates as a front-wheel drive until it senses a need for all four wheels. When the vehicle senses a need for all four wheels, it will engage the rear wheels. In a full-time all-wheel drive vehicle, power is transmitted to all four wheels all of the time. All-wheel drive is designed to handle slippery conditions on a paved highway, similar to driving in the four-wheel drive high setting. The disadvantage of an all-wheel drive vehicle is its increased weight, which translates to lower fuel economy.

Four-Wheel Drive

In a four-wheel drive (4WD) vehicle, power is transferred from the drive train to a transfer case. The transfer case then divides the power between the front and rear axles so that maximum torque is applied to all four wheels. If the rear wheels are stuck and have no traction, the front tires will still be able to provide grip and pull the vehicle out of a hazard.

There are full-time and part-time four-wheel drive vehicles. A part-time 4WD vehicle allows the driver to drive in two-wheel drive and only activate the 4WD when needed. Once 4WD is needed, the driver can decide to engage 4WD high and 4WD low.

Four-Wheel Low

Four-wheel low is designed to provide more torque (power) for off-road use. Four-wheel low should only be used at slow speeds when traversing rough terrain. Four-wheel low will provide more power to go up hills and provide increased engine braking to descend hills.

When switching to four-wheel drive low, it is important to know if the vehicle must slow to 2–3 mph, or come to a full stop. Keep in mind that switching to four-wheel drive low may cause noise from the transfer case. This is normal. Some vehicles may require the wheels to be locked. Again, this is something that must be determined ahead of time by reading the owner's manual.

Drivers must remember that they may need to stop or slow the vehicle when switching to four-wheel drive low. If the vehicle is in sand, mud, or snow, the vehicle may become bogged down if the driver has to stop and switch over. The driver should switch to four-wheel drive low before it is needed to prevent the vehicle from getting stuck or bogged down (fig. 18–1).

An important consideration when driving in four-wheel low is the vehicle's *crawl ratio*. Crawl ratio is the ability to travel over rough terrain in first gear at

FIGURE 18–1.   Think ahead and switch to four-wheel drive low before you need it. If you have to stop to switch, the vehicle may bog down (photo courtesy of TheDrivingCompany.com).

idle speed without having to depress the clutch pedal to keep the vehicle from stalling. A low crawl ratio is useful when descending a hill, as it allows the driver to maintain a slow and controllable speed without using the brakes. Applying the brakes on a loose or slippery surface could cause the wheels to lock and skid. A constant application of the brakes on a long, downhill grade could also cause the brakes to overheat and fade.

When using the brakes over long distances on loose terrain, it is better to use the retarding force of the engine instead of the service brakes. This will help prevent the wheels from locking. Drivers must also be wary of accelerating on loose terrain. Applying too much acceleration may cause the wheels to slip and lose traction. Know how to crawl the vehicle before it is needed.

Four-wheel low should not be used on paved roads during normal driving operations. It is also important to remember that four-wheel drive low will not provide more traction, only more torque. More torque can cause the tires to slip on snow, ice, or mud. The possibility of the tire slipping due to excess torque is another consideration when deciding on which four-wheel drive setting to use.

Four-Wheel High

In a part-time four-wheel drive vehicle, the driver can usually switch to four-wheel drive high while the vehicle is moving. This is known as "switch on the fly." Four-wheel drive high is designed for slippery conditions like loose dirt, gravel, or sand; it is not designed for rough off-road conditions. It is important to check the owner's manual to determine the vehicle's limitations before driving on a paved highway in four-wheel drive high. Some vehicles may have steering and braking issues if they are driven on a paved highway in four-wheel drive high.

Regardless of which type of four-wheel drive system the vehicle is equipped with, fire apparatus operators must remember that a four-wheel drive vehicle will not stop any quicker than a two-wheel drive vehicle. Four-wheel drive will only provide better rolling traction, which will help improve the performance of the vehicle when it is accelerating on slick roads or operating in off-road conditions. However, once the vehicle stops accelerating and begins to brake or corner, it will behave the same as every other vehicle on the highway. Hence, the large number of brand new 4 x 4 vehicles found flipped over in the median strip during winter driving conditions.

Drivers must be thoroughly versed in how and when to engage the four-wheel drive system, and when to use the high or low setting. Drivers must also remember to shut off the engine retarder during off-road situations so it does not engage on a loose surface and cause the wheels to lock-up. Understanding the abilities and limitations of the apparatus during off-road operations is a key facet of safe driving.

Understanding Differentials

When a vehicle enters a turn, the inside tires have to spin more slowly than the outside tires so the vehicle can pivot and make the turn. Consider a high school band as it attempts to make a turn on a football field. The band members on the outside of the drum line have to march faster, while the band members on the inside of the drum line march in place.

When a vehicle makes a turn, problems would arise if all of the tires rotated at the same speed. The inside tires would scuff and skid when the vehicle turned to the right or to the left. To solve this problem, manufacturers use *differentials*. Differentials allow the outside tires to spin quicker than the inside tires, preventing the inside tires from being dragged through the turn. In a tight turn, the inside tires may be barely moving, while the outside tires are moving quickly to allow the vehicle to make the turn.

Open Differentials

In a vehicle equipped with an *open differential*, torque is split equally between the wheels on the axle, regardless of the road conditions. Splitting the torque equally becomes a problem when one wheel is on a slick surface and the other is on a dry roadway. This is because the vehicle will only send enough torque to the wheels to make them spin, but not enough to make them slip.[1] If one tire is on dry asphalt and the other tire is on ice, it will take very little torque to make the tire on the ice spin. Because it does not take much torque to make the wheel spin on the ice, the tire that is still in contact with the pavement will not receive enough power to make it spin. Instead, the tire on the ice will be spinning freely, and the vehicle will not move. In an open differential, the wheels will receive equal torque but will have unequal rotational speed.[2]

Locking Differential

If the vehicle is equipped with a locking differential, the wheels on the axle are locked together and will both spin at the same speed when torque is applied, regardless of the traction. If one wheel is on a dry roadway and the other is on a slick surface, a locked differential will still provide power to the tire in contact with the dry road. There will be equal rotational speed and unequal torque at each wheel.[3] While locking differentials help maintain traction, differentials are not locked at all times because it would cause the tires to scuff during a turn. This explains the need for open differentials.

Limited-Slip Differential

In a *limited-slip differential*, engineers sought to combine the advantages of open and locked differentials. A limited-slip differential will allow the tires to rotate at different speeds, preventing wheel slip during turns. However, a limited-slip differential will also allow torque transfer between tires when needed. Many modern day vehicles are equipped with limited-slip differentials.

Off-Road Driving Considerations

Driving Surface

Is the vehicle driving on snow, gravel, sand, dirt, or mud? Remember that damp soil gives the best traction, while loose, dry soil can cause the tires to spin during acceleration or lock-up while braking. Wet or muddy soil can be dangerous, as it has a tendency to fill the tire tread and turn the tires into skis, reducing the vehicle's ability to stop and steer.

Terrain

Is the vehicle driving on hills? How steep are the hills? Is the vehicle driving up and down the hills, or driving on a cross slope? Are there rocks, stumps, or other ground obstructions that may cause the vehicle to hang up or blow a tire?

Vehicle

What type of vehicle is being driven—a Jeep or a large pumper? What is the rollover threshold of the vehicle? What type of tires does the vehicle have, and what are the tire pressures? What is the angle of approach and angle of departure? Does the vehicle have four-wheel drive? Does the vehicle have an automatic or manual transmission? Is the engine retarder turned off?

Off-Road Operations

Off-road vehicle operations are difficult and dangerous, especially when faced with rough terrain and an active fire. For this reason, fire departments must develop specific policies and procedures that outline the conditions during which a fire apparatus operator can drive the apparatus anywhere other than a paved- or hard-surface road. These conditions may include brush fires, technical rescues, or medical transports.

All fire apparatus operators must be well versed in off-road driving skills and consider the following points:

- Drive straight up or straight down a hill. Turning the wheel while driving up or down a hill may cause the vehicle to lose control and roll over.
- Driving across a cross-slope or down a hill at an angle may cause the vehicle to rollover. Fire apparatus are already prone to rollover on level ground. By introducing terrain that has roll to it, the safety margin is further reduced (fig. 18–2).
- If the vehicle begins to slide while traversing a cross-slope turn the tires in the direction of the slide and release the brakes and throttle. This may allow the vehicle to regain rolling traction and allow the driver to regain control of the vehicle. If the driver continues to allow the tires to slide sideways, it may induce a tripped rollover.
- Understand that the angle of approach and the angle of departure of the vehicle may cause it to get stuck as the vehicle tries to go up or down a hill.

FIGURE 18–2.   Most fire apparatus already have a low rollover threshold because of their high center of gravity. If the vehicle is driven on a cross-slope, the center of gravity will shift towards the outside of the vehicle, further reducing the vehicle's rollover threshold, making the vehicle even more unstable (photo courtesy of TheDrivingCompany.com).

- Know the hill-climbing limitations of the vehicle. How steep is too steep? If the vehicle can't make it, don't try.
- Understand that the slope of the hill may be too severe for the parking brake to hold. Set the brake and slowly release the brake pedal to see if it will hold. There may be times when a member has to sit in the cab with a foot on the brake. Under these circumstances, ask yourself if the vehicle should be there in the first place.
- Always use chocks when the vehicle is parked. At a minimum, the rear tires should be chocked in front and in back of the tire.
- When climbing a hill, the weight will shift towards the rear of the vehicle, causing the vehicle to sink onto the rear suspension. This may result in a loss of traction on the front tires. If the weight shift is severe enough, the front tires may actually lift off the ground and cause the vehicle to lose control.
- When climbing a hill, know what is on the other side. Beware of driving up a hill, only to reach the crest and discover a cliff or large rocks. Before ascending a hill, send a member to reconnoiter the area and ensure the vehicle will be able to safely continue after cresting the hill.
- If the vehicle stalls while climbing a hill, back down and try again. The driver may need to gather some speed before climbing the hill, depending on the ground conditions and the limitations of the vehicle and tires. Remember to account for the front and rear angles of deflection.
- Remember that if the vehicle has a manual transmission and the vehicle stalls while climbing a hill, depressing the clutch to restart the vehicle may cause the vehicle to roll backwards. Know how to start the vehicle safely in first gear without using the clutch or by using other means to keep the vehicle from rolling backwards.
- When going downhill, use the lowest gear possible and allow the engine compression to brake the vehicle. Manual transmissions typically have better engine braking than automatic transmissions. Don't ride the clutch.
- Applying the brakes while going downhill on loose ground may cause the vehicle to skid and lose control.
- If the vehicle becomes stuck in snow or loose ground, spinning the tires may cause the vehicle to dig in and sink further into the loose substance.
- When climbing or mounting obstacles, use low gear at idle speed. Only apply light brake or throttle if needed. Applying the brakes or throttle may lead to a skid or loss of control.

- Whenever possible, don't straddle rocks, stumps, or other ground obstructions. Drive over them instead. This will help prevent a ground obstacle from damaging the undercarriage of the vehicle.
- If the driver must straddle an obstacle, use a spotter and ensure that the vehicle does not strike the obstacle with any part of the undercarriage.
- Know the layout of the undercarriage and where the lowest points are located. A differential may not be located midline on the vehicle. The driver may be able to maneuver around the obstacle, keeping the lowest part of vehicle away from an obstacle.
- When crossing a ditch, know the angle of approach and the angle of departure of the vehicle. Beware of driving down into the ditch, only to become stuck or wedged inside of it.
- Before crossing water or mud
  - Be sure to check the depth of the obstacle before driving into it.
  - After crossing water or mud, remember that the braking efficiency may be reduced due to wet brake components. Lightly apply the brakes while driving at a slow speed to help heat them up and dry off the water.
  - Know where the air intake on the vehicle is located so you don't allow water into the engine and stall or destroy the engine.
  - Drive slowly through water. Remember that you can't see what's in front of the vehicle. If there is a wave forming in front of the vehicle, slow down.
- Beware of rocks that may become wedged in dual rear tires which may damage tire sidewalls and lead to a blowout.
- When driving in sand
  - Try to keep moving at all times so as not to become bogged down.
  - Reducing tire pressures will help balloon the tires, increasing the surface area that is in contact with the ground. The proper air pressure will depend on the consistency of the sand.
  - If the tire pressures are reduced, be aware that there is a higher likelihood that the tire bead may separate from the rim, resulting in a flat tire.
  - If the vehicle becomes stuck in the sand, try to wet the sand to increase traction.
  - Don't spin the tires on sand, as it may cause the vehicle to dig deeper into a hole.
  - If possible, roll to a stop. Applying the brakes may cause the vehicle to skid or dig into the sand.

- After deflating tires for off-road driving, re-inflate them immediately upon returning to the pavement. Driving on an underinflated tire is dangerous.

- Be very aware of the capabilities of the road. Avoid driving on unstable roadways which may collapse under the weight of the vehicle and cause it to rollover.

Off-Road Seatbelt Use

It should also be noted that NFPA 1500 addresses the use of seatbelts and members riding on the outside of fire apparatus during wildland firefighting operations.

> *NFPA 1500 A.6.3.1—The Fire Line Safety Committee (FLSC) of the National Wildfire Coordinating Group (NWCG) represents the U.S. Forest Service, Bureau of Land Management, Bureau of Indian Affairs, Fish and Wildlife Agency, National Park Service and National Association of State Foresters. Their position is that the practice of fire fighters riding on the outside of vehicles and fighting wildland fires from these positions is very dangerous, and they strongly recommend this not be allowed. One issue is the exposure to personnel in unprotected positions. Persons have been killed while performing this operation. Also, the vehicle driver's vision is impaired. The second issue is that this is not an effective way to extinguish the fire, as it can allow the vehicle to pass over or by areas not completely extinguished. Fire can then flare up underneath or behind the vehicle and could cut off escape routes. The FLSC and the NWCG strongly recommend that two firefighters, each with a hose line, walk ahead and aside of the vehicle's path, both firefighters on the same side of the vehicle (not one on each side), in clear view of the driver, with the vehicle being driven in uninvolved terrain. This allows the firefighters to operate in an unhurried manner, with a clear view of fire conditions and the success of the extinguishment. Areas not extinguished should not be by passed unless follow-up crews are operating behind the lead unit and there is no danger to escape routes or to personnel.*

Notes

1. Karim Nice, "How Differentials Work," *howstuffworks*, retrieved from http://auto .howstuffworks.com/differential3.htm.
2. "Locking Differential," *Wikipedia*, retrieved from https://en.wikipedia.org/wiki /Locking_differential
3. "Locking Differential," *Wikipedia*.

Off-Road Recovery Operations

Introduction

Despite our best efforts, there will be times when an off-road vehicle gets stuck. When this happens, a winch or recovery strap will help get the vehicle back on the road. Unfortunately, many departments fail to provide adequate training on the safe use of these devices.

Before operating in an off-road situation, ensure that the fire apparatus is equipped with properly rated straps, chains, hardware, and a winch. In order to ensure a safe recovery operation, the vehicle should be weighed, and the recovery equipment should have a safety factor of at least 2:1. The vehicle must also be equipped with properly rated recovery points which are securely mounted to the vehicle, such as tow hooks and brackets.

Fire departments should consider three possible scenarios in an off-road recovery situation:

1. A vehicle that is stuck off-road using its own winch to pull itself back onto the roadway
2. A vehicle that is stuck off-road being winched back onto the road by another vehicle
3. A vehicle that is stuck off-road being "snatched" back onto the roadway by another vehicle using a recovery strap instead of a winch

Each scenario will involve different equipment and recovery methods. Failure to use the proper equipment and recovery method can result in an unsafe operation that may cause severe injuries.

Tow and Recovery Points

Apparatus operators must have a thorough knowledge of the location and weight rating of the recovery points on each vehicle. Some stock vehicles that are converted to firefighting vehicles may be equipped with hooks and rings that appear appropriate for recovery operations. However, these hooks and rings are only meant to secure the vehicle to a dealership delivery truck. In these situations, fire departments should consider the aftermarket installation of properly rated recovery points for use in an off-road recovery. Check with the vehicle manufacturer to determine which anchor points are safe to use. Anchor points should be accessible from both the front and rear of the vehicle (fig. 19–1).

Winches

When selecting a winch for a vehicle, most manufacturers recommend multiplying the weight of the vehicle by 1.5. If the vehicle weighs 10,000 lbs, it should be equipped with a winch that has a capacity of at least 15,000 lbs. Remember that unwinding the winch rope affects the rated capacity of the winch.

A winch's maximum pulling capacity is rated with the first layer of rope wound on the drum. The pulling capacity is reduced as more layers wind onto the drum. On average, each layer of rope wrapped around the drum will reduce the pulling power by approximately 10–15%. Fire departments should create charts for the apparatus which provide the pull capacity of the winch based on how many layers of rope are wound on the drum. Exceeding the pull capacity of the winch will result in a failure of the system (fig.19–2).

If the pull is to be made extremely close to the load or anchor point, consider using snatch blocks to create a mechanical advantage system. This will increase the amount of rope that is unspooled from the winch drum while also adding a mechanical advantage to the recovery system. Also remember that the winch rope should be stretched before its first use. Refer to the manufacturer's recommendations on this procedure.

It is important to know if you have an electric winch or a PTO-driven hydraulic winch, as there are pros and cons to both systems. An advantage to the electric winch is that if the vehicle is stalled or disabled, the winch should still be able to run off of battery power, though not for very long. When running off of the battery, keep a close eye on the voltage gauges so the battery doesn't completely drain down. Also remember that using the winch will create heat inside the motor. During a long pull, it may be necessary to stop and allow the winch to cool down.

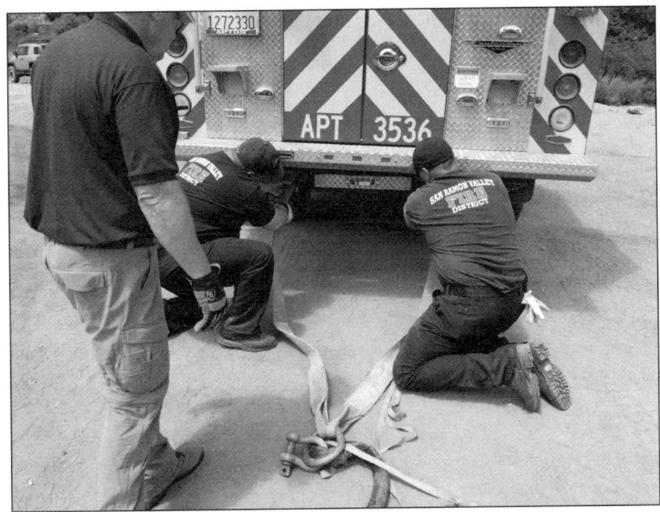

FIGURE 19–1. Fire apparatus should be equipped with properly rated recovery points. Do not assume that a hook or loop that is mounted on the vehicle is designed to be used for off-road recovery. It may simply be a secure point to lash a vehicle to a delivery truck (photo courtesy of TheDrivingCompany.com).

FIGURE 19–2. Fire apparatus operators must know the maximum pulling capacity of their winch. In this case, the manufacturer has listed the pulling capacity directly on the face of the winch. Notice how the capacity changes based on how many layers of rope are wrapped around the drum.

An advantage to a hydraulic winch is that it does not place as large a load on the vehicle's electrical system. Hydraulic winches are also not as prone to overheating. The downside to the hydraulic winch is that if the vehicle is stalled or bogged down, it might not function.

Anchor Straps and Recovery Straps

Once a properly rated winch has been mounted to the vehicle, ensure that the vehicle is equipped with anchor straps and recovery straps that are properly rated for the weight of the vehicle. As a rule of thumb, one inch of strap webbing equals 10,000 lbs of load: therefore, a 3-in. recovery strap would be rated at 30,000 lbs. Make sure to properly match the straps to the gross weight of the vehicle using a safety margin of at least 2:1. If the vehicle weighs 15,000 lbs, the vehicle should carry straps that are rated at 30,000 lbs.

There are different types of straps used for off-road recoveries. Fire apparatus operators must have a thorough understanding of when and where to use these different straps (fig. 19–3).

Tree Trunk Protector

A tree trunk protector, or "rigging strap," is made of nylon webbing and is non-elastic. Because the strap is nonelastic, it is safe to use as an anchor point for a winch pull. Tree trunk protectors should be wrapped around the anchor point and secured with a D-shackle or clevis. The winch hook is then attached to the D-shackle, and the winch pulls the vehicle free.

FIGURE 19–3. This fire department has constructed a rigging kit, complete with straps, chains, and D-shackles. These kits are a great idea as they keep all of the rigging equipment together in one place, making it easily deployable when needed.

Choker Chains

A choker chain can also be used as an anchor strap for winch operations. The advantage to the choker chain is that it can be used on sharp objects that may cut through a nylon tree trunk protector. The disadvantage is that a choker chain may damage the anchor point. Depending on the situation, this may not be an issue during fire department operations.

Recovery Straps

A recovery strap is used by one vehicle to pull or "snatch" another vehicle that is stuck off the road. These straps are often made of nylon and have elastic properties that act like a giant rubber band. The elastic properties of the recovery strap will allow it to stretch out, transferring the stretching energy to the trapped vehicle and pulling it back onto the road. Due to their elastic properties, these straps should never be used as anchor points for winch operations. These straps should only be used by one vehicle to snatch another vehicle free.

Tow Straps

A tow strap is used to pull one vehicle behind another when the wheels of the towed vehicle can move freely. Tow straps are typically made from polyester and are not elastic like a recovery strap. Tow straps should not be used for snatching another vehicle because the lack of elasticity may cause the strap to snap when the tow vehicle tries to pull the stuck vehicle free.

Using a Winch for Recovery Operations

There are two scenarios for a winch recovery:

1. The stuck vehicle is winching itself out.
2. The stuck vehicle is being winched out by another vehicle.

The first step in winch operations is safety. The anchor points selected for a winch operation should be secure and bombproof. Ensure that the winch and winch rope are not overloaded and are properly secured to the anchor points. If a winch rope or anchor point is incorrectly loaded, it may break or slip loose, resulting in a dangerous whip that could severely injure or kill someone nearby. At least five wraps of rope should always be left on the winch spool. This will ensure that the connection point between the rope and the drum does not take all of the strain from the pull.

During a winch operation, members should not be standing in the area of the winch rope because if the rope breaks it will have a tendency to snap back. Members should avoid touching a winch rope that is under tension and inspect the rope to ensure that there is no damage or frays before making a pull. Wear gloves at all times, as some ropes may develop burrs or slices that can severely cut your hands or fingers. Use a hook strap to hold the hook while spooling the winch cable in or out. Holding directly onto the hook could result in injury if the winch is accidentally engaged while the hook is being held in someone's hands.

Some experts recommend placing two mats or coats on the winch rope, one approximately 8–10 ft from the anchor point and another 8–10 ft from the winch drum. If the rope should snap, the mats will absorb some of the energy created by the whipping action of the broken rope. The snap of the rope will be dampened by the mats, or the rope may even be driven down towards the ground. It is also helpful to open the hood of the winching vehicle as it will create a shield to help protect the driver.

When using an electric winch, leave approximately 1 ft of slack in the rope before engaging the winch. This will give the winch time to wind up to full RPMs before it begins to pull on the load. This will ensure that the winch is at full power when it begins to pulling the load.

Anchor Points

A solid connection to a bombproof anchor point is very important. Only properly rated, nonelastic anchor straps or chains should be wrapped around the anchor point. Recovery straps should not be wrapped around an anchor point due to their elastic properties. At no time should the winch rope be wrapped around an anchor point and attached back onto itself, as this could result in serious damage and possible failure of the winch rope.

When selecting an anchor point, ensure that the winch rope is as straight as possible. When the winch rope goes to the right or to the left, it may have a tendency to cross over itself while wrapping back onto the drum. A straight pull will help ensure that the winch rope winds evenly back onto the drum and does not cross over itself. If it is difficult to make a straight pull, consider using a snatch block to adjust the path of the rope. By attaching a snatch block to another solid anchor point, the direction of the rope can be adjusted for a proper pull (fig. 19–4).

FIGURE 19–4. Consider the use of a snatch block to ensure a proper pull. The safe deployment of a winch is only limited by the skill and imagination of the firefighters deploying the tool (photo provided by Warn Industries Inc.).

Anchor points should be secure and bombproof. Avoid using dead trees, loose rocks, or other items that are not securely anchored into the ground. In the event that there are no natural anchor points, a second vehicle may have to act as an anchor point. In these cases, ensure that the parking brake is set, the vehicle is in neutral, and the wheels are properly chocked.

In a situation where a second vehicle is winching the stuck vehicle back onto the road, anchor straps should be securely fastened to the recovery points on the stuck vehicle. Do not attach the anchor straps to vehicle components such as struts, axles, or other parts of the undercarriage as these parts may break off during the pull.

Do not attach the anchor strap to a tow ball, as a tow ball is designed to handle the downward force of a trailer tongue, not the pulling force of a winch operation. If possible, use a strap that is designed to attach to two recovery points on the vehicle, as this will more evenly distribute the pulling force on the stuck vehicle reducing the force on a single recovery point.

The use of a snatch block will allow the operator to unspool more rope from the drum, increasing the pulling power of the winch. A snatch block can also be used as a pulley to increase mechanical advantage.

Calculating Winch Pulls

When planning a winch operation, remember that the terrain will affect how much load the winch will have to pull. A vehicle stuck in deep mud will require more pulling force than a vehicle that is stuck on wet grass. To calculate an approximate pulling force, multiply the weight of the vehicle by the following adjustment factors:[1]

Terrain	Ground Factor
Pavement	5%
Grass	10%
Wet sand or gravel	15%–20%
Dry sand	25%–30% +
Light, shallow mud	35% +
Heavy, deep mud	50%–60% +

loaded weight × ground factor = Winch Pull

In addition to the terrain, drivers must also account for the slope of the pull. To calculate the added pulling force required for a slope, the following formula is used:[2]

((slope in degrees)/60 × loaded weight)+loaded weight = Winch Pull

In order to calculate the total pulling force needed for a winch operation, drivers must account for both terrain conditions and slope. In these situations, drivers must first calculate the winch pull up the slope and then take into account the ground factor. Adding these two values together will give the total winch pull.

Using a Recovery Strap

There may be times when a recovery strap is used instead of a winch to free a stuck vehicle. When using a recovery strap, do not connect the two vehicles by

The concept of total winch pull is best explained using an example. Consider a 10,000-lb brush truck attempting to pull itself up a 20° slope in dry sand.

Step 1—Calculate the winch pull up the 20° slope

(slope in degrees/60) × loaded weight +loaded weight = Winch Pull

(20 degrees /60) × 10000 + 10000 = Winch Pull

(0.3333 × 10000) + 10000 = Winch Pull

3,333 + 10,000 = Winch Pull

13,333 = Winch Pull

Step 2—Calculate the terrain factor for dry sand

Loaded weight x terrain factor = winch pull

10,000 × 25% = winch pull

2,500 lbs = winch pull

Step 3—Combine the winch pull up the slope and the terrain factor

Winch pull up the slope + Terrain Factor = Total Load on the Winch

13,333 lbs + 2,500 lbs = 15,833 lbs

Step 4—The total load on the winch is 15,833 lbs. The slope and the terrain will add an additional 63% to the winch load of a 10,000-lb vehicle. The additional load will vary based on the slope and the terrain encountered

tying a knot in the strap and making a loop. Instead, make sure that the strap is choked on itself and secured to designated anchor points on both vehicles. Avoid attaching the recovery strap to vehicle components such as axles, tie rods, or steering racks. A vehicle component may fail during the recovery, causing the recovery strap to whip back and injure someone.

When the operation is ready to proceed, the recovering vehicle should gently drive forward in first gear. The recovery vehicle should avoid sudden acceleration, as this may jerk the stuck vehicle and cause damage or failure of the anchor point. Once the slack in the recovery strap has been taken up, the assisting vehicle can start to pull the stuck vehicle. If possible, the stuck vehicle can apply power and assist in the recovery.

Four-wheel drive situations are advanced and dangerous operations that must be left to highly trained and experienced fire apparatus operators. Driver training programs must include an extensive overview of a fire apparatus's off-road capabilities and limitations. A knowledge and understanding of off-road recovery operations, including the proper use of winches and recovery straps, is a must for fire departments with off-road vehicles in their fleets. Knowing this information ahead of time and drilling on it regularly will help improve vehicle safety in off-road situations.

Notes

1. Jim Allen, "Jeep 4x4 Performance Handbook," *Motorbooks International*, 1998, pg. 113.
2. Allen, "Jeep 4×4 Performance Handbook."

Backing the Apparatus

Best Practices

Backing crashes are a common occurrence in the fire service. They are typically considered high frequency/low severity crashes, meaning they occur on a frequent basis but typically cause minor damage. However, an examination of recent case studies indicates that a number of backing crashes have resulted in serious or fatal injuries. Almost all of these crashes could have been prevented.

Preventing a backing crash usually comes down to common sense. The use of spotters, a thorough walk-around of the apparatus, and defensive parking tactics will prevent most backing crashes. The best approach to prevent a backing crash is to think ahead and remove any need to put the vehicle in reverse.

Keep in mind that these theories apply to more than just putting the vehicle in reverse. There may be times when tight quarters, overhead obstacles, or other nearby hazards will require the driver to stop the vehicle, get out, and look around. If necessary, the driver may need to deploy spotters to assist with the situation.

While common sense is the best defense against a backing crash, there have been a number of suggestions and recommendations made throughout the trucking industry to help reduce backing crashes. Some of these suggestions include:

1. Park defensively. Whenever possible, pull the vehicle through a parking space so it is facing out. This will remove the need to back up. If time permits, back into a parked position when arriving at a scene, rather than backing out when it is time to leave. Taking a few extra moments to back the apparatus into a safe position at the beginning of an incident may put the vehicle in a better position to fight the fire; this is especially true for aerial devices or water tankers. There may be times when driving around the block will better position the vehicle to handle the incident and also make it safer to leave the scene.

2.   Learn to use the vehicle's mirrors and understand the vehicle's blind spots. When operating large apparatus, there will be blind spots. These blind spots should be mapped out ahead of time and made part of the driver training program for each apparatus.

3.   Ensure that before driving the vehicle, the mirrors are properly adjusted. Properly adjusted mirrors will help reduce blind spots. For career firefighters who may be assigned to drive the apparatus for an entire day, this is something that can be done at shift change. Volunteer firefighters will have to adjust their mirrors every time they get in the vehicle, as the last person to drive the vehicle may have changed the position of the mirrors. Follow these steps to properly adjust the mirrors:

   a.   Position your vehicle seat properly and lean towards the driver's window until your head just touches the glass. Adjust the mirror until you can just see the outer edge of the vehicle on the driver's side.

   b.   Now lean towards the passenger side window. Once in position, adjust the passenger side mirror until you can just see the outer edge of the vehicle on the passenger side.

   c.   If so equipped, adjust the rearview mirror to provide a clear view out of the rear window.

   d.   Make sure to practice and drill on mirror adjustments and blind spots by placing cones in the blind spots and adjusting the mirrors until they can be seen clearly.

4.   If there are blind spots, make sure to use spotters. If there are no spotters available, the driver should put the truck in park and check for themselves.

5.   When backing a vehicle without any spotters, make sure to conduct a walk-around **immediately** before putting the vehicle in motion. Do not conduct a walk-around, get back in the cab, do something else for a short time, and then initiate the backing maneuver. Things may have changed. Should this be the case, get out and check again.

6.   Remember that it is often easier to see with the driver's side mirror, rather than the passenger side mirror. Position the vehicle so that the driver's side mirror is the primary mirror for the backing maneuver.

7.   If the vehicle is so equipped, learn to use the backing camera. Backing cameras are excellent tools, but they are not a replacement for a spotter. Depending on the type of camera, it can give a false perspective and a lack of depth perception for the driver. Drivers should practice with the backing camera and understand its limitations. Also remember

that during inclement weather, the camera may become obscured by snow, rain, or dirt.

8. Use spotters whenever possible to assist with the backing maneuver. There are many recommendations for proper backing signals between the spotter and the driver. It does not matter what signals are used, as long as the spotter and the driver both understand them. A department-wide backing policy should include which hand signals should be utilized. Once these signals are agreed upon, they should be taught to all members of the department. If a member from another agency is acting as the vehicle spotter, ensure that the driver and spotter meet face-to-face and agree upon hand signals before conducting any maneuvers. This will help prevent confusion.

9. Make sure that the spotter does not become distracted by another task, such as a cell phone or a conversation with another person. A distracted spotter can be as dangerous as no spotter at all.

10. If the driver loses sight of the spotter, **stop immediately.** Many serious backing incidents were caused when the spotter tripped, slipped, or fell, causing the driver to back over top of the spotter. If you can't see the spotter in the mirrors, stop immediately until the spotter is back in view.

11. If using multiple spotters, ensure that only one spotter communicates with the driver. This will help avoid confusion. There may be times when there are spotters on both sides of the rig and at the front corners. This is a great use of resources, but make sure that it does not cause confusion. Select who the primary spotter will be and channel all communications through that person. Spotters should keep an eye on each other and communicate so no one slips, trips, or falls, resulting in an injury caused by the backing apparatus.

12. Ensure that spotters equip themselves with high-visibility clothing, a flashlight, and a radio. If a spotter should slip and fall, a radio will help them communicate with the driver. Make sure everyone is on the same radio frequency.

13. Spotters must avoid walking behind the apparatus, or anywhere in the apparatus's path of travel. With the advent of backing cameras, spotters have a tendency to stand directly behind the rig and communicate with the driver via the camera screen. This is a dangerous practice that must be discouraged. If the spotter trips and falls, they will be positioned directly behind the rig and could be run over.

14. Spotters should avoid walking backwards. It is easier to trip and fall when walking backwards. Find a safe position out of the way of the apparatus and within sight of the driver. If spotters should need to

reposition, ensure that the driver stops the vehicle until the spotter can find another safe position to resume their duties.

15. Be aware of overhead hazards and ground hazards. This is especially true for taller vehicles such as aerial devices and water tankers.

16. Remember that environmental conditions, such as smoke, fog, blowing snow, or condensing exhaust fumes, may create a sight obstruction between the driver and the spotter. In these situations, everyone must use extreme caution. It may also be necessary to find another way of communicating.

17. At night, use flashlights to direct the driver. Remember not to shine the flashlight into the driver's eyes. Make sure the department's standard hand signals translate to use with a flashlight in the darkness.

18. Before moving the apparatus, the driver and spotter should have a face-to-face meeting to discuss the driver's plan to maneuver the apparatus.

19. Be aware of sun glare when using mirrors to back the apparatus. There may be times when sunshine or lighting may create so much glare that the driver cannot see the spotter in the mirror. If this occurs, attempt to find another way to maneuver the apparatus. If it is not possible, the rear corner spotter can relay signals to the front corner spotter so the driver is not looking directly into the sun.

20. When backing into a fire station or other building, be cognizant of how changing light conditions will affect the driver's ability to see. If the driver is backing up from a bright and sunny concrete apron into a darkened engine bay, it will be very difficult to see inside the firehouse. It is a good idea to turn on the lights in the engine bay to improve visibility inside the darkened station (fig. 20–1).

21. Ensure that back-up alarms and emergency stop devices are working. If the back-up alarm is not working, members operating in the area may not be aware that a truck is backing up. If the vehicle is equipped with an automatic stopping device, ensure that it is tested regularly according to the manufacturer's recommendations. On a side note, I once taught a seminar at a department that had installed "old fashioned" bell alarms on all of their rear-axles. These bell alarms were not subject to failure if a wire or fuse went bad. The logic behind their installation was that they would always work, regardless of mechanical or electrical defects.

22. Keep apparatus bays clear of hazards and obstructions. Nothing irritates me more than having to back a fire truck into an engine bay and having to negotiate around the Shop-Vac that someone left out after they vacuumed a personal vehicle.

FIGURE 20–1. When backing into an apparatus bay during the daytime, drivers will be backing into a dark area while their eyes are adjusted to the sun. It may be helpful to turn the bay lights on to help with the backing maneuver.

Case Study—New Jersey

On January 2, 2009, a fire engine in New Jersey was dispatched to a working fire. At 0213 hours, the engine company arrived on-scene, with orders to conduct a reverse lay to supply an elevated master stream. In order for the apparatus driver to properly position himself, he was forced to back around a police car and a tow truck, both of which were positioned under a highway overpass.

During the course of this backing maneuver, the officer of the engine got out and began to spot the driver from the officer's side of the rig. As the driver was backing up, the tow truck drove around the engine, momentarily distracting the driver of the fire engine. During this brief period of distraction, the driver felt the vehicle run over something, which turned out to be the officer of the engine. It is believed that the engine officer was walking behind the vehicle in an attempt to get a better vantage point to assist in backing the engine. As the officer walked behind the apparatus, he may have been contacted by the rear of the engine which caused him to stumble and lose his balance.

Investigation revealed several contributing factors to this crash. The brief distraction of the engine driver was enough to take the driver's eyes off the mirrors when the officer slipped or fell to the ground. A lack of communication between the officer and the driver resulted in the driver not knowing that the officer had gone to the ground. Finally, while the vehicle was equipped with an automatic braking device in the event that something struck the rear bumper, this device had not been tested recently and may not have been working properly.

Driving in Hazardous Conditions: Nighttime and Weather

Driving at Night

Traffic fatalities occur three times more often at night than during the day.[1] Several factors influence this issue, including impaired drivers, drowsy drivers, headlight limitations, and nighttime visibility issues. Fire apparatus operators must understand the danger of operating a fire apparatus in the dark.

Seeing at Night

The overall level of lighting will influence the speed, accuracy, and distance from which a person can detect and identify an object.[2] At night, a lack of lighting makes it more difficult for a driver to detect and identify a hazard in the road. If the driver cannot detect and identify a hazard, they will not be able to properly react and avoid it.

At night, a driver's vision is also limited to the area in front of the vehicle which is illuminated by the vehicle's headlights. Drivers lose much of their peripheral vision to the darkness and can only see as far in front of the fire apparatus as the headlights will allow. Several factors affect a driver's ability to see a hazard in the road.

Conspicuity

The likelihood that an object will come to the attention of a driver is known as *conspicuity*.[3] An object which is more conspicuous may draw the driver's attention more than an object that is in the driver's direct field of view.[4] A digital billboard well off the side of the road may draw a driver's attention more than a pedestrian who is jogging just a few feet off the road.

Contrast

Contrast is defined as the characteristics which make something appear different from something else.[5] The contrast of an object will help determine how accurately and how quickly it can be seen.[6] The contrast of an object against its background and the amount of illumination will determine the visibility of an object.[7]

Consider a pedestrian who is standing on a dark road against a dark backdrop. A pedestrian who is wearing white clothing will be seen more readily than a pedestrian who is wearing dark clothing. This is because the pedestrian's white clothes contrast more with the dark backdrop. In order to be seen at night, an object must appear darker or lighter than the surrounding background.[8]

Glare

Glare from oncoming headlights can also affect a driver's ability to see at night. Light from oncoming headlights or bright lights near the road can scatter across a person's retina, reducing the contrast of other objects within the visual field. This reduced contrast will make it more difficult for the person to see.[9]

Once a driver passes an oncoming vehicle, it will take time for the driver to recover from the glare. The distance the vehicle travels while the driver recovers from the glare will depend on the speed of the vehicle and the age and condition of the driver. When traveling at highway speed, an older driver's vision may take several hundred feet to fully recover. During this recovery time, it will take that driver longer to detect and react to a hazard. This issue can also occur during the day when there is glare from the sun.

Age

Age plays a major role in how well a driver can see at night. Age causes a yellowing of the eye's lens, clouding of the vitreous body within the eye,[10] and a decreasing of the pupil's ability to constrict. These factors all reduce a person's visual acuity when driving at night. Older drivers must be especially aware of safety issues related to nighttime driving.

Understanding Headlights

As most roadways in our country are not equipped with street lights, headlights are the primary means to illuminate the road and provide forward visibility. The headlights on a vehicle also provide warning to other motorists that another vehicle is approaching. While more light provides more visibility for a driver, manufacturers must limit the light output of a headlight to reduce excess glare for oncoming drivers. As a result of this limited light output, headlights have a limited effective range.

On average, low beams project approximately 250 ft in front of the vehicle on a clear night, while high beams project approximately 500 ft on a clear night.[11] The effective range will depend on several factors, including the aim of the headlight, the age and condition of the headlight, weather conditions, mounting height, and the type of headlight installed on the fire apparatus. Fire departments should conduct drills to determine the effective range of each apparatus's headlights (see Chapter 25).

Most new apparatus are equipped with halogen or high intensity discharge (HID) headlights. A halogen headlight contains a tungsten filament inside an envelope filled with a halogen gas. When the filament heats up, it gives off light. The structure of the headlight assembly uses mirrors and prisms to direct the light and illuminate the area in front of the vehicle. Halogen headlights are not as efficient as an HID headlight. Instead, much of the headlight's light energy is wasted as heat.

HID headlights are more efficient and do a better job of illuminating the road. Instead of heating up a tungsten filament, electricity arcs between two electrodes producing the light which illuminates the road. The electrodes are enclosed in a quartz envelope filled with a combination of xenon and other gases selected to produce light with a similar color spectrum to daylight. As there is no filament to break, HID headlights will last longer while also producing more illumination. Due to the illumination created by HID headlights, it is important to make sure they are properly aimed. Improperly aimed HID headlights can create glare issues for oncoming drivers.

Higher-end vehicles are beginning to use LED headlights. In an LED headlight, electrical current flows through specially designed semiconductors. The current flow results in photons of ultraviolet radiation being given off. This UV radiation stimulates phosphors which give off visible light. While HID and LED headlights may give off similar amounts of light, the LED does so much more efficiently. Less electricity is needed for an LED bulb. They are also less influenced by vibration and shock.

Fire apparatus headlights are often overlooked when it comes time for vehicle maintenance and routine checks. Failing to keep a close eye on the condition of the headlights may result in further limiting the effective headlight range. Dirty or clouded lens covers and aged filaments can degrade the amount of light produced by the headlight assembly, making it more difficult to see when it's dark. One study concluded that dirty headlights can reduce the amount of light by 27%.[12] In addition to blocking useful light, dirty headlights may scatter the light and increase glare for oncoming drivers. Fire departments should consider European-style headlight cleaning systems for new apparatus.

A study conducted by NHTSA revealed that 2/3 of the vehicles examined had improperly aimed headlights.[13] Improperly aimed headlights will reduce the amount of light shining on the road ahead while also posing a potential glare issue for other drivers. Fire departments should make sure to have the aim of each headlight checked by qualified personnel on a routine basis.

It is also important to understand how vehicle loading can affect headlight aim. If the fire department loads the vehicle with equipment and causes the chassis to shift or settle on the suspension, the headlights may no longer be aimed properly. Once a fire apparatus has been outfitted and equipped, it should be loaded with personnel, and the headlights should be re-examined. Make sure the headlights are not shining too high now that the rear of the vehicle is loaded and has settled down on the suspension. Fire departments should consider purchasing vehicles with autoleveling headlight systems.

When examining the headlight aim on the fire apparatus, take note how the light is biased towards the right-hand side of the roadway. This is to help prevent excess glare for oncoming drivers. However, as a result of this right-hand bias, the left-hand side of the roadway will not be as well illuminated. This may cause issues, especially if a pedestrian, or other hazard, is crossing the road from the left to the right. Drivers must be aware of this reduced illumination and scan the road ahead accordingly (fig. 21–1).

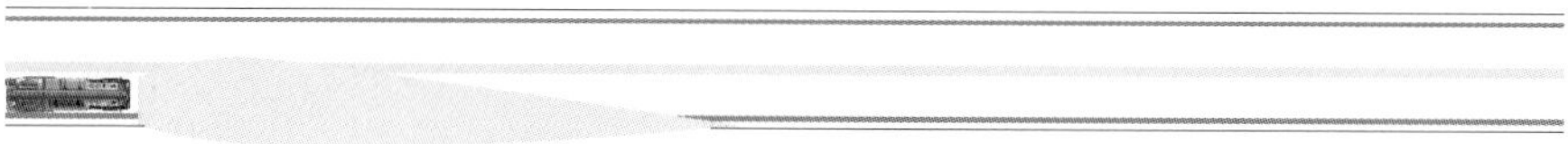

Low Beam illuminates approximately 250 feet in front of the apparatus.

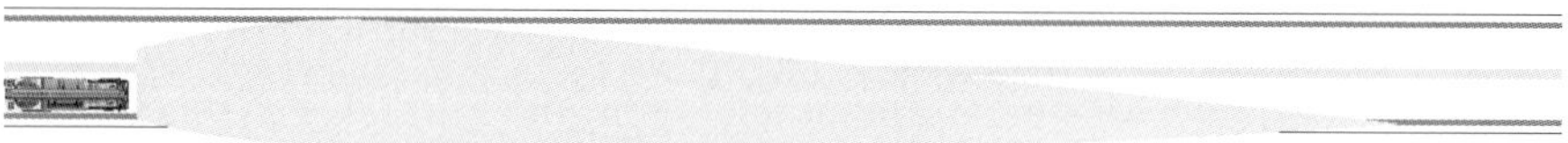

High Beam illuminates apporximately 500 feet in front of the apparatus.

FIGURE 21–1. Fire apparatus operators must be aware of their headlight range. Most high beams will project approximately twice as far as the low beams. Also take note of how the light pattern is biased towards the right. This is to help prevent excess glare for oncoming drivers. However, this bias can also create darker areas on the left side of the driver's approach.

In addition to headlights, fire apparatus operators must ensure that the apparatus windshield is clean and free from defects. Filming or fogging of the windshield or excessive chipping will contribute to reduced visibility. Routine vehicle checks should ensure that the windshield wiper fluid is full and properly working and the wiper blades have no grooves worn in them. Understanding how to properly clean and defrost the windshield is an important aspect of safe vehicle operations that is often taken for granted.

The issue of windshield visibility is especially important for volunteer firefighters. Volunteer firefighters who respond from home may not take the time to properly clean the windshield on their personal vehicles of frost, snow, or other debris. It is also not uncommon for a volunteer firefighter to get into a cold vehicle in the middle of the night to drive to the firehouse, only to have the windshield fog over from heavy breathing. During inclement weather, volunteers should try to park in a garage or use a tarp to cover the windshield. We will discuss the proper method of defogging the windshield later in this chapter.

Two-Lane Highways

The vast majority of fire departments operate on two-lane highways, otherwise known as local roads. These roads pose unique safety issues, especially when driving at night. The National Highway Traffic Safety Administration published a report to Congress entitled "Nighttime Glare and Driving Performance." This report addresses several issues specific to two-lane roadways. These issues are as follows:[14]

> Two-lane highways, as compared to multi-lane roadways and expressways, may present the "worst-case" scenario for nighttime glare. The reasons for this include:
>
> **Lower light levels:** Nighttime glare on unlighted roadways is more problematic because the visibility of an object is influenced by the overall light level. Since light levels tend to be lower on two-lane highways, objects along these roads are less visible in the presence of glare. In addition to the reduced visibility, the lower light levels on two lane highways increases discomfort glare because the relative brightness difference between headlamps and the roadway background is higher on these roads.
>
> **Oncoming traffic closer to driver's line of sight:** Because of the closer proximity of oncoming glare on two lane roadways, the beam pattern is directing more light at oncoming drivers, which produces more scattered light in their eyes.
>
> **Complex roadway geometry:** Two-lane highways often have complicated roadway geometries, including sharper curves and steeper

grades. Thus, drivers on two lane roadways may be exposed to a higher range of oncoming headlamp glare than on multilane roadways.

Less restricted roadway access: The greater potential for hazards to be found in any number of locations along the roadway increases the complexity of the driving task and makes glare more problematic and arguably less safe than it would be under easier driving situations.

Fewer roadway markings: Roadway markings in and of themselves have no direct impact on the amount of light that oncoming headlamps produce toward a driver's eyes. However, they make the driving task easier by providing visual guidance about the geometry of the roadways and other roadway conditions.

Closer proximity of pedestrians: Pedestrians are not always easily seen. While they may not be found frequently along two-lane roadways, they are in closer proximity when they are encountered on these kinds of roads. The difficulty of seeing them at night can increase the difficulty of nighttime driving, and therefore, increase the impact of glare on both visibility and comfort.

Expectancy

Expectancy is defined as the feeling that something is going to happen. The concept of expectancy plays a large role in driver safety. Often a fire apparatus operator's subconscious mind is expecting something to happen, even if the operator doesn't realize it. Consider the example of a firefighter who works in a large urban city. An attentive and well-trained driver will be constantly expecting such things as cross traffic and pedestrians. This is because the firefighter is used to driving in an urban environment and is well aware that there is traffic and pedestrian activity at all times of night. This firefighter will be keeping an eye on sidewalks and crosswalks to ensure no one walks or pulls out in front of the fire apparatus.

Now consider a firefighter who drives for a rural fire department that covers a large, sparsely populated district. This same firefighter often drives long distances to emergency calls and rarely sees another vehicle or pedestrian. In this case, the firefighter is driving to a call late at night and an intoxicated pedestrian suddenly walks in front of the fire apparatus on a dark, deserted road. The fire apparatus operator may strike the pedestrian because they weren't expecting a pedestrian to suddenly emerge from the darkness.

Fire apparatus operators must always expect the unexpected. Regardless of the district the driver protects, a professional fire apparatus operator should always drive in a defensive fashion, constantly scanning ahead and expecting the worst.

Adaption

Glare and visibility adaption is a common problem during nighttime driving. These issues can pose problems at other times as well. A fire apparatus operator who emerges from a tunnel or parking garage into bright daylight may face a painful glare that affects the ability to see. Another issue common to fire departments is backing or parking a fire apparatus in a dark firehouse after driving on a bright day. Fire apparatus operators must understand the need to pause and give their eyes time to adjust to the dark area of the firehouse. Having a fellow crew member get out and turn on the overhead lights in the apparatus bay may help reduce this issue.

Visual adaption may also be affected by streetlights and other ambient lighting for firefighters who operate in urban or suburban districts. Be wary of driving the apparatus on a busy thoroughfare that is lit with street lights and then suddenly turning into a dark, unlit housing complex. It will take time for the driver's eyes to adapt to the sudden darkness.

Roadway Geometry

The design and geometry of a roadway can also reduce headlight effectiveness. Keep in mind that headlights are aimed in a specific manner, designed to provide roadway illumination while not producing too much oncoming glare. Because headlights are aimed a certain way, there may be issues when a vehicle rounds a curve or approaches a rise or dip in the roadway.

The beam of a headlight is biased towards the right side of the road. This is to prevent too much glare from affecting an oncoming driver. If the vehicle is rounding a left-hand curve, drivers must remember that there will be less illumination towards the left side of the vehicle. Illumination bias may result in the driver failing to see a hazard.

Low beams are aimed closer to the ground to prevent excess glare for oncoming drivers. This may be an issue when the vehicle is negotiating a rise or dip in the road. Because the headlights are aimed a certain way, they may not illuminate a hazard or nonreflective road sign which is higher or lower than the road grade.

Be Careful Driving at Night

Driving at night is hazardous. Fire apparatus operators will not be able to see as well as they can typically see during the day. Reduced visibility may result in longer perception and reaction distances as the fire apparatus operator attempts to decipher the hazard they see on the road ahead. Because of this, fire apparatus operators must understand the need to slow down at night and drive accordingly. Figures 21–2 and 21–3 show what will happen when the driver uses the low beams instead of the high beams. Because the low beams only illuminate the hazard approximately 250 ft in front of the apparatus, the driver does not have enough time to perceive, react, and stop the apparatus before striking the hazard. However, a driver who has the high beams turned on will see the hazard much further away. As a result, the driver has enough time to perceive, react, and stop the truck. Unfortunately, in busy areas, with constant oncoming traffic, the fire apparatus operator is almost always driving with the low beams. Because of this, the fire apparatus operator must understand the need to slow down and not overdrive the headlights.

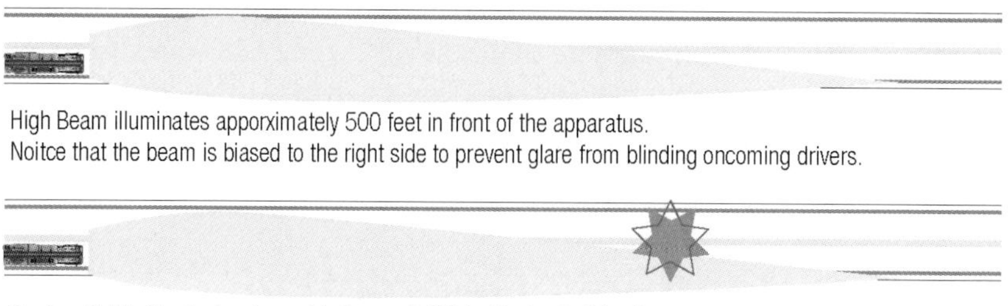

The headlights illuminate a hazard in the road 400 feet in front of the fire apparatus.

At 40 MPH, the apparatus will travel 93 feet as the driver perceives and reacts to the hazard.

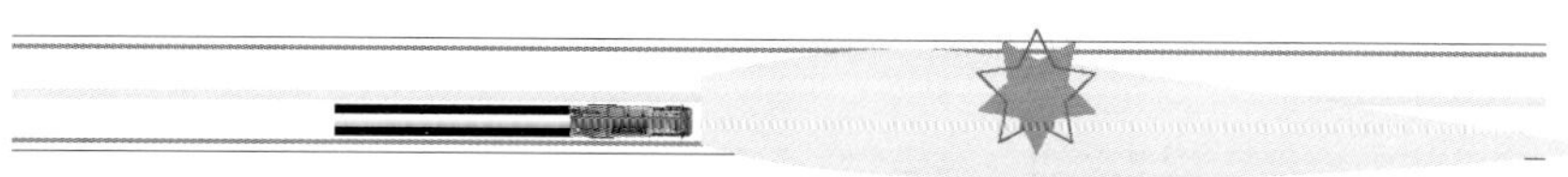

After the driver perceives and reacts, it will take the apparatus 109 feet to skid to a stop when the driver applies the brakes. Because the high beams were on, the driver was able to see the hazard 400 feet away. Becasue it takes 202 feet to perceive, react, and skid to a stop, the apparatus will stop before it strikes the hazard. The fire apparatus operator was not "overdriving" the headlights.

FIGURE 21–2. High beams headlight range. Drivers must understand the importance of headlight range and not overdrive their headlights.

Driving in Inclement Weather

Emergency responders do not have the luxury of sitting at home to enjoy a storm. In fact, a fire department's call volume will often increase during times of inclement weather. Fire apparatus operators must understand how poor weather will affect driving conditions.

Poor weather will affect a vehicle's handling characteristics in several ways. Reduced road friction will result in longer stopping distances and lower critical curve speeds. Visibility will be significantly affected as weather conditions will obscure forward sight distance and make it difficult to see around the vehicle. Poor weather may also have adverse effects on the mechanical condition of the vehicle, especially for fire departments operating old or poorly maintained fire apparatus.

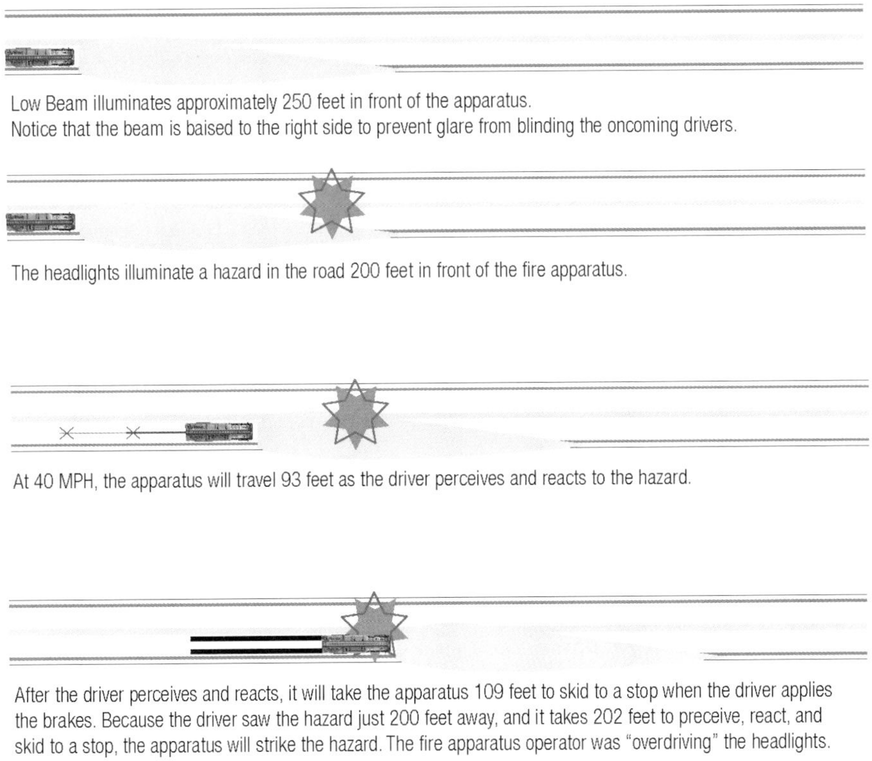

Low Beam illuminates approximately 250 feet in front of the apparatus.
Notice that the beam is baised to the right side to prevent glare from blinding the oncoming drivers.

The headlights illuminate a hazard in the road 200 feet in front of the fire apparatus.

At 40 MPH, the apparatus will travel 93 feet as the driver perceives and reacts to the hazard.

After the driver perceives and reacts, it will take the apparatus 109 feet to skid to a stop when the driver applies the brakes. Because the driver saw the hazard just 200 feet away, and it takes 202 feet to preceive, react, and skid to a stop, the apparatus will strike the hazard. The fire apparatus operator was "overdriving" the headlights.

FIGURE 21–3. Low beams headlight range.

Rain

Rain is the most common weather hazard faced by fire apparatus operators. While snow, rain, fog, and ice can all contribute to serious crashes, rain tends to be the most common cause of weather related fatalities. This may be due to the fact that drivers tend to consciously slow down for ice and snow. However, drivers tend to overlook the danger of driving in the rain and fail to change their driving habits accordingly.

Visibility

Rain can have a significant effect on visibility. Driving rain obscures forward sight distance, reducing visibility to sometimes just a few feet. Reduced visibility results in less time and distance to react to a hazard and bring the apparatus to a safe stop. Driving rain at night can create additional hazards as the effectiveness of the vehicle's headlights will be reduced. Raindrops will block the light from reaching in front of the vehicle and also reduce the amount of light returned to the driver's eyes.[15] The light will also have a tendency to strike the raindrops and scatter back towards the driver. This reflected light will illuminate the raindrop filled area directly in front of the apparatus, reducing contrast. Reduced contrast results in reduced visibility.

Rain covered roads will make it difficult to see the road markings. Rain creates a film of water on the road which acts like a mirror. This mirror-like effect will project headlight illumination forward instead of back towards the driver. This forward reflection will cause the road to appear dark, while also increasing glare for oncoming drivers (fig. 21–4).

Rain can also cover and obscure the glass beads used as a reflective material in many road markings. As a result, road markings such as double yellow lines and white fog lines may become invisible to a driver. If road markings are obscured by a film of water, it can become difficult for drivers to properly position themselves in their lane of travel.

Friction

Rain can have a significant effect on the coefficient of friction of the roadway. Reduced friction will result in longer total stopping distances and reduced critical curve speeds. While a driver may have been able to negotiate a curve at a particular speed on a dry day, they may find themselves spinning off the road at the same speed on a wet day. These issues are especially true within the first 30 minutes of a rainfall. At the beginning of a rainfall, the rain will cause oils and debris to float to the surface of the roadway further reducing the coefficient of friction.

FIGURE 21–4. Rain can create significant glare and visibility issues, especially at night.

Hydroplaning

Rain slicked roads can also lead to hydroplaning. A vehicle's tire tread is designed to push the water out of the way and allow the rubber tire to grip the road. If the vehicle is traveling too quickly, the driver may not give the tread enough time to move the water out of the way. As a result, the tire may ride up on the water surface and begin to hydroplane. Drivers must remember to slow down and give the tire tread enough time to do its job during wet weather.

Windshield Wipers

When driving the apparatus in the rain, operators must remember to use the vehicle's headlights and windshield wipers. While this may seem like common sense, many drivers forget to do this, especially when they are distracted by other tasks. Windshield wipers will help the driver see the road, while the headlights will help other drivers see the fire apparatus.

Poorly maintained windshield wipers will also result in reduced visibility. If the windshield wipers are old and worn out, they will not effectively clean the windshield. Instead of wiping the water away from the windshield, the wipers may streak or leave large patches of rain which will make it difficult for the driver to see.

While an effective tool, windshield wipers are fragile. Wiper blades are made of thin rubber and are mounted on the outside of the vehicle where they are exposed to ultraviolet light, dust, debris, and other damaging elements. These elements can damage the wiper blade and render them ineffective.

When the rubber wiper blade dries out or becomes contaminated by sap, oils, or tar, it will begin to leave streaks on the windshield. Streaks are also caused when the edge of the wiper blade becomes rounded from excess use. If the windshield wiper begins to skip on the windshield, the blade most likely became curved and misshapen from lack of use. Fire apparatus operators should look for these issues during routine vehicle checks. Operators should also wipe down the wiper blade with a damp towel to ensure the blades are clean and ready for use.

Window Fog

Interior windows will fog as a result of moist air inside the vehicle. Just as moisture will form on the outside of a cold soda can, moisture will also form on the inside of a cold windshield. This moisture can come from breathing, wet fire gear, or anything else that can introduce moisture into the vehicle. Once the moist air contacts a cold windshield, the water will condense onto the windshield, obscuring the driver's view. The fastest method for defogging a windshield is as follows:

1. Turn the temperature setting to high heat.
2. Turn on the defroster to full fan.
3. Turn on the air conditioning.
4. Turn off the vehicle recirculation and use fresh air from outside.
5. If you need to do it quickly, open the windows.

Snow

Snow reduces forward visibility, obscures signs, and covers roadway markings. At night, snow will block the headlight beams, decreasing their effective range.

Snow can also reflect light from the headlight back towards the driver's eyes, causing glare and lack of contrast.

Even after a snow storm has passed, the fallen snow will have an effect on roadway visibility. Fallen snow will obscure road markings and cover street signs. Without effective road markings, drivers will have a difficult time maintaining their lane of travel, or knowing where to stop on the road (fig. 21–5).

FIGURE 21–5.   Snow can pose unique driving hazards. In this case, it has completely obscured a stop sign. A motorist who is unfamiliar with the area may drive through the stop sign unaware that it is a controlled intersection.

Deep snow can create ruts in the road that can trap a vehicle and make it difficult to turn or take evasive action. Snow can also fill the tread of all-season tires, turning the tires into skis. This is why it is important to switch to snow tires if the vehicle will be driven in deep snow (see Chapter 11).

Snow will reduce the coefficient of friction of the roadway, increasing total stopping distance and reducing critical curve speeds. Any braking, accelerating, or turning movements made during snowy conditions should be done slowly and smoothly to account for the reduced friction of the roadway. Slamming on the brakes, turning the wheel suddenly, or stomping on the accelerator will cause the tires to break traction with the road. This fact is also true for all-wheel drive and 4x4 vehicles. While these vehicles will not get stuck as easily as a two-wheel drive vehicle, they will still stop and turn like a two-wheel drive.

When driving in deep snow, the driver should only stop when necessary, especially when traveling uphill. Stopping in deep snow may cause the vehicle to bog down and get stuck. Maintaining forward momentum will help the vehicle to push through the snow and keep the tires rolling, which will help maintain traction.

In severe storms it may be best to have a snow plow accompany the apparatus on calls. It is better to take a few moments and give the plow time to clear the road rather than risk getting the apparatus stuck. There may also be times when a roadway just isn't passable and other response procedures should be implemented.

Fog

Fog is a cloud that has formed close to the ground. Fog consists of small water droplets or ice crystals. There are several types of fog, but the most common fog is formed when moist air condenses over cool ground. If fog condenses or freezes on the road surface, it can reduce the coefficient of friction of the roadway.

Fog can affect a driver's depth perception and cause them to misjudge following distance or closing speed. Fog also obscures the road and restricts forward sight distance, making it difficult for a driver to see a hazard on the road ahead. When fog reduces forward visibility, there is little a driver can do to counter this. Using the vehicle's high beams will only exacerbate the issue as the fog will reflect the headlights back towards the driver, making it more difficult to see.

Many vehicles are equipped with fog lights, but fog lights do little to increase the driver's forward sight distance. The purpose of a fog lamp is to light the area directly around the vehicle allowing drivers to see the road markings and maintain their position on the road. Ironically, helping the driver to feel comfortable about road positioning may give them a false sense of security. This false sense of security can lead to higher speeds.

Furthermore, using the fog lights will illuminate the area directly in front of the vehicle and cause the driver's pupils to constrict. When the driver's pupils constrict, they will not be able to see as far down the road. Fog lamps should only be used when needed and at low speeds.

One of the causes of fog-related crashes is the fact that the vehicles driving in the fog often slow down to different speeds. One vehicle may have slowed down to account for the fog, while the vehicle behind them did not slow down as much. As a result, the faster vehicle drives into the back of the slower vehicle. Many chain reaction crashes are started this way.

Fire apparatus operators must understand the need to slow down for foggy conditions and account for the limited forward sight distance. If the fire apparatus operator is overdriving the forward sight distance, they may not be able to react in time when a slow moving vehicle suddenly emerges from the fog. As a result, the fire apparatus may strike the back of the slow moving vehicle and start a chain reaction crash.

Fire apparatus operators must also be wary of being rear-ended by a vehicle that is traveling behind them in the fog. Anything that can help increase the conspicuity of the fire apparatus should be turned on, including emergency warning lights and hazard flashers. Hazard flashers will be more readily seen through the fog than just taillights. The same holds true when driving a personal vehicle. Just keep in mind that it may be necessary to balance the use of these warning lights with the possibility of creating more glare that will reduce the fire apparatus operator's ability to see.

Ice

Ice covered roads are treacherous. An ice covered road can increase stopping distance as much as four times. If the road is ice covered and includes a steep grade, there may be no way for the fire apparatus operator to bring the vehicle to a stop.

Drivers must be aware of potential ice locations, especially areas of black ice. Black ice is a thin glaze of clear ice which covers the road surface. Black ice will form when the air temperature near the road surface is less than 32°F and there is rain falling. Black ice can also form as a result of melting snow, water run-off, or at fire scenes where hoses and pumps are in use.

Black ice is more common in shaded areas and on bridges and underpasses where air is able to freely circulate under the elevated roadway. While black ice is more common in the early morning and early evening, it can form at any time of the day. Drivers must be especially aware of black ice, as it is difficult to see and often forms on an otherwise dry road. A fire apparatus operator may be driving on a dry road when they suddenly encounter an invisible area of black ice. Remember that black ice is not black; it's clear. If the ice is on a concrete or light surface, it will appear the same color as the underlying surface.

Black ice can also form as a result of condensation from vehicle exhaust. Fire apparatus operators should be especially wary of black ice as they approach an intersection where vehicles may have been stopped and waiting for a traffic light to change. Condensation from the waiting vehicle's exhaust pipes may have formed an area of black ice near the intersection.

Black ice can also form when the air temperature is above 32°F. If the weather should warm suddenly while the road surface is still below freezing, any water deposited on the road may still turn to ice. This scenario can be especially deceiving as drivers are not expecting icy road conditions because of the warm weather.

Fire apparatus should be equipped with thermometers so the apparatus operator knows the outside temperature. However, apparatus operators must remember that the thermometer may not always be accurate. If the vehicle thermometer is mounted high above the road surface, or close to the engine, the thermometer may give readings that are higher than the actual temperature at the road surface. Fire apparatus operators should make a habit of walking outside with a heat gun or thermal imaging camera to keep an eye on the actual temperature of the road surface.

In the event that the apparatus drives over an area of black ice, the driver should avoid applying the brakes or turning the wheel. The driver should let off of the accelerator and allow the wheels to continue rolling freely while trying to keep the vehicle on a straight path. If it becomes absolutely necessary to steer or brake, do so in a smooth and gentle fashion to avoid locking the tires or inducing a loss of control. If it is safe to do so, try to steer towards an area of increased traction, such as dry asphalt, or even snow.

If there is a chance of encountering black ice, fire apparatus operators must be prepared. Know where black ice may form and know how to spot it. While black ice is often invisible, there may be times when it gives off clues, such as a shiny area on the road ahead. If you see the black ice approaching, do whatever you can to avoid it. Once a vehicle drives on black ice, it will be difficult to keep it under control.

Sleet and Freezing Rain

Sleet and freezing rain are often used interchangeably; however, sleet and freezing rain are different. Freezing rain occurs when raindrops fall through a thin layer of cold air. The raindrops do not spend enough time in the cold air layer to freeze and remain in liquid form. However, when the raindrop lands on a cold surface, it immediately freezes and creates a layer of ice. Freezing rain is a common cause of black ice.

When sleet forms, the raindrop travels through a thicker layer of cold air. Because the raindrop spends more time traveling through the thick layer of cold air, the drop has time to freeze into an ice pellet before landing on the ground.

Sleet and freezing rain will cause slick roads that reduce the coefficient of friction. A reduced coefficient of friction will increase stopping distances and lower critical curve speeds. Sleet and freezing rain will also create visibility problems, just like rain and snow.

Fire apparatus operators should also be aware of issues related to falling trees and branches and low hanging wires during periods of freezing rain. Freezing rain will form a layer of heavy ice on trees and branches. If the branch or tree reaches a point where it can no longer support the weight of the ice, it will break. Broken branches and fallen trees can cause unique hazards for fire apparatus operators. When driving in ice conditions, fire apparatus operators must constantly scan the road ahead, looking for fallen trees and limbs.

Wind

Many drivers are surprised at the effect wind can have on a vehicle. This is especially true for drivers who are operating large vehicles or trailers. The large size and cross section of a fire apparatus can make the vehicle especially susceptible to high winds.

Sustained winds can make it difficult for a vehicle to maintain its lane of travel. In areas where high winds are common, most drivers are able to anticipate and compensate for the effect of the wind on the behavior of the vehicle. However, sudden gusts of wind can catch a driver off-guard, resulting in a loss of control and subsequent crash. Apparatus operators must be prepared for a gust of wind when responding during heavy storms.

Fire apparatus operators should be familiar with their districts and anticipate potential areas of high wind. These areas include bridges and overpasses. Drivers must also anticipate how a sudden wind gust from a passing tractor trailer can affect their vehicle. This is especially true for drivers who may be pulling a trailer.

When driving in high winds, leave plenty of space around the vehicle. Also keep an eye out for other large vehicles that may be blown out of their travel lane by a gust of wind. Make sure to give these vehicles plenty of room to maneuver. Drivers should also remember to watch for branches, trees, and other debris that may be blown onto the road by windy weather.

Weather and Stopping Distance

Snow, ice, and wet road conditions will reduce the coefficient of friction of the roadway. A reduced coefficient of friction will result in longer total stopping distances. These issues are compounded when the vehicle is poorly maintained or is operating on roadways that have downhill grades. Table 21–1 provides the total stopping distance of a civilian vehicle traveling 35 mph on different road surfaces:

A fire apparatus will not have 100% braking efficiency like a civilian vehicle. Stiffer tires and air brakes will reduce the braking efficiency of the vehicle to approximately 65% that of a civilian vehicle. As a result, the total stopping distance will be more affected by changing road conditions. Table 21–2 provides the total stopping distance of a fire apparatus traveling 35 mph on different road surfaces.

Tables 21–1 and 21–2 demonstrate the significant increase in stopping distance as the coefficient of friction of the roadway decreases. A fire apparatus traveling on icy roads will require nearly twice as much distance to come to a stop. Fire apparatus operators must constantly consider this issue and slow down accordingly when driving in foul weather.

TABLE 21–1. Total stopping distance when driving 35 mph, assuming a 1.6 second perception/reaction time and 100% braking efficiency.

Conditions	Coefficient of Friction	Braking Efficiency	Perception Reaction Distance	Skid Distance	Total Stopping Distance
Dry Road	0.75	100%	82 ft	54 ft	136 ft
Wet Road	0.50	100%	82 ft	81 ft	163 ft
Icy Road	0.20	100%	82 ft	204 ft	286 ft

TABLE 21–2. Total stopping distance when driving 35 mph, assuming a 1.6 second perception/reaction time and 65% braking efficiency.

Conditions	Coefficient of Friction	Braking Efficiency	Perception Reaction Distance	Skid Distance	Total Stopping Distance
Dry Road	0.75	65%	82 ft	83 ft	165 ft
Wet Road	0.50	65%	82 ft	125 ft	207 ft
Icy Road	0.20	65%	82 ft	314 ft	396 ft

Fire apparatus operators must also consider how an increased total stopping distance will affect the approach to an emergency scene. Nothing is more embarrassing than sliding past a house fire, or worse yet, sliding into a vehicle crash that the apparatus is responding to. Even though the adrenaline will be flowing when the fire apparatus rounds a curve and the crew sees a working fire, the driver must take a deep breath and slow down well in advance of the incident scene.

Compounding the stopping distance issue is the reduced sight distance caused by the weather. Depending on the speed, a fire apparatus may require nearly 400 ft to perceive, react, and skid to a stop on an icy road. If the icy weather conditions have reduced the forward visibility to 200 ft, there will be no way for the fire apparatus operator to see a hazard, react to the hazard, and stop the vehicle before crashing into it.

There may be times when it is too dangerous to drive the fire apparatus close to an incident scene. This is especially true when responding to vehicle crashes that have occurred at the base of a steep hill or on a slick curve. In these instances, it may be best to use the apparatus to block the roadway and stop any more vehicles from driving on the road. Protecting the scene and walking to the incident may be better than risking the entire apparatus.

Weather and Critical Curve Speed

The reduced coefficient of friction of the road that is associated with inclement weather will affect a vehicle's ability to safely round a curve. As the coefficient of friction goes down, so does the critical curve speed. A vehicle will not be able to drive through a curve on a wet day as it normally would on a dry day. Drivers must remember to slow down and account for the weather conditions. Failure to do so could result in a loss of control. Table 21–3 provides an examination of critical curve speeds for a curve with a 300 ft radius, while table 21–4 provides the critical curve speeds for a curve with a 150 ft radius.

Case Study—Virginia

On February 13, 2012, a firefighter was killed during an apparatus crash that involved black ice. The victim firefighter was assisting with a tanker shuttle at a working fire. After assisting the driver by engaging the dump valve at the dump site, the firefighter became trapped on the tailboard when the fire apparatus operator drove away without ensuring the firefighter had gotten off of the truck. As the apparatus was driving to a fill site, the vehicle drove over an area of black ice that had been created by the leaking fill valve from another water tanker involved in the shuttle. The water tanker spun around and threw the firefighter off of the tailboard, fatally injuring him.

TABLE 21–3.   Critical curve speeds for a curve with a 300 ft radius.

Condition	Coefficient of Friction	Curve Radius	Critical Curve Speed
Dry Road	0.75	300 ft	57 mph
Wet Road	0.50	300 ft	47 mph
Icy Road	0.20	300 ft	29 mph

TABLE 21–4.   Critical curve speeds for a curve with a 150 ft radius.

Condition	Coefficient of Friction	Curve Radius	Critical Curve Speed
Dry Road	0.75	150 ft	40 mph
Wet Road	0.50	150 ft	33 mph
Icy Road	0.20	150 ft	21 mph

This incident was an incredible mix of bad luck. However, one of the causative factors in the crash was the fact that the apparatus drove over an area of black ice and lost control. Whenever fire apparatus operators spot black ice, they should communicate the location of the ice to other apparatus operators. This will allow the other operators to take the appropriate precautions to avoid driving over the ice and losing control.

Notes

1. Based on NHTSA statistics
2. M. S. Rea and M. J. Ouellette, "Relative visual performance: A basis for application." Lighting Research andTechnology 23, no. 3, 1991, pg. 135–144.
3. Paul L. Olson, "Visibility Problems in Nighttime Driving." SAE Technical Paper 870600, February, 1987.
4. Olson, "Visibility Problems in Nighttime Driving."
5. Olson, "Visibility Problems in Nighttime Driving."
6. Rea MS, Ouellette MJ. 1991. Relative visual performance: A basis for application. Lighting Research and Technology 23(3): 135-144.
7. Olson, "Visibility Problems in Nighttime Driving
8. Olson, "Visibility Problems in Nighttime Driving
9. Olson, "Visibility Problems in Nighttime Driving
10. This is the fluid inside a person's eyeball.
11. Thanks to James Sobek, P.E. for providing these values based on his real-life testing.
12. M. Flanagan, M. Sivak, and B. Schoettle. "Benefits of Headlamps and Cleaning for Current U.S. Low Beams." The University of Michigan Transportation Research Institute. Ann Arbor MI. November, 2007.
13. Lighting Research Center, Rensselaer Polytechnic Institute. 2005. Assessment of Headlamp Glare and Potential Countermeasures: New Methodology for the Assessment of Headlamp Aim [report submitted to NHTSA]. Troy, NY: Rensselaer Polytechnic Institute
14. NHTSA. "Nighttime Glare and Driving Performance—Report to Congress," February 2007.
15. Olson, "Visibility Problems in Nighttime Driving."

Distracted, Sleepy, and Fatigued Driving

Distracted Driving

While many drivers understand the dangers associated with drunk driving, they often underestimate the danger of distracted driving. There are three primary types of distracted driving: visual, manual, and cognitive.

Visual Distractions

Visual distractions cause drivers to take their eyes off the road. Examples of a visual distraction include reading a text, looking at the radio, or looking down to find something that has fallen on the floor. When drivers think of distracted driving, visual distractions are most commonly envisioned.

The amount of time drivers take their eyes off the road will depend on the distraction they are dealing with. As an example, studies have shown that reading a text will cause drivers to take their eyes off the road for an average of 3–5 seconds.[1] If there is a more complex issue in the vehicle, drivers may take their eyes off the road for even longer. Depending on the speed of the vehicle, drivers may have traveled the distance of an entire football field without looking at the road. Imagine what could happen in that distance if the driver isn't paying attention.

Drivers must remember to keep their eyes on the road and continually scan their surroundings. Looking somewhere other than the road is an invitation to disaster. If traffic should suddenly slow down, or a small child should run in front of the vehicle, a distracted driver will not see the hazard and react accordingly. Table 22–1 gives the distance a vehicle will travel at a given speed if drivers take their eyes off the road for 5 seconds.

TABLE 22–1. When drivers take their eyes off the road for just a few seconds, the vehicle will travel a considerable distance without the driver looking at the road ahead. This table demonstrates the distance traveled at different speeds if drivers take their eyes off the road for just 5 seconds.

Speed of the Vehicle	Eyes Off the Road for ...	Distance Traveled Without Looking
25 mph	5 seconds	183 ft
35 mph	5 seconds	256 ft
45 mph	5 seconds	329 ft
55 mph	5 seconds	403 ft
65 mph	5 seconds	476 ft

Manual Distractions

A manual distraction is a distraction that causes drivers to take their hands off the wheel. Manual distractions include eating, applying makeup, or manipulating a handheld electronic device. If a driver's hands aren't on the wheel, they will be unable to maneuver the vehicle should they encounter a hazard in the road. Furthermore, if drivers suddenly grab the wheel in attempt to maneuver around a hazard, they may grab the wheel too hard and cause the vehicle to lose control.

When drivers take their hands off the wheel, the vehicle may have a tendency to drift outside of the travel lane. If a large fire apparatus drifts off of the right side of the road, it may drop off the road and rollover. If the apparatus drifts to the left side of the road, it may encroach into the oncoming traffic lane and crash head-on into another vehicle. Drivers must keep both hands on the wheel at all times to prevent the vehicle from leaving its lane of travel.

There is a strong temptation for fire apparatus operators to perform other tasks while driving the vehicle. These tasks include sounding the air horns, working the siren, or manipulating a mobile radio. Whenever possible, fire officers should tend to these tasks and allow fire apparatus operators to keep both hands on the wheel. The fire officer should act as a co-pilot, doing as much as possible to prevent the fire apparatus operator from becoming manually or visually distracted.

Cognitive Distractions

While visual and manual distractions are easily recognized, cognitive distractions are often overlooked. A cognitive distraction will take the driver's mind off the road. While many drivers recognize the danger of taking their eyes off the road or their hands off the wheel, they often underestimate the danger of not paying attention. In reality, cognitive distractions are the leading cause of distraction crashes. One of the most common cognitive distractions is talking on a phone while driving a vehicle.

Many states have banned the use of hand-held cellular phones while driving. The mistaken belief was that the danger of using a cell phone was directly related to drivers not keeping their hands on the wheel. Studies have now revealed that there is little difference in the number of crashes caused by handheld cell phones and hands-free cell phones. It is not holding the phone that causes the distraction. The issue is paying attention to the cell phone conversation instead of the driving task. This is because humans are not able to multitask.

The term multitasking is commonly used to refer to a person's ability to do more than one activity at the same time. While many people believe they are able to multitask effectively, the truth is that the human brain is unable to do more than one task at a time. Instead, the brain prioritizes the tasks and then switches back and forth from one task to the other.[2] As the brain becomes more and more overloaded with information, it will begin to filter out certain information. Because a driver may not be able to consciously decide what information is filtered out, critical information related to the driving task may be filtered out in favor of the cell phone conversation.

The driver's inability to multitask may lead to *inattention blindness.* Inattention blindness occurs when a person looks at something but does not see it. This happens because at any given time the human brain is overwhelmed with sensory information. As a result, the brain must filter all of the information and decide which information is important. If the brain is focused on multiple tasks, it may deem valuable information as unimportant, and the person will fail to see it. As an example, how many times have you been focused on reading a book or doing a crossword puzzle, and you don't realize someone has walked into the room and is standing in front of you? You fell victim to inattention blindness. If you are talking on a cell phone while driving, your brain may fail to see a hazard because it deemed the cell phone conversation more important. Now imagine a fire apparatus en route to a house fire. The driver is blowing the air horns with one hand and listening to the chief's orders on the radio, all while driving the apparatus at night. If a pedestrian suddenly emerges from the darkness, will the apparatus operator see the pedestrian and swerve to avoid him? Or will the driver hit the pedestrian because his mind was focused somewhere else?

A study conducted by Carnegie Mellon University discovered that when a person listens to spoken sentences, activity in the portion of the brain that controls spatial processing decreases by 37%, as compared to someone who was performing an undisturbed driving task. When a driver attempts to comprehend language and drive at the same time, mental resources are taken away from the task of driving the vehicle. When these resources are taken away, driving performance deteriorates.[3] This deterioration occurs regardless of whether the driver is using a hands-free or hand-held phone.

Distracted driving will also result in increased perception and reaction times. A person who is talking on a cell phone will need to switch focus from the phone conversation to the driving task if they encounter a hazard in the road. While it may only take a few microseconds to switch focus from one task to another, this delay will increase the driver's perception and reaction time and distance.

Tunnel vision is another side-effect of distracted driving. A driver should constantly scan the road ahead, looking for unexpected hazards. However, studies have demonstrated that a driver who is cognitively distracted will tend to focus on a smaller viewing area directly in front of the vehicle.[4] Due to this limited viewing area, the driver may fail to see a hazard encroaching from the side of the road.

It is interesting to note that cognitive distractions will decrease the variability of lane positioning, while visual distractions increase the variability of lane positioning.[5] In other words, when drivers take their eyes off the road, they tend to drift in and out of their lane of travel. However, when drivers are thinking or talking on the phone, they tend to zone out and do not maneuver in and out of the lane when needed.

While many states ban civilians from using a hand-held phone while driving, many of these same states provide exemptions for emergency responders. It should be noted that an emergency responder's brain works the same way as a civilian's brain. Using a hand-held or hands-free device will still result in cognitive distractions for emergency responders. Emergency agencies should implement policies that prohibit the use of cellular phones while operating an emergency vehicle.

Keep in mind that most of the current studies related to cognitive distractions revolve around the use of cellular phones. However, the results of these studies also apply to anything that distracts a person's attention while driving. This could include conversations with other passengers, listening to the radio, or simply thinking about something other than the driving task. These distractions are even more numerous for a fire apparatus operator who is driving an emergency vehicle to an incident.

Driving the fire apparatus should be the operator's primary task, while all other tasks should be left to the fire officer riding in the passenger seat. A fire apparatus operator who begins to do other things while driving the vehicle may fall victim to the effects of distracted driving and fail to perceive and react to a hazard in the roadway.

Fatigued and Sleepy Driving

Fatigued driving is a dangerous practice that often goes unrecognized. Little evidence is left behind to indicate that a crash was caused by a fatigued driver. This lack of evidence may cause investigators to overlook fatigue as the cause of a crash.

While the terms "sleepy" and "fatigued" are used interchangeably, their definitions are different. When we feel sleepy, we want to go to sleep. When we feel fatigued, we want to stop what we're doing. Studies have demonstrated that sleepiness and fatigue both result in decreased mental and physical capacity.[6] This decrease in mental and physical capacity is especially dangerous for a driver behind the wheel.

> A publication by the European Road Safety Observatory provides the following explanation: "Sleepiness is the drive for sleep, while fatigue can be seen as a signal from the body that we should end the ongoing activity, whether it is physical activity, mental activity, or just being awake."[13]

Sleepiness or fatigue are usually not an issue for a fire apparatus operator who is responding lights and sirens to an emergency call. Most emergencies are located just minutes from the firehouse, and the adrenaline and excitement of an emergency response will overwhelm the feeling of sleepiness or fatigue. However, there may be situations when fatigue becomes a problem. These situations include long drives to training schools and conferences, or a long-distance response to a large incident.

Another common issue in the fire service is a long drive home from work. It is not uncommon for firefighters to drive several hours as they return home after a 24-hour shift. Firefighters who make these long drives must be especially wary of fatigue. One study examined the effects of fatigue on nightshift workers who were driving home from work. The study revealed an increase in the number of times the vehicle drifted outside its lane of travel, increased drifting within the lane, increases in the number of times the driver closed his eyes, and an increase in subjective feelings of sleepiness.[7] This issue may be especially true for firefighters who work in busy companies and aren't able to sleep restfully through the night.

Fatigued driving can manifest itself in many ways. Fatigued drivers may suffer slow reaction times, lack of vigilance, poor attention, and reduced coordination.[8] Fatigued drivers may also demonstrate diminished steering performance and difficulty maintaining a proper following distance.[9] These issues have been shown to be more prevalent under the following circumstances:

1. During late night, early morning, and midafternoon hours
2. On a straight, high-speed roadway
3. When the driver is young and alone in the vehicle

Causes of Fatigue

Acute Lack of Sleep

A lack of sleep, or constantly interrupted sleep, is a common cause of fatigue. An acute lack of sleep is defined as little or no sleep within the past 24 hours. This scenario is not uncommon for firefighters and emergency responders who work long shifts in busy companies. Even if firefighters are able to lay down to sleep, their sleep pattern may be interrupted by an emergency call. Interrupted sleep is just as bad as no sleep.

Studies have shown that an acute lack of sleep may result in symptoms similar to drunk driving. One study demonstrated that a driver who stays awake for 17 hours will have impairment similar to that of a driver who operates a vehicle with a blood alcohol concentration of 0.05%. A driver who has stayed awake for 24-hours will have impairment similar to someone who operates a vehicle with a blood alcohol concentration of 0.10%. [10]Acute lack of sleep manifests itself behind the wheel in the same manner as drunk driving.

Chronic Lack of Sleep

Chronic lack of sleep occurs when the normal rest cycle is not long enough. Most people require 7–9 hours of sleep each night in order to feel fully rested. If this rest cycle is shortened by a few hours each night, the person will begin to feel a chronic sense of fatigue over the course of several days. The only way to reduce this sense of fatigue is to catch up on sleep. It is not unusual for a person to require several full nights of sleep to lose the feeling of chronic fatigue.

Unfortunately, today's economic demands often require firefighters and emergency responders to work multiple jobs to make ends meet. When a firefighter is working at multiple fire stations, or working back-to-back overtime shifts, chronic fatigue may become an issue. It is important for fire department administrators to recognize this issue and ensure that emergency responders are getting enough rest to safely perform their job. In situations where a lack of pay is causing firefighters to have to work multiple jobs, it may be helpful for union administrators to highlight the issue of chronic fatigue in hopes of securing increased pay.

Medical Conditions

Medical conditions such a sleep apnea, narcolepsy, and age-related conditions can all affect the quality of a person's sleep. Poor quality sleep can lead to reduced sleep hours. If a person suffers poor sleep over the course of several weeks, it can result in chronic fatigue. Managing medical disorders and ensuring that they do not affect the quality of a person's sleep is important to overall health and physical performance.

If a firefighter is experiencing a medically related sleep disorder, fire officers and administrators should examine the problem and attempt to rectify the situation. Solutions may include alternative sleeping arrangements or shift alterations to account for the firefighter's medical condition.

Circadian Rhythm

The human body follows a predictable cycle which revolves around a 24-hour circadian clock. The circadian clock helps determine a person's daily cycle, such as when they will wake up, eat, or feel sleepy. Most humans adapt to a normal circadian rhythm, which tends to follow the cycle of the sun. At night, humans sleep. During the day, humans are awake. Unfortunately for the fire service, fires and emergencies occur at all hours, which means firefighters must work 24-hour shifts. This around-the-clock schedule can lead to sleep disorders and fatigue related problems. Circadian rhythms can also make it difficult for a night-shift worker to sleep during the day.

Humans are not nocturnal and feel the need to sleep at night. Humans also have an increased need for sleep at certain times of the day, usually midnight to 0400 hours, and 1400–1600 hours. Many firefighters refer to these times as "the witching hours," as these hours are when fatigue tends to be greatest. The need to sleep at particular times of the day can cause periods of extreme fatigue during a shift.

Monotonous and Lengthy Tasks

Long and boring tasks can increase fatigue. A lack of mental or visual stimulation, combined with a long and monotonous road, may cause a driver to zone out and not pay attention. During a long and monotonous drive, steering performance may deteriorate as fatigue sets in.[11] As fatigue sets in, the driver may fall asleep or experience a period of microsleep. Microsleep occurs when a person nods off for a few seconds. Microsleep is common in a person who is extremely tired, which is why a fatigued or sleepy driver on a long trip must be vigilant to avoid succumbing to a microsleep.

While fatigued or sleepy driving may not be an issue during an emergency response, firefighters must remember to take frequent breaks and rotate drivers during long rides. When making a long response to a neighboring state for a large incident, it is important to ensure that more than one qualified driver is on the rig. This will allow the drivers to take turns and help alleviate safety issues related to fatigue.

Medications

Some prescription and over-the-counter medications cause drowsiness. It is important for firefighters to discuss these medications with their doctors before operating a vehicle. If the medication states "do not operate a vehicle or heavy machinery while taking this medication," take heed. If a firefighter is involved in a crash and investigators determine that there was a sleep inducing medication in the firefighter's blood, the firefighter may face legal sanctions. Anytime a firefighter is prescribed medication, it is important to notify the shift commander and determine if the firefighter is fit for duty. This is especially true for fire apparatus operators.

Countermeasures to Fatigued or Sleepy Driving

Firefighters and fire apparatus operators must take measures to ensure that they do not succumb to the ill effects of fatigue. Firefighters must ensure that they get plenty of rest before reporting for duty. Firefighters must also be wary of taking medications which could result in drowsiness or reduced mental capacity.

For firefighters who may be driving long distances, it is important to take breaks and constantly rotate drivers. Fire departments should consider policies which require multiple drivers if the fire apparatus is going to be driven longer than a few hours. Having more than one driver on the apparatus will allow the driver to pull over and give someone else a turn to drive.

Most experts recommend some type of break every 2 hours or every 100 miles of travel.

For firefighters who face long rides home after work, it may be best to take a nap or rest before driving home. While there is an understandable rush to finish work and get home, it may be best to take an extra hour or two to rest. For firefighters working overnight and driving home at 8 am, this rest period may actually benefit the firefighter by giving rush hour traffic some time to subside. If needed, drink a caffeinated beverage before hitting the road. Researchers at the University of Britain at Loughborough conducted studies to determine the most effective methods for fighting fatigue while driving. One of the most effective methods was the "Caffeine Nap."[12] This method involves drinking a cup of coffee and then immediately taking a 15–20 minute nap. Because it will take time for caffeine to work its way through the body and into the brain, there is a short window of time to take a nap. A caffeine nap should last no more than 15–20 minutes, as anything longer will result in the a deeper stage of sleep that will take time for a person to recover from. If a person sleeps for 15 minutes, they will wake up just as the caffeine is reaching the brain and beginning to block the chemical which causes the feeling of tiredness. The person will then feel more alert.

Case Study—Michigan

A wildland fire apparatus was responding from Michigan's Upper Peninsula to a wildfire in Utah. As the vehicle traveled through Minnesota, it drifted off the right side of the road, struck a cable barrier, and rolled over. As a result of the crash, two firefighters were killed.

Investigation revealed that the driver of the fire apparatus may have fallen asleep behind the wheel. The firefighter told investigators that he remembered waking up to one of the firefighters yelling at him as the fire apparatus left the roadway. The fire apparatus operator was unable to regain control of the vehicle and it crashed.

During the investigation, the driver admitted that he had been awake for almost 28 hours before the crash. The driver further admitted to having smoked marijuana the morning of the crash and ingesting cocaine two days prior. As a result of the investigation, the fire apparatus operator was charged with two counts of vehicular homicide.

Notes

1. NHTSA, "Distracted Driving," retrieved from https://www.distraction.gov/stats-research-laws/faq.html,accessed January 28, 2017.
2. "Understanding the Distracted Brain: Why Driving While Using Hands-Free Cell Phones is Risky Behavior." National Safety Council White Paper. April, 2012.
3. M. A. Just, T. A. Keller, and J. Cynkar, "A decrease in brain activation associated with driving when listening to someone speak." *Brain Research 1205*, pg. 70–80, 2008.
4. Harbuck, J.L., Noy, Y.I., Trbovich, P.L., & Eizenman, M. (2007). An on-road assessment of cognitive distraction: Impacts on driver's visual behavior and braking performance. Accident Analysis & Prevention, 372–379.
5. Cooper, Mederios-Ward & Strayer, in press...found in "Measuring Cognitive Distraction in the Automobile," AAA Foundation for Traffic Safety. June 2013.
6. European Road Safety Observatory, "Fatigue," (2006), retrieved May 9, 2008 from http://www.erso.eu
7. T. Akerstedt, B. Peters, A. Anund, and G. Kecklund, "Impaired alertness and performance driving home from the night shift: a driving simulator study." *Journal of Sleep Research, 14*, pg. 17–20, 2005.

8. NCSDR.NHSTA Expert Panel on Driver Fatigue and Sleepiness, "Drowsy Driving and Automobile Crashes," retrieved from https://one.nhtsa.gov/people/injury/drowsy_driving1/Drowsy.html

9. European Road Safety Observatory, "Fatigue."

10. A. M. Williamsonand A. M. Feyer, "Moderate sleep deprivation produces comprehensive cognitive and motor performance impairments equivalent to legally prescribed levels of alcohol intoxication." *Occupational and Environmental Medicine, 57*, pg. 649–655, 2000.

11. M. van der Hulst, T. Meijman, and T. Rothengatter, "Maintaining task set under fatigue: a study of time-on-task effects in simulated driving." *Transportation Research* Part F, 4 (2), pg. 103–118, 2001.

12. https://www.ncbi.nlm.nih.gov/pubmed/9401427

13. European Road Safety Observatory, "Fatigue."

Hose Loads, Wheel Chocks, and Electronic Recording Devices

First arriving fire apparatus should be laying a supply line whenever possible. Due to today's highly combustible fire loads, it is important to ensure a reliable water source before initiating interior operations. Unfortunately, there are times when a supply line may block the path for next arriving units. If the road is blocked by a charged or uncharged hose line, fire apparatus operators should do whatever they can to avoid driving over it. Hose can get tangled around a wheel or axle, resulting in damage to the hose, the apparatus, or the hydrant. A dragging or burst hose line may result in serious injuries to firefighters and civilians standing in the area. Taking a deep breath and a few extra seconds in the beginning of the firefight can save a great deal of headaches in the long run.

One person can easily move an uncharged hose line. Taking the time to stop the apparatus and slide an uncharged line out of the way is much safer than driving over the hose and risking damage. When faced with this scenario, and if manpower permits, members should dismount the truck and help the hydrant person move the supply line to the side of the road before it is charged.

Driving over a charged hose line could cause it to burst or create a dangerous water hammer, which will destroy the water supply to the members fighting the fire. A ruptured hose line can also create a dangerous missile that could injure anyone in the vicinity. When the pressure of a multi-ton vehicle is applied to a hose line that is resting on an asphalt road, curb, or other sharp surface, it may cause the hose to slice or rupture. While a charged hose line is more difficult to move, it can still be done with the proper methods and enough manpower. Hose ramps can also be used to avoid driving over charged hose lines. Ensure that the hose ramp is tall enough to allow the apparatus to mount the hose line without striking or rubbing it.

Most manufacturers strongly discourage driving over a hose line under any circumstances.[1] However, there will be times when driving over a hose line is absolutely necessary. If a hose line must be driven over, attempt to drive over it when the line is charged. More damage is likely to occur when driving over an uncharged hose line than a charged hose line.[2] Keep in mind that if the hose is not charged, it would be much easier to have someone jump out of the rig and move it out of the way.

When absolutely necessary, hose lines should only be driven over when there is sufficient ground clearance under the apparatus. Avoid driving on couplings and cross the hose at no less than 45° to help prevent the hose from becoming tangled between the rear wheels.[3] Drive over the hose line slowly and do not apply the brakes or turn the wheels. Turning the steering wheel or applying the brakes while there is hose under the tires will significantly increase the chance of damaging both charged and uncharged hose lines.

Driving over a fire hose should be avoided at all costs. By slowing down, taking a deep breath, and planning ahead, apparatus operators can avoid this problem all together. Planning how to lay the line, or having someone take a few seconds to move the line out of the street (a great job for all the cops standing around), will prevent the need to drive over a hose in the first place. In the rare event that a hose line must be driven over, ensure that it is done properly and safely.

Loading Fire Hose

After laying heavy supply lines, many fire departments will load the supply line back onto a moving fire apparatus rather than carry the hose to the apparatus. While this technique is expedient and reduces heavy lifting by firefighters, fire apparatus operators must understand the dangers associated with having crew members standing on and around a moving fire apparatus. This is especially important considering members may be tired and exhausted from a lengthy fire-ground operation.

NFPA 1451 provides guidance on how hose loading operations should take place. Loading hose back onto a fire apparatus is the only time firefighters are allowed to stand on the tail step, side steps, or running boards while the vehicle is in motion. However, firefighters should only be allowed to stand on the moving apparatus after several conditions have been met. The first condition is that the fire department has a written procedure which outlines the safety requirements for loading hose onto a moving fire apparatus. All of the members involved in a hose loading operation must be trained in these procedures.

When loading hose back onto the fire apparatus, one member must be designated as a safety observer. The safety observer must have a clear and unobstructed view of the hose loading operation and maintain visual and voice communication with the fire apparatus operator. The safety observer must be able to immediately stop the operation in the event that someone falls off of the truck or there is an imminent safety hazard.

All civilian traffic must be excluded from the area surrounding the hose loading operation. If civilian traffic must be allowed near the operation, it must be under the direct control of someone authorized to control traffic, such as a police officer or fire-police officer. This will prevent a vehicle from striking any of the members who are involved in the hose loading operation.

During the hose loading operation, the apparatus should never be operated any faster than 5 mph. While members are allowed in the hose bed during a hose loading operation, they are not allowed to stand. Standing will increase the chance of a member losing their balance, while also making them more susceptible to being hit by low hanging obstructions such as wires and branches.

Every hose loading operation must be evaluated to ensure that the members are complying with all safety procedures. In the event that the members can't adhere to these requirements, or if the operation is deemed unsafe, hose must not be loaded onto the moving fire apparatus. Instead, other methods of hose loading must be utilized.

Securing Hose Loads

It is important that the hose loads on every apparatus are properly secured by a retention system so the hose does not inadvertently deploy on the highway. NFPA requires all apparatus to have a hose storage area equipped with a positive means to prevent the unintentional deployment of hose from the top, sides, front, and rear of a hose storage area when the vehicle is in motion (fig. 23–1).[4] Many firefighters are shocked to discover how many civilian injuries and fatalities are caused by hose deploying off an apparatus as it responds to or from an emergency. If the hose should deploy, it will create a dangerous whip that can seriously injure anyone in the vicinity and cause a great deal of damage to anything in its path. Drivers must inspect each hose load to make sure it is properly secured and also keep an eye on the mirrors while driving to or from an emergency. If it appears that the hose is about to deploy, or has already deployed, stop immediately and take corrective action.

Understanding Wheel Chocks

For most fire apparatus operators, a wheel chock is something they may or may not remember to deploy at a fire scene. Wheel chocks are not optional. They are

FIGURE 23–1. NFPA requires all apparatus to have a positive means of preventing the unintentional deployment of hose. This is one example of such a system installed for a crosslay.

an integral part of vehicle safety and crash prevention. Though they are taken for granted because of their simplicity, there is actually a great deal of science involved in wheel chock design. In addition to fire departments, airlines, railroads, and private citizens who tow trailers or keep cars in their garage also use wheel chocks. The height, width, and composition of a wheel chock play important roles in defining the size of the vehicle that a wheel chock can safely hold in place.

A wheel chock is an inclined plane designed to prevent a vehicle from rolling forward. Because tires have difficulty rolling uphill, a wheel chock creates an artificial hill. To move in the direction of the road grade, the tire must climb the artificial hill. If the vehicle should start to roll, the wheel will press the front base of the chock into the ground. The base of the chock is covered with metal teeth or some type of friction surface which digs into the ground and plants the chock as the vehicle rolls on top of it. As long as the grade is not too steep, or the tire too large, the vehicle will run up against the planted chock and stop rolling forward.

The size of the wheel chock depends on the size of the tire. Most sources state that the height of a wheel chock should be ¼ the height of the tire. If the tire on the fire apparatus is 36 in. high, the wheel chock should be at least 9 in. high.

Another consideration when selecting a wheel chock is the type of surface the vehicle is parked on. Dry, solid surfaces work best, while loose or wet surfaces cause issues. A loose or wet surface will cause the chock to slide instead of plant itself into the ground. This is especially true when using a chock in off-road situations, such as brush fires or technical rescues, where the vehicle may be parked on loose or wet soil. For vehicles that operate off-road, a chock with a large footprint is a better choice. A chock with a large footprint will spread the load of the vehicle across a larger surface area. As the weight of the vehicle is spread over a larger surface area, the chock will better resist sliding or sinking into soft or loose ground.

On extreme grades, more than one wheel chock may need to be deployed, or the chock may need to be higher than ¼ the height of the tire. There may be times when a member may actually have to stay with the vehicle with a foot on the brake to keep the vehicle from rolling forward. Keep in mind that if this is the case, the vehicle is probably at the upper limit of its operating capabilities and the driver should question if the vehicle should be there in the first place.

Types of Wheel Chocks

Each type of chock has advantages and disadvantages. A contoured chock fits better than a pyramid chock, as the tire will nest better along the chock face (fig. 23–2). Aluminum chocks tend to have better gripping abilities in all weather conditions and will not compress under heavy weight. Rubber or plastic chocks are lighter but do not have very good gripping qualities on cold, icy, or slippery surface conditions. Rubber or plastic chocks may also compress or lose shape over time, especially if they are placed under heavy loads.

In July, 2000, a study was conducted by Turtle Plastics to compare the performance of aluminum and plastic wheel chocks. The study, entitled "Field Testing and Standard Evaluation of Wheel Chocks Manufactured from Recycled

FIGURE 23–2. In this case, a trailer is chocked in the apparatus bay using a rubber pyramid chock. Notice the gap between the chock and the tire. This is why contoured chocks tend to fit better against the tire face.

High-Density Polyethylene," was designed to examine the effectiveness of plastic wheel chocks versus aluminum wheel chocks under both static (parked/standing still) and dynamic (rolling vehicle) conditions. The study examined conditions on both a 3-degree and 30-degree slope. Three styles of wheel chocks were tested during the study and the details are provided in table 23–1.

Static Testing—3-Degree Incline

The vehicle was parked on a 3-degree incline, the chock was placed in contact with tire, and the parking brake was released. All three of the chocks held firmly. In the case of the aluminum chock, the vehicle rolled forward slightly until the tire fully engaged the chock, and the vehicle came to a stop.

Static Testing—30-Degree Incline

The vehicle was parked on a 30-degree incline, the chock was placed in contact with the tire, and the parking brake was released. The aluminum chock moved approximately 1–3 in. After this slight movement, the aluminum teeth on the bottom of the chock dug into the road surface and stopped the truck.

The plastic chock shot out from under the tire when the parking brake was released, and the driver had to engage the brakes to stop the vehicle. According to the researchers, it appeared that the plastic teeth were unable to grip the road surface and stop the vehicle.

TABLE 23–1.   Description of wheel chocks tested. The study used a 2000 Pierce ladder truck, equipped with 44.5 in. tires.[8]

	Plastic	Aluminum (Non-folding)	Aluminum (Folding)
Material	100% Recycled HDPE	Aluminum	Aluminum
Height	11.1 in.	12.0 in.	12.0 in.
Width	14.5 in.	11.3 in.	10.7 in.
Length	15.0 in.	21.0 in.	20.9 in.
Average Weight	18.2 lbs	17.0 lbs	21.8 lbs
Tooth Configuration (Note: "Front" surface refers to the bottom surface that fits furthers under the tire)	Bottom surface has three rows, running the lengthwise direction, with 5 teeth per row; teeth are consistent in size.	Teeth around entire bottom surface perimeter; 22 small teeth in front, two large teeth on sides, and five large teeth on back.	Bottom surface is comprised of two separate aluminum pieces. Front surface has 34 small teeth and back surface has five large teeth.

Dynamic Testing—3-Degree Incline

The vehicle was parked on a 3-degree incline, and the chock was placed 4 in. in front of the tire. The truck was allowed to roll forward 4 in. before engaging the chock. All three of the wheel chocks held firmly and stopped the vehicle from rolling any further.

Dynamic Testing—30-Degree Incline

The vehicle was parked on a 30-degree incline and the chock was placed 3 in. in front of the tire. The truck was allowed to roll forward 2 in. before engaging the chock. In this case, the truck rolled approximately 10–15 in. with the aluminum chock wedged in front of the tire before the chock dug into the road surface and stopped the vehicle. Due to the poor performance of the plastic chock during the 30-degree static testing, the plastic chock was not tested on a 30-degree incline.

As is evident from this study, the aluminum wheel chock performed better on the 30-degree incline than the plastic wheel chock. While there was significant movement of the aluminum chock on the 30-degree incline when the vehicle was allowed to roll, the aluminum wheel chock stopped the vehicle from rolling any further than 15 in.

Proper Use of a Wheel Chock

The previous study demonstrates the importance of making sure that the wheel chock is fully engaged against the tire. There should be no space between the chock and the tire face. Any free space will allow the vehicle to roll forward and gather momentum which will make it more difficult for the chock to stop the vehicle. When using wheel chocks, the chock should be pressed firmly against the face of the tire and in the center of the tire. Depending on the situation, drivers may need to use more than one wheel chock to properly secure the vehicle.

Many drivers question why modern day fire apparatus require the use of wheel chocks. In a vehicle equipped with air brakes, a loss of air pressure should automatically engage the spring brake (parking brake). While it is true that a loss of air pressure in the system should cause the parking brake to engage, the parking brake may fail for several reasons. These reasons include the following:

Operator Error

Operators may forget to engage the parking brake, or they may not completely engage the actuator valve. The brake release may also be knocked accidentally, causing the parking brake to release. In the case of a PTO pump, if the driver does not engage the pump correctly, throttling the vehicle to flow water may cause the vehicle to lurch or drive forward.

Mechanical Failure

The most important part of the parking brake system is a large and powerful spring. To drive the vehicle, one side of the brake chamber is charged with air, which compresses the large spring. When the spring is compressed, the wheel can spin freely. When the parking brake is engaged, the air is released from the brake chamber, allowing the spring to expand. When the spring expands, it engages the parking brake. If this spring becomes damaged or broken, it may result in a failure of the parking brake.

Some brake chambers allow contaminants such as water, salt, and de-icing chemicals to enter the chamber. When this happens, the power spring can become contaminated and start to corrode. While some manufacturers coat their power springs with anticorrosion materials, these protective coatings can wear off over time, resulting in unprotected parts of the spring that may be susceptible to corrosion. A corroded spring may break, causing a loss of the parking brake.

Another issue that may result in a loss of parking brake power is *sag loss*.[5] As the vehicle and brake chambers age, the repeated compression and expansion of the power spring can result in a loss of parking brake power. While the parking brake may not show any issues when it is holding the vehicle on flat ground, the vehicle may start to move on its own if it is parked on a steep grade. This is another reason why regular tests of the brake system should be mandatory.

Don't Forget the Chock!

It is common for drivers to forget to take up the wheel chocks before driving away. When this occurs, the vehicle could run over the chock resulting in significant damage or even injuries. If the chock finds its way between two dual tires, it will be extremely difficult to remove the chock without further damage to the apparatus.

Chocks should be painted a bright, fluorescent color that is immediately noticeable to a driver. Drivers should conduct a walk-around of the apparatus anytime they are about to move it. During the walk-around, a brightly painted wheel chock will be more noticeable than a dark metal or black rubber chock. It is also helpful to place a sign or sticker somewhere at eye level when the driver enters the cab that says "CHOCK!" If the driver fails to pick up the chock during the walk-around, a reminder sign will prompt them to pick it up.

It is important to remember that wheel chocks are not a substitute for using a parking brake. However, if the parking brake should fail for some reason, the wheel chock will provide a backup system to prevent the vehicle from rolling down a grade resulting in damage or injury.

Vehicle Data Recorders

Modern vehicles are equipped with event data recorders, commonly referred to as a "black box." While some may question the privacy issues related to these devices, an event data recorder can be an extremely useful teaching and investigative tool. In fact, this type of technology is not new to commercial motor vehicles. Older fire apparatus are already recording this information in the engine control module. The only difference is the data recorded by the engine control module is not as readily accessible as the data recorded in the black box of a modern fire apparatus. As a crash reconstructionist, I have been downloading commercial motor vehicle engine control modules for years; however, each download required special software that was specific to the engine manufacturer. The NFPA 1901 standard requires that the data be readily accessible to the fire department without the need for expensive, proprietary software.

NFPA 1901 requires that all new fire apparatus be equipped with a vehicle data recorder. The vehicle data recorder constantly records information which can be readily accessed by anyone who has the appropriate computer software to download the data. The NFPA standard requires these devices to record the following information at least once every second and store the information in a 48-hour loop:[6]

Data	Unit Of Measure
Vehicle Speed	mph
Acceleration (from speedometer)	mph/sec
Deceleration (from speedometer)	mph/sec
Engine Speed	rpm
Antilock Braking System Event	on/off
Seat Occupied Status	Occupied: yes/no by position
Master Optical Warning Device Switch	on/off
Time	24-hour Clock
Date	Year/Month/Day

While the data contained in the vehicle data recorder can be used for crash investigation purposes, it can also be used as part of a comprehensive driver training program. Many departments pay for the event data recorder and never use them. This is a huge mistake. Data should be downloaded after every run, or in the case of a career department, after every shift. This data should then be used as a teaching tool for the driver of the apparatus. Just remember that NFPA 1500 requires any fire department that uses a data recorder to have a policy that requires a written procedure for uploading, monitoring, and reviewing the data.[7]

Information from the event data recorder will also allow a fire officer or crash investigator to objectively examine the operation of the fire apparatus in the event that a complaint is received. As an investigative tool, it removes warped perceptions that civilians may have when they see a fire truck screaming to a call. Then again, perhaps the complaint of unsafe driving was warranted. Regardless, a vehicle data recorder will remove guesswork and speculation from an investigation and provide objective data that can be shown to both the complainant and the apparatus operator to prove or disprove an allegation.

In addition to resolving complaints, a vehicle data recorder is a useful training tool. Data recorders can help identify issues such as hard braking, sudden decelerations, and excess speed. Driver training programs can then direct training towards the root causes of these issues, such as following too closely, driving too fast for conditions, and aggressive driving tactics. These are common causes of hard braking or sudden decelerations.

Keep in mind that vehicle data recorders are installed in more than just fire apparatus. Civilian vehicles have had such technology since the early 1990s. These devices, commonly referred to as "Event Data Recorders," are found on all modern vehicles. The devices are used to deploy the air bags and crash avoidance equipment. When there is enough of a jerk, or if the airbags deploy, the event data recorder will record up to 5 seconds of pre-impact data. This data is recorded in the memory of the device and can be easily retrieved by investigators after the appropriate search warrants or subpoenas are obtained. Few people realize that their own vehicles, in addition to their fire apparatus, are equipped with such technology.

On-Board Video Systems

Video systems provide an even better option than event data recorders for identifying and resolving issues related to unsafe driving. These systems are also excellent tools for training drivers and reducing liability. On-board video provides an objective, first-hand account of how an emergency vehicle driver is operating the apparatus. Fire officers and investigators can check the tape to determine if an alleged violation should be substantiated or dismissed.

Many fire apparatus operators get caught up in the moment and fall victim to the adrenaline rush that accompanies an emergency response. They often don't realize the risks they are taking as they navigate tight streets filled with vehicles and pedestrians. Having the ability to go back and review an emergency response with video evidence is an excellent tool for driver training instructors.

The use of video systems can also provide valuable evidence to refute claims of unsafe driving. It is not unusual for an intersection crash to involve the simple fact of who had the red light. Both the apparatus operator and the driver of the civilian vehicle may claim to have had a green light. Video evidence will provide irrefutable proof of who had the right-of-way before the crash occurred. During my days as a chief officer, I can't tell you how many times I handled a complaint about unsafe driving from a member of the public or from a firefighter riding or following a rig. Almost every time it boiled down to a "he said—she said" situation. I wish I had access to a vehicle data recorder for each and every one of those complaints.

To those who worry about liability issues related to vehicle data recorders and video systems, I ask "what are you hiding?" If a fire department is worried about recording an emergency response in a multi-ton fire apparatus, they have bigger issues to examine. If a fire department is worried that the data recorded by these devices may paint apparatus operators in a negative light, it is definitely time for the fire department to examine its vehicle operations program.

For departments that have a strict vehicle operations program, digital recording technology can provide valuable evidence to help protect the driver and the fire department should a motor vehicle crash occur. Having irrefutable proof that the fire apparatus operator was not at fault can often put a quick end to a lawsuit brought against a fire department. Rather than waste countless dollars and many years trying to defend a baseless lawsuit, digital evidence can send the lawyers packing.

Some drivers will take issue with "Big Brother" watching them via black box technology. Whether they like it or not, all vehicles are equipped with such technology, and it is here to stay. Fire departments must embrace the technology as a useful teaching and investigative tool that will increase the safety and credibility of a vehicle operations program.

Notes

1. "Fire Hose Care Dos & Don'ts." North American Fire Hose Corporation. https://www.nafhc.com/hose-maintenance-and-care
2. Arthur R. Couvillon, *Fire Captain Written Exam Study Guide*.
3. Bill Gustin, "Working with Large-Diameter Fire Hose Part 1." *Fire Engineering Magazine*, January 1, 2000.
4. NFPA 1500—Section 6.1.8
5. "The Consequences of Using Low-Quality, Low-Cost Spring Brakes: A Look at the Critical Issues". Bendix Spice Foundation Brake, LLC.
6. NFPA 1901, Section 4.11
7. NFPA 1500, Section 6.1.2.1
8. "Field testing and standard evaluation of wheel chocks manufactured from recycled high-density polyethylene." Prepared for the CWC by Turtle Plastics. July, 2000.

Developing Driver Training Programs and Selecting Drivers

A fire department vehicle safety program is only as good as the driver training program. Understanding how to properly develop a driver training program that combines effective classroom training with hands-on training is imperative for fire department administrators and driver training coordinators. Unfortunately, many driver training programs do not thoroughly address vehicle dynamics and crash causation. While a typical low-speed cone course provides valuable practice for precision driving, this training does not provide in-depth knowledge of how a vehicle will handle under real-world conditions. Just as a doctor needs to learn anatomy and physiology to treat disease, a fire apparatus operator must understand the dynamics of a moving vehicle to become a safe and effective driver.

The thought of managing a driver training program may seem like a daunting task. While many fire departments already have a program in place, other departments are struggling to find enough qualified members to fight fires, let alone train drivers. Taking the time to design a thorough driver training program will save years of future headaches and liability. This chapter is designed for fire departments starting a brand new driver training program or seeking to improve upon an already existing program. The goal of this chapter is to provide ideas for new or current driver training instructors.

Remember that only qualified drivers should operate a fire apparatus under any circumstances. If the driver is in training, they should not operate the vehicle without another qualified driver by their side. Furthermore, fire departments should strive to develop individual certification programs for each apparatus, as different vehicles may have different design and handling characteristics. Training and certification programs should also include members of the fire department who operate non-emergency vehicles during routine operations, as well as members who may operate specialized or unconventional vehicles. These vehicles may include bulldozers, bicycles, and other unusual vehicles. Training should include policies and procedures that justify when these vehicles can respond to the scene of an emergency.

National Standards

When designing, modifying, or evaluating a current driver training program, it is important to research the National Fire Protection Association (NFPA) standards. Anyone involved in the supervision or development of a driver training program should purchase the following standards:

- *NFPA 1002—Standard for Fire Department Driver / Operator Professional Qualifications*
- *NFPA 1451—Standard for a Fire Service Vehicle Operations Training Program*
- *NFPA 1500—Standard on Fire Department Occupational Health and Safety Program*

While driver training and vehicle operations are referenced in other NFPA standards, NFPA 1002, 1451, and 1500 are the definitive standards for driver training and vehicle operations. Purchase them, know them, and follow them. All driver training programs should be designed to meet the requirements set forth in each of these standards, as the fire department will be judged against these standards should an apparatus ever be involved in a crash.

Standard Operating Policies and Procedures

A fire department must establish written policies and procedures for safely driving, riding within, or operating a fire department vehicle during an emergency response, nonemergency response, or routine driving. Policies and procedures provide the guidance necessary to ensure that a fire department delivers quality service in a safe, efficient, and consistent fashion. This is especially true for vehicle operations, which is one of the greatest liabilities a fire department may face.

Fire departments must consider the health and safety of its members and the public as the primary concern when developing any policies or procedures that deal with the purchase, operation, or maintenance of a fire department vehicle. Vehicle operations policies should emphasize that the fire department's first priority is the safe arrival at the incident scene of the fire apparatus and emergency responders. At no time should the fire department vehicle operation procedures be less restrictive than state motor vehicle laws.

While the terms "policy" and "procedure" are often used interchangeably, they are different. A policy refers to what the department wants the members to do. A procedure lays out guidelines for how it should be done.[1] Some experts believe that if a member deviates from a written procedure in the slightest manner, it may result in legal action should something go wrong. For this reason, many fire departments use the term "guideline" instead of "procedure." Fire departments should check with qualified legal counsel to ensure that their policies and procedures allow members to effectively do their job without exposing them to unnecessary liability.

Driver training programs should have strict policies and procedures dictating who can drive and strict criteria regarding the tests and examinations used to qualify drivers. Keeping things in black and white will leave little room for the chronic complainers who allege that one person is allowed to drive a truck because they are "in the loop" while another person is not allowed to drive because they are considered an "outcast." Removing the often heard allegation of "it's not fair" and holding everyone to the same driver training standards will increase the morale and overall professionalism of a fire department.

Vehicle operation procedures should clearly identify how the vehicle is to be driven safely when responding to or from an emergency and during routine driving. For those agencies that allow members to drive their personal vehicles to the firehouse or incident scene, the fire department must have policies and procedures in place to address personal vehicle operations. The policy and procedure manual should also include policies that address circumstances that may require variations from the policy, such as natural disasters, civil unrest, and mass disorder.

Policies and procedures are a means for fire department administrators to manage and direct even when they are not present. While many members feel that policies and procedures are a means of controlling members and limiting their decision making abilities, it should be explained that this is not the case. As long as members stay within the boundaries of the policy, members will have the freedom to make decisions and act on their decisions even in the absence of an administrator or chief officer.

Members must be held accountable to the fire department's policies and procedures. If a member fails to follow a policy or procedure, it is the front line supervisor's responsibility to immediately address the issue. A thorough investigation should be conducted to determine the facts of the incident. If evidence suggests that a policy was not followed, the driver should be made to answer. Administrative staff officers must support the front line supervisor's decision and follow through with recommended disciplinary actions. Keep in mind that there may be times when the fire officer is also culpable for the unsafe operation of a fire apparatus. In that case, the officer's direct superior should conduct the investigation.

Because a policy infraction may result in disciplinary action, the fire department must ensure that a policy is in place for progressive discipline. Regardless of whether the department is career or volunteer, the ramifications of a policy infraction must be clearly spelled out before an incident takes place. Use of a fair and unbiased investigative and disciplinary process will help reinforce the fact that the fire department administration takes safe vehicle operations very seriously and those who fail to drive in a safe manner will be held accountable.

Writing a Policy and Procedure Manual

While writing a thorough set of policies and procedures may seem like a daunting task, several organizations provide guidance that will make the job much easier. The International Association of Fire Chiefs (IAFC), the International Association of Firefighters, and numerous insurance companies have all created easy to follow templates that can be used to implement and maintain a driver training program.

Keep in mind that these initiatives are templates. While these templates are a great place to start, they may not address all of the issues associated with an individual fire department. Each policy and procedure should be thoroughly examined and customized. Do not cut and paste the template directly into the fire department policy manual.

When developing a policy and procedure manual, it is important to solicit input from the members. There will be much better buy-in if the members have input regarding how policies are written and implemented. Firefighters may become resistant to change when policies are arbitrarily dictated from on-high. Allowing the members to review and give input before implementing a policy will result in better acceptance and adherence by the rank-and-file. This is especially true for fire departments that deal with labor unions, as collective bargaining agreements may require the union to approve a policy or procedure before it is implemented.

Also remember that a policy and procedure manual is a living document and should be evaluated every year. Do not fall victim to the age old fire service adage of "that's how we've always done it." If a policy or procedure is no longer relevant or is outdated, it should be modified or removed from the manual. Keeping the policy manual up to date will demonstrate that the administration takes the process seriously and is dedicated to maintaining a progressive outlook about the job.

Selecting Instructors

After implementing vehicle operation policies and procedures, the fire department must select qualified instructors. Instructors must be able to effectively convey the important message of safe vehicle operations to driver trainees. NFPA 1451 states that only qualified persons should be assigned as driver trainers and these persons should, at a minimum, meet the qualifications for Instructor I, as specified in *"NFPA 1041—Standard for Fire Service Instructor Professional Qualifications."*

So what makes a driver training instructor qualified? The answer will set the tone for the entire vehicle operations program. Qualified driving instructors should be members with a long and proven safety record who are able to remain calm and cool behind the wheel and don't take chances. A driver training instructor should not be selected because they can drive the rig to an emergency faster than anyone else. Speed demons should not be used to train young and impressionable drivers.

There is no perfect template for the ideal driver trainer candidate. What I can say from years of visiting firehouses and speaking to drivers is that I have found there tends to be a certain personality that I would strive for in a driver trainer. Usually, these are the folks who are calm, soft-spoken, and drive as professionals for a living. In career departments, it will be easy to identify instructor candidates, as there will be a pool of full-time fire apparatus operators to choose from. In volunteer departments, I have found that some of the most responsible fire apparatus drivers are those who drive large trucks as part of their day job. Long-haul tractor trailer drivers, fuel oil delivery drivers, and public works equipment operators usually make excellent instructors.

This is especially true in states that do not require fire apparatus operators to have a commercial driver's license (CDL). Driving full-time with a CDL can be very helpful when teaching someone how to drive a fire apparatus equipped with air brakes. However, some of these CDL drivers may think that possessing a CDL, or driving in their full-time career, gives them super powers behind the wheel of a fire apparatus. Avoid these types of over-confident or aggressive drivers when selecting instructors.

Once instructors are selected, it is helpful to have them certified by an outside agency. Many state agencies and insurance companies run programs that certify driver training instructors. These programs teach driving instructors the importance of lesson plans, how to set up driver training courses, and how to identify and correct issues related to unsafe driving.

A fire department should use certified instructors because it will help protect the driver training program in the event of a lawsuit. Being able to say a department's driving instructors have been officially trained and certified by a legitimate outside agency will lend credibility to the driver training program should the fire department find itself in a court of law.

Lesson Plans

Once the instructors are certified, the department must develop lesson plans. Lesson plans ensure that fire apparatus operators are trained in a safe and consistent fashion in accordance with the fire department's written policies. Should a fire department find itself involved in a lawsuit, the attorneys will ask to see all of the lesson plans to determine what was taught to the apparatus drivers—or more importantly, what was **not** taught.

Lesson plans should address both classroom and hands-on training. Successful driver training programs start with a classroom program and then slowly evolve into behind-the-wheel exercises. Classroom sessions should discuss vehicle dynamics, crash causation, and basic operation of the apparatus. Fire apparatus operators must have thorough understanding of how a vehicle is able to maneuver on a roadway, and more importantly, why things go bad.

Selecting the Fire Apparatus Operator

The safe operation of a fire apparatus is the direct responsibility of the fire apparatus operator. While a professional fire apparatus operator must possess many positive traits, the most important trait is their attitude. A driver with a calm, professional demeanor will make a safer and more effective driver than a reckless cowboy.

Remember that the same holds true for the fire officer who is riding the apparatus, as they also assume responsibility for the driver's actions. If the officer or members in the back seats have an irresponsible attitude towards vehicle operations and safety, they may influence the behavior and actions of the fire apparatus operator. I have seen the most professional fire apparatus operators fall victim to the peer pressure placed on them by other members of the crew. For this reason, it is important to understand the concept of the driving team. All members should be trained in safe apparatus operations and have a thorough understanding of how the apparatus should be operated to and from emergencies.

The fire apparatus operator is the quarterback of the emergency response. When the apparatus is in motion, the apparatus operator is in full control of how the vehicle will respond to the call. Fire apparatus operators must be senior members of the department who demonstrate a high level of maturity and authority. The apparatus operator must be strong-willed so they do not fold to the peer pressure of other crew members, who may only be focused on getting there first. The fire apparatus operator must be able to set this pressure aside and do what is right for the safety of the public and the crew.

The fire officer must also understand and respect the fire apparatus operator's point of view. Some officers get caught up in the moment and urge the apparatus operator to drive in an unsafe manner. This can create a difficult situation for a fire apparatus operator, who may be outranked by the apparatus officer. The apparatus operator will have to choose between driving safely and ignoring a direct order from a superior officer. Fire officers must be trained in the same manner as a fire apparatus operator so the officer can understand the danger of operating a heavy vehicle.

The ideal fire apparatus operator should have a clean driving record, a clean disciplinary record, and the ability to remain calm under pressure. Keep in mind that driver training programs should be designed to weed out those members that do not fit these criteria. It is much safer to wash out an ill-suited driver than deal with the long-term ramifications of putting that person behind the wheel.

Motor Vehicle Record Checks

The fire department must ensure that the fire apparatus operator has a valid license that is appropriate for the class of vehicle that the driver will operate. The department must also determine if the state requires a commercial driver's license or any special endorsements. While some states do not require a CDL, they do have other requirements, such as an exemption card issued by the fire chief. Make sure that each driver meets the license requirements for your respective state.

Once a fire apparatus operator is qualified to drive the apparatus, many fire departments fail to conduct on-going checks on the status of the member's driver's license. Failure to conduct these checks could be an invitation to disaster. In several recent case studies, the fire apparatus operator did not have a valid driver's license at the time of the crash. Allowing a member to drive with an invalid or suspended driver's license may result in a significant lawsuit against the fire department.

NFPA requires fire apparatus operators to notify the fire department if their license is no longer valid. However, if the member fails to notify the fire department that their license is suspended or revoked, the fire department will have no way of knowing. A member's driver's license could be suspended or revoked for many reasons. Too many traffic citations, a drunk driving arrest, or perhaps a medical related issue, are all grounds to suspend or revoke a driver's license. For this reason, fire departments should conduct routine driver's license checks on every member at least twice a year.

To make it easier for a fire department to conduct regular license checks, each fire apparatus operator should provide their driver's license information to the driver training coordinator. The coordinator can then take the appropriate steps to check the member's driver's license. In some cases, the fire department's insurance company will conduct the checks. Regardless of who runs the check, ensure that it is done at least twice a year.

Medical Surveillance

Many fire departments do not have a medical surveillance program for their members. There are many reasons for this, including cost prohibitions and concerns that members will be kept from running calls due to medical concerns. While these concerns are understandable, failing to ensure that a member is healthy enough to perform the duties of a firefighter is unacceptable.

Medical surveillance of fire apparatus operators is especially important. Several recent crashes were the result of a fire apparatus operator suffering some type of medical emergency when they were behind the wheel. Medical emergencies are hazardous enough, let alone if they occur while the member is in control of a large fire apparatus.

Fire apparatus operators should receive annual physical exams. These exams should include checks of members' vision, blood pressure, blood sugar, and any other issue that could affect the ability to safely operate a vehicle. This is also an excellent time for members to discuss any prescription medications that they may be taking, which may have an effect on their ability to operate a vehicle.

Training Records

Fire departments must maintain meticulous training records, especially for fire apparatus operators. Each driver should have an individual file which documents which apparatus the operator is certified to drive, initial training, annual training, driver's license status and information, and annual medical reports. These files must contain the date(s) of training, the subject covered, and whether or not the member successfully completed the training session and was issued a certificate. Fire departments should also consider documenting each driver's time behind the wheel by instituting apparatus trip logs. This will allow each driver to track their experience operating each apparatus. Should the driver ever find themselves involved in a crash, this information will prove invaluable.

Training the Driver

Once the driver trainee is selected and undergoes basic classroom training to understand the anatomy of driving, it is time to learn the rig. Fire apparatus operators must have a thorough knowledge of every gauge, lever, button and moving part inside the fire apparatus. I must admit that when I was a 21-year-old driver trainee, all I cared about was getting behind the wheel. I had no idea what the air pressure gauges, the oil pressure gauges, or anything else in the cab meant to the safety of myself and the crew. Vehicle operation programs must teach drivers all of these important facts, including how to conduct basic tests, inspections, and servicing functions on each apparatus. This training must include the hazards of conducting these tests and inspections, as there may be dangers associated with batteries, flammable liquids, and extremely hot surfaces. Before a driver trainee puts the rig in gear, they must understand how the vehicle works and be able to recognize when things are going wrong. Drivers who are going to drive newer apparatus must also be familiar with how to properly conduct the engine regeneration process.

Driver trainees must also have an understanding of local, state, and federal DOT laws and regulations related to the operation of an emergency vehicle. Trainees must understand the rules and exemptions for emergency vehicles in their state and all department policies and procedures related to vehicle operations. This portion of the driver training program will have to be customized, as the rules and regulations will vary among states and departments.

While discussing rules and regulations, the fire department must also inform the trainee in writing of any conditions and limitations related to the trainee's personal and civil liability in the event of a crash. While the fire department must provide worker's compensation and liability insurance that protects against potential financial losses that may result from a crash, an act of gross negligence by the driver may void this protection.

Once the driver trainee has a thorough understanding of how the vehicle works and the principles of safe driving, it is time to put the truck in gear. Before driving on a major roadway, it is important to practice in a controlled environment. A basic cone course will give the trainee a chance to get familiar with the vehicle and how it handles. Keep in mind that for many driver trainees, this is the first time they have ever been behind the wheel of a large vehicle. It will take time to learn and understand the different issues associated with accelerating, braking, and turning a large vehicle. This is especially true when learning on specialized equipment such as a ladder truck or water tender.

Once the trainee is comfortable with how the vehicle will handle at low speeds, it is time to hit the road. It is best to limit new drivers' exposure to traffic until they are comfortable with driving the vehicle on a real road. Take advantage of Sunday mornings and late evenings when traffic is light. Allow the trainee to get the feel of the rig on a real road, but without the pressure of oncoming vehicles.

As trainees grow more comfortable behind the wheel, it is time for them to drive with other traffic. Allow driver trainees to take the rig out during busier times of the day and grow more confident in their ability. The trainee will learn how to turn and adjust the lane position of the apparatus when other vehicles are on the road. Trainees will also grow more confident in judging the acceleration and braking capabilities of the apparatus.

Once the trainee has extensive experience behind the wheel, it is time to begin emergency driving. At no time should a new driver be cut loose on his own before they have made several emergency responses with the assistance of a qualified instructor. I have seen countless driver trainees who are fine when they are driving under nonemergency circumstances but lose their mind when the lights and sirens turn on. Having an instructor or senior officer nearby will help keep the trainee calm as they grow more experienced in emergency vehicle operations.

Hopefully, these procedures look familiar. Unfortunately, not all fire departments follow these guidelines. Some departments will allow their members to get behind the wheel with extremely limited training, or "field baptize" someone over the radio because they need a truck to get to the scene and no one else is around to drive. These types of scenarios are an invitation for disaster.

Understanding Stress

Driving an emergency vehicle can be a stressful exercise. Operating a vehicle is difficult enough, and responding to an emergency is even more difficult. Fire apparatus operators must understand how stress can affect their ability to operate a vehicle, and also cause issues such as tunnel vision and "sirencide."

The Federal Emergency Management Agency (FEMA) conducted an in-depth analysis of a crash which involved a fire apparatus and a train. While this report was authored in 1989, and may seem dated, the conclusions are still valuable to this day. Among the many recommendations the FEMA report made for the fire service, the report touched upon issues related to the stress of driving a fire apparatus.

In September, 1989, a fire department in Virginia was dispatched to a vehicle fire. While en route to the vehicle fire, the apparatus became lost and had difficulty making its way to the incident. To compound the matter, the apparatus had left the station without a fire officer, in violation of department policy. As the apparatus arrived at the driveway where the fire was located, the driver drove past the driveway and had to back up to properly enter the driveway.

All of these factors are believed to have had a significant impact on the driver's stress level. As the stress increased, the driver may have fallen victim to tunnel vision as he attempted to redeem himself and arrive at the fire.[2] According to the operator of the train that struck the apparatus, neither the fire officer nor fire apparatus operator looked at the train as the fire truck crossed in front of it. More than likely, the apparatus operator and officer were so intent on getting to the scene quickly, they failed to see the large train approaching. As a result, the train and fire apparatus collided.

Chiefs, officers, and fire apparatus operators must understand how stress will affect the driver's ability to safely handle a vehicle. Tunnel vision, sirencide, and poor judgment may all result from the unique stress related to driving an emergency vehicle. Because of this, the fire apparatus operator, the fire officer, and all those riding the rig must work as a team to ensure that the apparatus is driven in a safe and responsible manner. If the driver is falling victim to tunnel vision and driving too fast or too dangerously for conditions, someone must speak up. Keep in mind that the fire apparatus operator is holding the lives of the entire crew in his hands. Failing to speak up and point out unsafe driving can lead to a serious crash.

The most important piece of the vehicle safety puzzle is the operator. Fire departments must focus on selecting the right person to drive and operate the rig. Only those members with a reputation of being calm, cool, and professional should be allowed behind the wheel. Furthermore, the entire crew of the apparatus must work as a team to ensure that the apparatus arrives at an incident safely. Encouraging unsafe driving from the officer's seat or jump seats is a practice that must be discouraged. Every fire apparatus operator must remember that if the apparatus crashes, the driver will be held responsible in a court of law. Proper selection and training of the fire apparatus operator will help prevent crashes and reduce liability.

On-going Training

Fire apparatus operators must complete a separate driver training program for each piece of apparatus they are expected to drive. This will provide the

NFPA makes mention of a "Challenge and Response" procedure when a fire apparatus is responding to or from a call. When a fire apparatus operator is approaching a hazard in the road, such as a traffic signal, railroad crossing, or other road-related hazard, the fire officer should ask the driver what the driver intends to do to deal with the hazard. This procedure helps ensure that there are more than one set of eyes scanning and looking at the road ahead. The procedure also helps prevent a fire apparatus operator from falling victim to tunnel vision and not seeing an impending hazard. Implementing a "Challenge and Response" procedure will also help solidify the apparatus operator and apparatus officer as a driving team.

apparatus operator with the opportunity to learn and understand the capabilities, limitations, and unique issues related to each individual rig. If members have not been trained on an apparatus, they should not be driving it.

If the fire department purchases a new apparatus, or there is a significant change to the operational procedures or technology on the current apparatus, drivers must be retrained. Fire departments must ensure that each driver receives formal training on a newly arrived apparatus from the apparatus manufacturer. This training must include vehicle limitations, manufacturer's operating recommendations, and any differences between the old and new rig. Fire departments must also take this opportunity to reevaluate their current policies and procedures to see if anything will have to change because of the new rig.

Once a member is qualified to drive an apparatus, the fire department must ensure that the member participates in on-going driver training. According to NFPA standards, drivers must complete formal hands-on training at least twice a year. In addition, fire apparatus operators must receive annual reauthorization to drive each rig. This will help keep the authorized driver's list clean and prevent someone who is no longer certified to drive the rig. I should hope the department allows its drivers to train more than **twice** a year.

Driver Training Exercises

NFPA 1451 states that the fire department safety officer must monitor the driver training program and ensure that all rules and procedures are properly followed. Before conducting a driver training activity, the safety officer must review the lesson plans and location of the exercise and each exercise must be supervised by a qualified driver training instructor. If more than one vehicle is being used in the exercise, an instructor must be assigned to each vehicle.

During the driver training exercise, the safety officer must ensure the use of all safety equipment and notify the driver training instructor of any unsafe situations. In the event that there is an unsafe condition that may result in a crash or injury, the safety officer or his designee has the authority to immediately stop the driver training operation.

When selecting a location to conduct driver training exercises, it is important to find a safe location for the members, the apparatus, and bystanders. Large parking lots that can be easily secured or remote training locations with little traffic are ideal locations. Once the location has been selected, the department must implement safety procedures before conducting the exercise. These procedures should include:

1. Separating the vehicles into safe areas if more than one vehicle will be used
2. Agreement on hand signals that will be used during backing exercises
3. Control of all members and vehicles that will operate in the training area
4. Ensuring an adequate number of instructors and safety personnel for each exercise
5. A pre-trip inspection of each vehicle that will be used for the training

Any vehicle that is not involved in the training must be safely parked outside of the training area. At no time should any of the exercises extend beyond the secured area, and any personnel not participating in the training must be prohibited from entering the secured area.

Special Hazard Training

NFPA 1451 requires fire apparatus operators to have training on specific hazards related to fire apparatus operations. The list of these hazards is quite lengthy and is included in Appendix B of the NFPA 1451 standard. Fire department safety officers and driver training instructors should examine the list and design the fire department training program to include all of the hazards.

In addition to the hazards found in Appendix B, NFPA 1451 makes reference to several other hazards directly in the standard.[3] These hazards are as follows:

1. Personnel must be trained to inspect and identify all pinch and crush points on each apparatus. These points include such things as the aerial ladder, hydraulic ladder racks, and telescoping booms and towers.

2. Once identified, the fire department must implement procedures to enable personnel to perform their job while avoiding these pinch and crush points.

3. Personnel who climb or operate an aerial ladder must be trained to recognize and avoid the danger of climbing or standing on a ladder while it's being extended or retracted.

4. Fire apparatus operators must be trained to recognize the location of the air intake for the vehicle motor and understand how burning embers may be ingested into the air intake and cause a vehicle fire.[4]

5. Fire apparatus operators must understand the danger of operating a diesel engine around fuel vapors. This is because a diesel engine does not require a spark to ignite. If the vehicle is operated in an atmosphere that has a high concentration of fuel vapors, the engine may increase speed uncontrollably. If this should occur, the only way to shut down the motor is to shut off the air intake to the engine. The battery and ignition switch will not shut down the motor under these conditions.

6. Fire apparatus operators who are assigned to a vehicle that pumps water must be trained on the proper methods of eliminating pressure in a hose line before removing inlet or discharge caps.[5]

7. Fire apparatus operators who are assigned to an aerial device must be trained in the safe operation of the device as directed by the operator's manual.[6]

8. Members must be trained to safely enter and exit the cab of the vehicle and to ascend and descend any stairs, steps, or ladders on the vehicle. Members must face the apparatus at all time and maintain three points of contact.[7]

9. Fire departments must train members, including junior members, with training that addresses the hazards of responding in a privately owned vehicle or unconventional means of transportation.[8] The fire department must also have a policy that requires the use of personal protective equipment and appropriate clothing when responding on an unconventional means of transportation.

Risk Management Programs

The NFPA 1451 standard also requires each fire department to implement a risk management program for vehicle operations. The purpose of a risk management program is to identify and evaluate vehicle related risks and implement strategies to control these risks. During the risk evaluation, fire departments must include all of the vehicles used by the members, including nonemergency and unconventional vehicles such as bicycles and trailers. The risk management program must address issues that relate to the administration of the vehicle operations program, driver training, vehicle operations, personal protective equipment and clothing, operations at emergency and nonemergency incidents, and other vehicle related activities. Once the risks have been identified and evaluated, control strategies should be developed and incorporated into the driver training program. Once the risk management and control plan has been implemented, the fire department must evaluate the effectiveness of the program a minimum of once every three years and submit a report of the findings to the chief and members of the health and safety committee.

Case Study—Texas

In February of 2014, an 18-year-old volunteer firefighter crashed a water tanker while responding to a fire. Investigation determined that the apparatus operator did not have a valid license and was not officially qualified to operate the vehicle. Instead, the fire chief called him over the radio and authorized the firefighter to drive the rig due to short staffing. The firefighter had only driven the rig a short distance when the vehicle overturned. While the fire chief stated to news reporters that the incident was "just an accident," lack of training and a lack of a valid driver's license were most certainly causative factors in this crash.

Notes

1. Craig McQuate, Penning Effective Policies, Security Management, Dec 2002.
2. USFA-TR-048 / September 1989. "Fire Apparatus Train Collision". FEMA.
3. NFPA 1451 Section 8.2.10
4. NFPA 1451 Section 8.2.9
5. NFPA 1451 Section 8.2.15
6. NFPA 1451 Section 8.2.16
7. NFPA 1451 Section 8.3.7
8. NFPA 1451 Section 4.3.4 Note: this includes, bicycles, skateboards, Segways, etc.

Training Ideas

G-Meters

It is one thing to sit in a classroom and discuss how a fire apparatus may rollover if it experiences 0.4 g of lateral acceleration, but what does that feel like? How does it feel in the seat of the driver's pants when the vehicle decelerates at a normal rate of 0.1 g compared to slamming on the brakes and skidding at 0.6 g? How does it feel to take a corner at 5 mph versus 10 mph? What is the corresponding difference in g-force? By relating the physical sensations experienced during normal driving to an actual g-force reading, a driver will have a better understanding of the vehicle dynamics learned in the classroom.

Teaching the practical aspects of g-force can be accomplished with the purchase of a g-meter[1] (fig. 25–1). By sitting in the officer's seat and watching the reading on a g-meter, the trainee will see how the g-force changes as the speed of the vehicle and the radius of the curve go up and down. Once the trainee has had the opportunity to experience g-forces first-hand, it is time to return to the classroom and reinforce the calculations discussed in Chapter 3.

This type of training will teach drivers to fully appreciate the forces placed on the vehicle as a result of their driving behavior. Drivers will come to understand that slowing down just 10 mph can make a huge difference on the g-force placed on the vehicle during a driving maneuver. A thorough understanding of g-force and lateral acceleration will give the driver a better understanding of vehicle control and rollover prevention.

Just keep in mind that at no time should driver trainees or instructors place an excessive amount of lateral g-force on the vehicle. To do so may result in a rollover crash, as evidenced by several recent training accidents throughout the fire service. If g-force training is taking place in a high center-of-gravity vehicle, students and instructors should strive to minimize the amount of lateral g-force placed on the vehicle. Anything in excess of 0.20–0.25 lateral g's could risk

FIGURE 25–1. This is an example of a g-meter. The meter can be mounted on the inside windshield to give students an idea of how lateral g-force will increase and decrease with speed and severity of a curve.

disaster. If instructors wish to demonstrate higher lateral g-force to their students, a low center-of-gravity vehicle should be used and the demonstration should take place on a controlled course

Speed

A fire apparatus operator must be able to accurately sense the speed of the vehicle without watching the speedometer. To demonstrate this point, the instructor should mount a GPS device which displays speed on the officer's side of the apparatus. The trainee will then cover the speedometer with a piece of paper and drive for a period of time. At random times, the trainee should be asked how fast they think they are going. The instructor will be able to tell the trainee the actual speed of the apparatus by looking at the speed readout on the GPS. Many students are shocked at how they underestimate the actual speed of the apparatus. During this type of training or during an actual emergency response, instructors must remember to keep a close eye on the actual speed of the apparatus and tell the driver to slow down if he begins to drive too fast.

This exercise is especially relevant during emergency calls. Drivers may fall victim to "sirencide" and drive faster than they think they are driving. At random times during an emergency response, the officer or instructor should ask the driver how fast they think they are going. Without looking at the speedometer, the driver must identify the speed of the apparatus.[2] Understanding speed control is an important concept for the fire apparatus operator.

Stopping Distance

As discussed in Chapter 1, a commercial motor vehicle with air brakes and truck tires will not stop as effectively as a vehicle equipped with hydraulic brakes. To demonstrate this, find a large open space with enough distance to safely stop a fire truck and a passenger car. Drive the fire truck at a given speed and have the driver bring the vehicle to an emergency stop. Using spray paint or spray chalk, measure the total stopping distance. Repeat the exercise using a standard sized car and measure the total stopping distance for the car. What is the difference in stopping distance between the two vehicles? How is this difference important as it relates to following other vehicles too closely, especially in wet or inclement weather?

Once the difference in stopping distance has been demonstrated, examine the stopping distance as speed changes. Create a chart or spreadsheet which will allow students to record the stopping distance of each vehicle at a particular speed. By increasing speed, the students will be able to see how the stopping distance rises accordingly.

While some administrators may be hesitant to allow a fire apparatus to emergency stop in a parking lot, this type of demonstration is invaluable. Just make sure there is enough room to safely stop the vehicle, the area is completely closed off to cars and pedestrians, and the apparatus is sent out to be checked once the training is complete. Remember to conduct straight-line stopping exercises, as turning the wheel too sharply may cause the vehicle to rollover. If the department is hesitant to use a full-size apparatus, using a smaller vehicle will still provide an excellent training demonstration.

Night Driving and Visibility

Fire apparatus operators must never overdrive the headlights. To demonstrate this issue, park the fire apparatus in a dark parking lot. Have a member pretend to be a pedestrian and walk away from the apparatus until they can no longer be clearly seen. How far away is the pedestrian when the driver can no longer clearly see them? This distance is the effective range of the headlights.

This drill should be repeated with both the high beams and the low beams. The drill should also be repeated for each apparatus, as some apparatus headlights will work better than others. Keep in mind that this is not a true experiment. In this training scenario, the driver knows what they are looking for. In reality, the driver may see something ahead of the apparatus, but not be able to perceive it as a pedestrian until it is too late.

Make sure to play fair so the driver gets an accurate portrayal of the effective range of the headlights. Do not have the pedestrian walk away from the apparatus to a point where only the pedestrian's sneaker can be seen in the headlights. In real life, will the driver really be able to perceive the sneakers and understand what they are? The pedestrian should stop walking away from the apparatus when they are no longer clearly perceived as a pedestrian.

Once the effective headlight distance has been established, compare it to the total stopping distance of the apparatus at various speeds. As an example, assume the pedestrian cannot be seen any further than 100 ft from the apparatus when the low beams are in use. If the total stopping distance of the fire apparatus is 165 ft at 35 mph, will the apparatus operator be able to perceive, react, and skid the apparatus to a stop before striking the pedestrian?

Blind Spots

It is important for fire apparatus operators to understand what they can and cannot see. An effective method for locating blind spots is to park the apparatus in a dark parking lot. Once the vehicle is parked, hang a clear light bulb over the driver's seat at the eye level of the driver and turn on the light bulb.[3] Now have the students walk around the apparatus and take note of how the light falls on the ground. Anywhere the parking lot is lit up at ground level is where the driver will be able to see all the way to the ground while sitting in the driver's seat. The areas cast in shadow represent a blind spot for anyone sitting in the driver's seat.[4]

After identifying the areas of the vehicle that are affected by a blind spot, the next step is to determine the height of the blind spot. This can be accomplished by having a member walk around the outside of the apparatus and stand in the areas of shadow with a white index card. The member should raise the index card off the ground until it is illuminated by the light bulb. When the index card is illuminated by the light bulb, it will represent the height of the blind spot. If the index card does not illuminate until it is 4 ft from the ground, anything shorter than 4 ft cannot be seen at this location. This is important when dealing with smaller obstacles, such as children, pylons, or curbs.

Inclement Weather Driving

Many driver training programs fail to address adverse weather conditions. Due to concerns about damaging or dirtying the fire apparatus, many driver training sessions only take place on sunny afternoons. This does a great disservice to the overall efficacy of a driver training program and the safety of those who ride the apparatus.

Driver trainees must understand the unique issues related to driving in poor weather. New drivers must learn how adverse weather and wet roadways will affect stopping distance, curve speed, and visibility. These differences should be learned in a nonemergency environment under the supervision of a qualified driver training instructor. Providing driver trainees with the opportunity to drive in bad weather under nonemergency conditions allows them to gain confidence in the proper operation of the emergency vehicle during inclement weather or poor visibility.

Particular attention must be paid to the proper use of manual tire chains, automatic chain systems, and auxiliary braking devices. It is not uncommon for a driver trainee to complete an entire qualification course having never driven with the auxiliary braking device deactivated. The first time they respond to an emergency during foul weather, they turn off the auxiliary braking device as instructed, only to be shocked by the sudden decrease in stopping power. Driver training programs must include a specified number of driving hours in which the auxiliary braking device is not used. This will give the driver trainee a solid feel for the additional stopping distance required when driving with the auxiliary braking device deactivated.

I would also recommend driver training in the snow and rain. While most chiefs would consider this a bad idea for fear of risking damage to the apparatus, I ask what these chiefs expect to happen when an emergency call is received? An emergency response should not be the first time a new driver gets behind the wheel during snow or rain conditions. New drivers must have experience driving in the snow or rain so they will be ready to drive in a real emergency.

Because snow storms are rare events for most of our country, it is important to take advantage of the storm and send apparatus operators out for training. Provide drivers with the opportunity to drive using automatic chain systems or manual tire chains so they learn to understand the limitations and restrictions they must place upon themselves when driving under these conditions. Drivers should also get comfortable with downshifting the apparatus to control speed and descend hills. Because most modern apparatus have automatic transmissions, few drivers have a great deal of experience downshifting the vehicle for speed control, especially during foul weather.

Sight Distance and Stopping Distance

Driver training programs should include exercises that help provide a practical understanding of sight stopping distance. While this concept may sound complex, it is not hard to teach. First, calculate the stopping distance of the apparatus at various speeds on a dry road and a wet road using the lessons learned in the Chapter 1. Write down these distances and identify an area in the first-due district that has a sight distance issue, such as the crest of a hill or a sharp curve. Take the trainees to this location and after securing permission from the proper authorities, shut down the road for safety. In busier regions where this may not be possible, you may have to be creative. Use a driveway that goes around the back of the firehouse or an empty corporate park driveway on a weekend morning.

Once the training area is secure, have the trainees examine the actual sight distance of the roadway. If the obstruction is a curve in the road, park the apparatus at some point in the curve and have a member walk away from the apparatus and into the curve. Have a trainee sit in the driver's seat of the apparatus and take note of when the member walks out of sight and can no longer be seen. When the member disappears from view, call them on the radio and have them stop walking. Measure the distance between the front of the apparatus and where the member stopped walking. This represents the forward sight distance into the curve. Place a line of cones at this location.

Having identified the forward sight distance into the curve, now go back to the apparatus and measure the total stopping distance of the apparatus that was previously calculated in the classroom. For instance, if the total stopping distance was calculated at 25, 35, and 45 mph, there will be three separate stopping distance values on the chart. Measure off the three calculated stopping distances using different colored cones to identify where the apparatus will skid to a stop.

How does the total stopping distance of the apparatus compare to the forward sight distance of the curve? Would a driver be able to safely skid the truck to a stop if they can only see 100 ft into the curve? If not, they are overdriving the sight distance of the curve. The driver would not be able to see a stopped car, react to the stopped car, and skid the apparatus to a stop before striking the rear of a stopped vehicle. This is because the curve is blocking the forward sight distance.

This type of exercise will demonstrate to trainees the need to slow down when approaching an area with a sight distance issue. If done properly, this exercise can identify the safest speed a driver can approach the area and still be able to safely stop. Not only will this exercise teach students, but the information learned in the exercise can be put to use during future emergency responses. Just remember that the chart you are using will assume the driver is SKIDDING to a stop. This is not an ideal situation. You should also examine the stopping distance of the vehicle if the driver were slowing in a more controlled fashion. In this case, an *f* of around 0.4 would be used in the skid to stop formula, as discussed in Chapter 1.

Intersections and Sight Distance

Emergency vehicle operators must provide civilian drivers with proper *notice of approach*. Notice of approach is defined as giving a civilian driver enough time to see or hear the fire apparatus, react to the fire apparatus, and yield the right of way. Notice of approach is especially important when the emergency vehicle is approaching a negative right-of-way intersection such as a red light or stop sign.

Many fire apparatus operators fail to come to a complete stop at a negative right-of-way intersection. Instead, they briefly slow down and roll the intersection. This is a dangerous practice, especially at intersections that have sight obstructions on the corners. Sight obstructions, such as buildings, signs, or vegetation, may prevent the fire apparatus operator from being able to see a civilian vehicle that is approaching on a cross street. If the driver fails to see an approaching vehicle, the fire apparatus operator may proceed into the intersection directly into the path of the approaching vehicle.

A fire apparatus operator cannot pull into an intersection when there is an approaching civilian vehicle that is "close enough to constitute a hazard." This means the fire apparatus operator cannot enter the intersection if the civilian vehicle is so close that the civilian driver will not have time to perceive, react, and bring the vehicle to a stop before colliding with the fire apparatus.

The term "close enough to constitute a hazard" defines the distance between the civilian vehicle and the encroaching fire apparatus. This distance will depend on the speed limit of the roadway and on the weather conditions. If the road is wet, it will take the civilian vehicle longer to skid to a stop. Keep in mind that fire apparatus operators should not be driving in such a way that causes civilian vehicles to skid to a stop. Instead, fire apparatus operators should be giving civilian vehicles enough time to slow down in a safe and controlled fashion.

The purpose of this drill is to demonstrate the need for a fire apparatus operator to come to a complete stop at a negative right-of-way intersection. The first step in this drill is to determine the distance it will take a civilian driver to perceive, react, and skid to a stop once they see the fire apparatus. This should be completed as a class room exercise. Calculate the total stopping distance of the posted speed limit, write it down, and bring it out to the field. Keep in mind that this assumes the civilian is traveling at the speed limit. If the civilian is speeding, the total stopping distance will be longer.

The calculated total stopping distance of the civilian will determine the civilian's "point of no escape." As an example, assume the total stopping distance of the civilian is 100 ft. If the fire apparatus pulls into the intersection when the civilian vehicle is any closer than 100 ft, the civilian driver will not have enough time and distance to perceive, react, and skid to a stop. Instead, the two vehicles will collide.

After calculating the point of no escape for the civilian vehicle, measure the distance from the intersection and mark this point in the roadway with traffic cones. The cones will signify the last point on the roadway where the civilian vehicle will be able to avoid a collision with an encroaching fire apparatus. If the civilian vehicle is any closer than the traffic cones, the vehicle will collide with the encroaching emergency apparatus. Having done this, the question becomes: where can the fire apparatus operator see this point of no escape?

Having marked the point of no escape on the cross street, fire apparatus operators should return to the intersection and look down the road. Can they see the cones when they stand at the intersection? Now walk backwards 10 ft and look down the roadway. Can the fire apparatus operator still see the cones? 20 ft back? 40 ft back? At what point can the fire apparatus operator no longer see the cones in the roadway?

Once it is determined where the fire apparatus operator can no longer see the cones, measure the distance from that point to the intersection. If the fire apparatus operator loses sight of the cones 30 ft from the intersection, is 30 ft enough distance to perceive, react, and skid the fire truck to a stop before colliding with the civilian car? How does this distance change based on the speed of the fire apparatus?

After completing this drill, the fire apparatus operator should have an understanding of why it is so important to come to a complete stop and not roll the intersection. This is especially important at intersections where there is limited sight distance due to an obstruction on the corner.

Tire Blowouts

Most fire chiefs would take issue if an instructor exploded a tire as part of a training demonstration. However, a tire blowout can be simulated during driver training. As discussed in Chapter 11, the worst thing a driver can do after experiencing a tire blowout is apply the brakes. Instead, the driver should apply power and accelerate to regain the vehicle's forward momentum. When the vehicle regains its forward momentum, it will be easier to bring the apparatus back under control. Once the vehicle has been brought back under control, the driver can gently decrease speed and come to a controlled stop.

For this drill, the driver trainee should drive on an open stretch of straight roadway. At some point during the exercise, the instructor should pop a balloon or simply yell "bang" to simulate a tire blowout. When the trainee hears the bang, he should accelerate to simulate regaining control of the vehicle. Once the apparatus has accelerated, the trainee can gently come to a controlled stop. Ideally, this type of training should be done in a large parking lot or an empty road which can be shut down for training with the permission of the local authorities. Remind the driver trainee not to drive at the governed speed as it will give no extra room to accelerate the vehicle when needed.

Antilock Brakes

Before the invention of antilock brakes, a vehicle would lose steering control when the wheels locked during a panic stop. The purpose of antilock brakes is to allow the driver to maintain steering control, even during a panic stop. However, few drivers are provided with adequate training on how to properly use the ABS system. This exercise will ensure that the fire apparatus operator has a thorough understanding of how to maintain steering control during a panic stop.

In an ABS training exercise, cones are placed in the middle of a long, open straight-away. The driver of the emergency vehicle will drive towards the cones, at which point the instructor will yell "left" or "right." The driver trainee should firmly apply the brakes and steer around the cones. It is recommended that an entry gate and an exit gate be placed before and after the cones. The driver should enter through the first gate, steer around the hazard, and attempt to exit the exercise through the second gate. During this exercise, driver trainees must remember to **stomp** their foot on the brake pedal, **keep** their foot on the brake pedal, and **steer** around the hazard.

Remember that a vehicle with a high center of gravity may roll over if there is a sharp and sudden steering input by the driver. This provides a catch-22 when it comes time to train fire apparatus operators. While drivers should be given the opportunity to practice ABS steering maneuvers, there are concerns regarding the possibility of a rollover if the maneuver is not performed correctly. Therefore, large fire apparatus should only be used for this training if they are equipped with an after-market skid training system and the exercise is overseen by a qualified instructor. If this type of training device is not available, low center of gravity vehicles should be used, such as a police vehicle. As many police EVOC instructors are well versed in this type of training, consulting with one of them before attempting this type of training is highly recommended.

Siren Audibility

Fire apparatus operators must understand the limited effective range of the emergency siren. To demonstrate this point, an emergency vehicle equipped with a siren should be positioned at a 90-degree angle to a roadway. The driver trainee should then drive towards the siren in a second vehicle and take note of when they can hear the siren. This exercise should be completed multiple times at different speeds. Driver trainees should take note that as the speed of their vehicle increases, they must be closer to the siren to hear it effectively. The car stereo and HVAC fan should also be turned on at different levels to demonstrate their effect on siren audibility. Remember to use a large parking lot or a closed (or little-used) roadway for this exercise so as not to confuse the public with the siren.

It is also beneficial to teach driver trainees how difficult it is for a civilian driver to localize the siren once it is heard. To demonstrate this, find a large, open parking lot. Have the driver trainee sit inside a vehicle while blindfolded, with the radio and HVAC system turned on. An emergency vehicle should then be driven to different points around the vehicle at a distance of approximately 150–200 ft. At each point, the siren should be sounded and the blindfolded trainee should try to identify what quadrant of the vehicle the siren is coming from. Do not be surprised if the trainee has difficulty locating the siren. As discussed in Chapter 6, drivers have difficulty localizing sirens because the structure of the civilian vehicle will distribute the sound waves across the surface of the vehicle, making it difficult to determine the location of the siren.

Backing Cameras

Many modern emergency vehicles are equipped with backup cameras. While excellent tools, backup cameras can create problems. One of the more common problems is the fact that drivers develop an overdependence on the camera while not having a full understanding of what the camera is telling them. Every driver trainee should be given the opportunity to calibrate the backup camera so they know exactly what they are seeing (fig. 25–2).

FIGURE 25–2. Most modern fire apparatus are equipped with backup cameras. Do your drivers understand how to use them properly? What do the lines on the screen mean? It is important that every driver have the opportunity to calibrate the monitor so that they understand how each line on the screen relates to distance.

Most camera screens provide a visual indication to designate the path of the apparatus, or red, yellow, and green lines to show how close the apparatus is to an obstacle. It is important to mark these lines with cones and allow the driver trainees to walk to the rear of the apparatus and see the distance from the rear of the vehicle that each colored line represents. How far back is the green line? How far back is the yellow line? Providing a sense of reference will teach drivers not to become fixated on a small screen and instead to conceptualize just how far away an object or obstruction is.

Tire Pressure Charts

Understanding proper tire inflation pressure is an important skill for a fire apparatus operator. Drivers must know how to determine the proper air pressure for each tire using the tire pressure charts provided by the tire manufacturer. Using these charts to determine the proper air pressure will ensure that the tires are properly inflated and not at risk to overheat and blow out.

The first step in understanding how to use a tire pressure chart is to weigh the vehicle. Ideally, each tire should be weighed individually to determine exactly how much weight is resting on the tire. Most large police departments or state DOT inspectors have a set of portable scales and should be more than willing to assist you.

Having determined how much weight is resting on each tire, use the manufacturer's tire inflation chart to determine the proper air pressure. How does the recommended air pressure differ from the maximum air pressure that is printed on the sidewall of the tire? Does it differ from the manufacturer's plate on the inside of the cab? Once this information has been recorded during the training exercise, it can be set aside and used for future tire inspections. Not only is this drill an excellent training tool, it is a practical exercise as well.

Cone Courses

Understanding how to design and set up a cone course is an integral part of any driver training program. At a minimum, a driver training program should include the driving exercises that are included in the NFPA standards. This will ensure that the driver training program is NFPA compliant, should the driver training program need to be defended in a court of law.

While the NFPA standards are a good place to start, there are also other resources that provide cone course layouts. An internet search will reveal countless online documents related to emergency vehicle driving, many of which include cone course dimensions and instructions for building the courses.

However, instructors should not limit themselves to cone courses that are already designed by others. Every driver training instructor should learn how to design a cone course, especially a course that relates directly to the fire department and the obstacles the apparatus operators may encounter. I once created a cone course for a police department that was modeled after the sally port the police officers must pull into when dropping off a prisoner. With a tape measure, a pencil, and some paper, it is easy to design your own course. If you design a cone course, make sure to add it to the written lesson plans.

Creating a cone course is no different from drawing a scale diagram of a building. Use the engine bay as an example. By measuring the width of the door and the depth of the bay, and including any obstacles (other apparatus or gear racks), an instructor can develop a cone course that will reflect a scenario that the apparatus operators will encounter every time they ride out.

Explaining Skids

Fire apparatus operators must understand how locked tires or loss of traction will cause a vehicle's tires to skid. To demonstrate how a vehicle will skid, use a small wooden racing car that is commonly used by scout organizations for track races. Drill a hole in each tire so that the tire can be locked in place with a small nail. Using a wood or plastic ramp, the instructor can lock different tires and show how the locked tire will affect the vehicle. If the front tires are locked in place, the vehicle should skid forward in a straight line. If the rear tires are locked, the vehicle should spin around. This type of demonstration is an excellent way to show students how locked tires affect vehicle dynamics.

Vehicle Inspections

Fire apparatus operators must understand the importance of a thorough vehicle inspection. A vehicle inspection checklist should be developed by the fire department in accordance with NFPA and federal motor carrier guidelines regarding pre-trip inspections. Instructors can hide business cards or Post-It notes in various spots on the apparatus to ensure that trainees are completing a thorough inspection. As an example, a Post-It note stuck to the inside of a dual tire which says "nail" would demonstrate the importance of properly inspecting all tires. If the student fails to find one of the mock issues, they would fail the drill.

Drunk Driving

It is important to understand how alcohol and controlled substances can affect a driver's ability to safely operate a motor vehicle. This concept is difficult to demonstrate in real life because it's unsafe to have driver trainees consume alcohol and drive. However, these scenarios can be simulated.

There are several devices on the market which allow instructors to simulate intoxication. These devices, affectionately referred to as "beer goggles," simulate some of the effects of alcohol. Remember that alcohol affects a person's coordination, balance, and ability to follow directions. Alcohol will also affect a person's fine motor skills and ability to multitask. Because a person using beer goggles will not actually be drunk, the effects are not as drastic. The person should still be able to multitask and follow directions. However, beer goggles do an excellent job of affecting a person's balance and coordination during a training scenario.

There are several ways to use beer goggles as a demonstrative tool for instruction. For fire departments with access to a driving simulator, allow each student to operate the simulator while wearing beer goggles. Another option is to rent a golf cart and set up a mini emergency vehicle operator course (EVOC). Allow students to try and drive the EVOC in the golf cart while wearing the goggles. Just make sure a second person rides on the golf cart and can take over if the student gets too out of control.

On-Board Cameras and Event Data Recorders

Vehicle based video systems and event data recorders provide driving instructors with an excellent tool to review the driving habits of fire apparatus operators. While these reviews can be done one-on-one, individual coaching sessions can sometimes be awkward or confrontational depending on the student's attitude. This is especially true when conducting a review of a member who has been driving for several years. It may be more beneficial to use these devices in a group setting. By removing any personal identifiers and making the videos anonymous, you can create an entire drill which reviews and critiques real-life driving scenarios. Relating specific vehicle safety topics to actual scenarios can be a valuable teaching tool for the entire fire department.

Critical Curve Speed and Lateral G-Force

Fire apparatus operators can learn a great deal about critical curve speed and the friction circle with the simple use of a large, open parking lot or a specially designed skid pan. Instructors can create a large circle with the radius of their choosing by tacking a string in the center of the circle and tracing the circle with traffic cones. The radius should be large enough to not cause dizziness while driving, but small enough to allow safe space for a vehicle to spin out of the circle. Ensure that there are no obstacles and that there is enough room for the vehicle to run off. Many larger fire academies will have a specially designed pad for this purpose.

Using the formula learned in Chapter 3, students can calculate the critical speed of the curve on a white board before driving the circle. The students will then have a rough idea of the speed at which the vehicle will begin to break traction. Using a civilian style vehicle with a low center of gravity (so the vehicle doesn't rollover), students should drive around the circle at increasing speeds. Eventually, the student will reach a speed where the vehicle begins to lose traction. This speed should correspond closely to the calculated speed.

If the fire department has purchased a g-force meter for training, this is an excellent time to put the meter to use. By riding with a partner, one trainee can drive the vehicle while the other trainee watches the increase in g-force as the vehicle increases speed. This drill will demonstrate the importance of slowing down for curves and the dangers related to excessive g-force. Also remember to have the trainee apply the brakes as they grow closer to the critical curve speed. This will demonstrate the effects of combined g-force, as discussed in Chapter 5

Off-Road Driving

Fire apparatus operators must practice off-road driving techniques before driving in an actual emergency. Trainees must understand the abilities and limitations of the apparatus so they do not over-extend the vehicle during a firefight. If there are off-road areas where a driver may commonly respond, it will be helpful for the trainee to go to these areas and practice driving under nonemergency conditions. Preplan these areas and identify hazards, safe routes, and other obstacles. This information will be extremely helpful during an emergency call.

Off-road driver training will depend on the type of terrain that a fire apparatus operator will face. Does the department respond to wooded areas, where stumps, creek beds, and trees could cause trouble? Or does the department respond to a desert or beach area where sand and water are the major obstacles? Driver training instructors must identify the common issues a driver trainee will face in real life and design the training program accordingly.

Winch Operations

Recovering a vehicle that is stuck off-road can be a hazardous operation if the fire apparatus operator is not properly prepared. Training on winch operations should include topics such as the rated capacities of the winch and rigging, the use of snatch blocks for mechanical advantage, and the proper operation of the PTO or electric-driven winch. Consult with local towing companies, as they have a great deal of experience in these matters and may be willing to assist with a drill.

Winch drills should include a review of anchor straps, recovery straps, and the use of bomb-proof anchors. Review the rated anchor points of the vehicle and explain where anchor straps should not be attached to a vehicle. Simulated exercises should be designed so the driver trainee has the opportunity to use the winch and rigging to recover a stuck vehicle. Conduct this training during daylight and at night, and also consider training in various weather conditions.

Backing Maneuvers

There can never be too much time spent practicing backing maneuvers. Backing drills can be as simple as a cone course, or they can include exercises which incorporate spotters, hand signals, and proper radio communication. Backing practice does not have to be overly complicated. Simply giving each driver ample time to back the apparatus around a parking lot is still an effective method of training.

Field Trips

Fire apparatus operators must have a thorough knowledge of how different parts of the vehicle work. There is no better way to learn this than to see these parts in action. Most fire departments have a truck depot or diesel mechanic somewhere in the first-due district. Arrange a meeting with a qualified diesel mechanic who is willing to give a demonstration to fire apparatus operators on the inner workings of a large truck. As an example, pulling a wheel off and watching the drum brakes engage is a valuable teaching tool for air brakes. The same can be said for inspection procedures and out-of-service criteria.

There are also online training programs which provide excellent training tools for fire apparatus operators. Some of these courses may cost money, but they are well worth the cost.

Conclusion

The only limit to your driver training program is your imagination. Most driver training instructors have found ways to increase student interest by developing driver training exercises that are directly related to the regions in which their students will operate emergency vehicles. By introducing real-life examples and scenarios, driver training students will come to understand the value of safe driving and the consequences of their unsafe acts.

Notes

1.   Also called an accelerometer.
2.   During a real call, do not cover the speedometer. This should only be done under controlled circumstances in a training environment.
3.   Use a clear light bulb, not a diffused light bulb to get an accurate depiction of the blind spots.
4.   Thanks to Kevin Johnson of Crash Consulting Services LLC, who showed me this method while teaching at the World Reconstruction Conference in Orlando, Florida.

Emergency Vehicle Crash Investigation

The Crash

A motor vehicle crash is a violent and unexpected event. Immediately following the crash, the apparatus crew may find themselves dazed and confused, struggling to figure out what happened. Despite this fact, firefighters involved in a crash must remember their first priority: the care and safety of those involved.

The crew's first priority is to triage and treat anyone who was injured during the crash. The apparatus officer must conduct an immediate survey of the scene to determine the number of patients, the severity of the injuries, and the presence of entrapped victims. Most importantly, the officer must determine if the scene is stable and secure.

If someone is injured or trapped, help must be summoned immediately. While assistance is typically requested via radio or a mobile data terminal, some crashes may be so severe that this means of communication is no longer operational. In the event of a damaged communications system, members should consider alternative means of communication such as cell phones, nearby landlines, or bystanders.

If the crash occurs in an unsafe location and both vehicles are operable, it is imperative that the vehicles be moved off of the travel lanes and onto a nearby shoulder. Any delay in moving the vehicles from the travel lanes will increase the chances of a second and potentially more severe crash. If time permits, and it is safe to do so, snapping a quick picture of the crash scene before moving the vehicles can prove quite useful to the crash investigation.

In addition to caring for any victims, proper preservation of crash-related evidence will go a long way in assisting the investigation. Valuable evidence such as broken vehicle parts, skid marks, fluid spills, and roadway evidence must be secured in place for later documentation. If conditions permit, the scene should be taped off and entry prohibited by anyone who is not directly involved in patient-care or the crash investigation. Fire apparatus operators must remember that any motor vehicle crash could result in criminal charges. Therefore, the crash scene should be treated no different from a crime scene.

Once the scene has been secured, the proper authorities must be notified. In some states, drivers are only required to notify the police department when someone is injured or when a vehicle needs to be towed. These crashes are referred to as "reportable" crashes. In cases where the crash is minor and there is only cosmetic damage to the vehicles, some states only require the drivers to exchange information instead of notifying the police department. These crashes are referred to as "nonreportable" crashes.

As a best practice, the police department should be called to every crash scene, regardless of how minor. This is due to the potential liability posed by a crash. While the driver of a civilian vehicle may say they are uninjured and only wish to exchange information, it is not uncommon for this same civilian driver to suddenly develop neck pain several days later and seek legal advice which may lead to a lawsuit. The possibility of a future lawsuit is the reason fire departments should request a police report, even for minor, nonreportable crashes. Fire departments should also have members trained to identify, gather, and secure evidence should it be needed to defend a future lawsuit.

In addition to a police response, members must also know who to notify in their immediate chain-of-command. In large departments, the apparatus officer may only have to notify their battalion chief. In small, rural departments, the apparatus officer may be required to contact the fire chief or a local representative. Fire departments must outline the notification procedure long before a crash occurs. Standing on the roadside at 2 a.m. is not the time to wonder what should be done.

The Scene

After notifying the proper authorities, the fire department must secure the crash scene and begin an investigation. It is best not to allow any members involved in the crash to lead or conduct the investigation. Due to the possibility of criminal or internal charges, there may be a conflict of interest if an involved member is allowed access to a sensitive investigation. Having said this, the members onboard the apparatus at the time of the crash are valuable assets for determining the root cause of the crash.

Any members involved in the crash should walk the investigator through the crash scene, pointing out what they saw at the time of the crash and what they feel is important. Once a member provides this initial walk-through, it is best to have them transported to a safe location, such as a fire station or police station, so that they can give a formal statement.

Once the investigator has been given an overview of the crash, he must begin to identify and document evidence. In the case of a serious or fatal crash, much of this documentation will be conducted by police officers. However, in the event of a minor or nonreportable crash, it may fall upon a fire officer to conduct the investigation. Speak with the on-scene police commander to determine who will be responsible for the crash investigation. Depending on the relationship between the fire and police department, the investigation may be done concurrently. Concurrent investigations further impress the need to pre-plan how these incidents will be handled. If these issues aren't discussed ahead of time, the fire department investigator may face serious allegations if he begins to paint, move, or collect evidence before authorized to do so by the police.

Provided the fire officer has been authorized by the police investigator to do so, evidence should be identified, photographed, and marked. While the purpose of this chapter is not to train the fire department member to act as a crash reconstructionist, the proper identification and documentation of crash evidence may prove valuable in later litigation. A motor vehicle crash will typically leave behind different types of evidence. Some of this evidence is transitory and should be the first evidence marked and secured. Such transitory evidence may include the location of broken vehicle parts, or loose objects that may have been thrown from the vehicle during a crash. The location and distance that these items have traveled could prove useful in determining pre-impact speeds and the heading and direction of the vehicle both before and after the crash occurred. While it is human nature for emergency responders to wander a crash scene picking up pieces and returning them to the vehicle, it is important that all members understand not to do this. In the event that the police department is handling the full investigation, it is the duty of the fire officer and fire apparatus operator to advise all fire department and emergency medical members not to touch or disturb this type of evidence.

Evidence

The following section will identify several different types of evidence that an investigator will likely find at a crash scene. This evidence should be photographed, measured, and documented for later use in the investigation.

Gouge

A gouge is a deep scar or chop in the roadway. Gouges are often used to identify where the collision occurred and can be very useful in determining the position of a vehicle at the time of the collision. When two vehicles collide, they will reach a point of *maximum engagement*. When this occurs, it is common for the vehicles to be driven into the roadway and leave a deep scar. Gouges are very useful for determining a definitive point of impact.

Scratches

A scratch is not as deep as a gouge. Scratches are typically shallow and are caused by parts of the vehicle which come in contact with the road during a crash. Scratches can be helpful, especially in determining the postimpact path of a vehicle. Postimpact paths can be used to determine a number of different things, including speed at impact.

Skid Marks

A skid mark is created by a tire sliding across the road surface and can be made before or after impact. Skid marks should be documented and measured, as the length of a skid can be helpful in determining speed. While a skid mark will typically last several days after a crash, it is important to identify and measure the marks as soon as possible. Often, there are very faint "shadow skids" which precede a dark skid mark. These shadow skids are useful for speed calculations but are very transitory in nature. As soon as a roadway is reopened, traffic traveling on the roadway will tend to destroy this evidence. This is especially true if the skidding vehicle was equipped with antilock brakes.

Offset Skid Marks

It is also important to identify any offset skid marks. Most skids marks are straight. At times, they may drift to the right or to the left due to the crown of the roadway. If the investigator should notice a straight skid that suddenly changes direction to the right or to the left, this is an example of an offset skid mark. The change in direction corresponds to the point where the skidding vehicle struck another object, causing it to change direction. Offset skid marks are valuable in identifying the area of impact and should be documented accordingly.

Yaw Marks

A simple definition of a yaw mark is a tire mark that is left on the roadway as a vehicle begins to spin around its center of mass. Yaw marks are often seen after a vehicle is struck on the front or rear quarter panel and spins out. Yaw marks are also seen when a vehicle is traveling too fast as it maneuvers or travels through a curve, causing it to lose control and spin out.

Crash reconstructionists use a special method for measuring and recording yaw marks. These measurements are useful for calculating a pre-impact speed. If a yaw mark is found on the roadway, it would be beneficial for the fire department investigator to locate a trained crash reconstructionist who will be able to properly measure and document the mark.

Fluid and Debris

The location of debris, such as glass and vehicle parts, and the location of fluid spills can be useful evidence in crash investigations. Fluid trails can help identify the postimpact path of travel of a vehicle. Depending on the type of fluid or debris, this evidence can be transitory in nature and should be photographed and documented as soon as possible.

Furrows

When a vehicle leaves the roadway, it will often leave marks on the soft shoulder or road side. These marks are known as furrows. When a vehicle rolls over, it often leaves deep impressions or disturbances in the roadside. These marks are very useful in determining postimpact speeds and paths of travel.

Coefficient of Friction

In addition to marks and evidence left on the road, it is important to examine and document the condition of the roadway at the time of the crash. Is the road dry or wet? Are there any foreign substances, gravel, or ice that may have caused the vehicle to skid?

The stickiness of a roadway is referred to as the *coefficient of friction* or *drag factor*. Drag factor plays an important role in almost all of the calculations used to determine a vehicle's pre-impact speed. The simplest way to document the drag factor is to subjectively note the road conditions and then measure the grade of the roadway. A trained investigator can then approximate a drag factor based on existing charts and studies.

A more accurate method for determining drag factors is to use a scientific instrument to measure and record this information. Most police crash reconstruction teams are equipped with an accelerometer or drag sled. Whenever possible, ask the police department to use this equipment to properly measure the road conditions. Once the drag factor is measured and recorded, it can be turned over to a qualified person who is able to perform speed calculations.

In the event that the drag factor cannot be measured at the time of the crash, the investigator should document the ambient conditions that exist at the time of the crash. It may be possible to return to the crash scene at a later time and measure an accurate drag factor once the investigator locates the proper equipment.

Photographs

It is important to preserve the events of a crash scene with photographs and video. Investigators should thoroughly photograph the scene, documenting the location of evidence, final rest positions of the vehicles, approach paths, sight distances, etc.

Photographs should first be taken with no paint markings on the road. In the event of future litigation, unmarked photographs will provide an unadulterated view of the crash scene for jurors, attorneys, or judges. Once the scene has been properly photographed with no markings, the investigator should mark the location of evidence with spray paint. Marking evidence with paint will allow other investigators to return at a later date to complete a follow-up investigation. Once the scene is painted, the investigator should take another set of pictures. The painted marks will help show evidence in the photographs that was previously difficult to see.

When photographing the scene, the investigator should walk around the scene in a circle, taking photographs from a distance to provide context. Evidence such as gouges, scrapes, and marks on the road should then be photographed at close range with a ruler laid next to the mark. Using a ruler as a point of reference will assist with later analysis of the mark and its significance.

When photographing tire marks such as skids and yaw marks, the investigator should start photographing from far away to provide context and perspective. The investigator should then take close-up photographs of the marks to assist in their identification. With today's modern digital cameras, it is almost impossible to take too many pictures. Also don't forget the importance of an aerial photo to assist in the investigation. As many fire departments now have access to a drone, use of drone photography will help preserve and memorialize important evidence that can be later analyzed by a qualified crash investigator.

Statements

Following a crash, the apparatus operator, the apparatus officer, and all crew members who were riding the apparatus should be interviewed as soon as possible. Keep in mind that in a serious crash this process may be complicated if law enforcement is involved. In cases where criminal charges could be filed against the driver of a fire apparatus, certain procedures must be followed to ensure that the apparatus operator's right to due process is not violated. These procedures could include Miranda warnings (the right to remain silent) and representation by an attorney. In cases where law enforcement is not involved but crew members may face internal discipline, union representation may be warranted. As with any crash investigation, these issues must be worked out and agreed upon before the crash happens. Fire departments should contact their local police departments, prosecutor's office, and union delegates to discuss these issues before an incident.

Once the policy for gathering statements has been agreed upon by law enforcement, the fire department, and the union, crash investigators may begin this investigative process. It is best to record interviews so there can be no future controversies involving "he said/she said" situations. Bear in mind that in some states, the investigator must secure permission from the person being interviewed to allow voice recording.

Once permission to record the statement has been granted, the interview may proceed. It is best to find a quiet, controlled area to conduct these interviews. Regardless of where the crew members are brought for an interview, it is best to remove them from the immediate scene to avoid distractions. Keep in mind that the crew members should remain separated so as not to taint each other's statements with false memories. By gathering each member's version of events, small details can be identified and corroborated which may become important in the crash reconstruction.

The types of questions asked during an interview are limitless and far beyond the scope of this chapter. That said, keep in mind important details such as the following when it comes time to conduct an interview: What direction was the vehicle traveling? What lane was the vehicle in? How fast do you think you were going? Where was the other car (if applicable)? Did you see the other car? The more thorough the interview, the more accurate the crash investigation.

It is also important to interview the drivers of any other vehicles involved. Keep in mind that it is common for a civilian vehicle to have fault in a fire apparatus crash. It is important to consult with local law enforcement to ensure that the statement provided by the other driver is legally obtained and admissible in court.

Keep in mind that statements should not be limited to persons directly involved in the crash. One of the first duties of a crash investigator is to identify potential witnesses. Identifying third party witnesses is an essential part of a thorough crash reconstruction. Once a witness is identified, it is important to determine where they were when the crash occurred. How well could they see the crash scene? Did they see the actual crash or events leading up to the crash? Did they arrive shortly after the crash? If so, did any of the people involved in the crash make statements or comments about what happened? These are just some of the important answers a witness can provide.

Video and Electronic Data

In addition to human witnesses, there are often video and electronic witnesses to a crash. Whether it's an ATM surveillance camera or a factory installed black box, most crashes are recorded on some form of electronic media. Identifying and securing this evidence is a crucial part of any investigation.

Investigators should conduct a thorough canvas of the area to identify potential camera locations. These camera locations could be anywhere: gas stations, automatic banking machines, a bank drive-thru, or even a private citizen's house. By going door to door and speaking to property owners, potential video evidence can be located and secured.

Video evidence is crucial to proving the truthfulness of a witness statement. Video evidence is especially important in crashes involving issues that leave little physical evidence, such the color of a traffic signal or the location of a pedestrian when he was struck by a vehicle. A picture is worth a thousand words.

In addition to video evidence, most modern day fire apparatus are equipped with an electronic data recorder or black box. Data recorders can be found inside the vehicle motor or as an aftermarket device installed by the vehicle manufacturer. Regardless of what type of data recorder the vehicle is equipped with, it is imperative that this data be properly preserved and analyzed.

Depending on the type of recorder, the fire department may have direct access to the data. If the vehicle is older and only equipped with an electronic control module on the motor, a qualified expert with special equipment may be required to conduct a download. In a case where the fire department is able to download the data, make sure that law enforcement is advised of this fact and is present during the download. It does not take much for an investigator to accidentally damage or delete this evidence. It is important that there is no chance that anyone can be accused of misconduct. In a situation where the crash is serious and digital evidence is present, an impartial third party should be used to recover on-board data.

Siren Audibility and Warning Lights

As discussed in Chapter 6, many fire apparatus crashes are the direct result of a fire apparatus operator overdriving the vehicle's emergency warning devices. The issue of when a civilian vehicle could see or hear the emergency vehicle may become an integral part of the crash investigation. Investigators should identify where the civilian could see the warning lights and hear the siren.

To determine the effective range of the siren, investigators should purchase a sound level meter. The sound level meter will allow the investigator to map the sound levels of the siren and superimpose the map onto a crash diagram. If possible, sound level readings should be taken at the scene of the crash to account for reflectivity and localization issues caused by the surrounding environment. It is also important for the investigator to interview the fire apparatus operator and fire officer to determine what siren settings were in use at the time of the crash.

Having created a sound level map of the siren, investigators must determine the interior sound level and insertion loss of the civilian vehicle. Investigators should use an exemplar vehicle to ensure that the ambient noise and sound insulating properties of the vehicle are properly accounted for. It is also important that the crash investigator attempt to determine the volume setting of the radio, the speed of the vehicle, the HVAC fan setting, and whether the windows were up or down. These factors will have a substantial effect on the ambient noise inside the civilian vehicle at the time of the crash. The ambient noise inside the vehicle should be measured at different speeds to determine an accurate range.

To calculate insertion loss, sound level readings should be taken at various measurement points from the siren while standing outside of the civilian vehicle (remember to use proper hearing protection). The civilian vehicle should then be parked at the same points and sound level readings should be taken while the investigator sits inside the vehicle. The difference in the two values is the insertion loss.

Once the ambient noise and insertion loss are calculated, the investigator can begin to reconstruct where the civilian driver could have heard the fire apparatus. Studies have demonstrated that a siren must rise to approximately 10 dB above ambient noise to effectively warn a driver. Using this referenced data, the investigator can determine how much siren sound must arrive outside the driver's window to effectively warn the driver. The required sound level can then be compared to the sound level map to determine where the civilian driver could have heard the siren.

> Ambient Noise + 10 dB + Insertion Loss = Required Siren Sound to Warn Driver

Vehicle Records and Inspections

It is not unusual for a fire department to face a crash-related lawsuit due to a mechanical defect on the vehicle. For this reason, it is important to maintain detailed records of all maintenance and repairs that are performed on a vehicle during its service life. These records should include the original specification data, any history of modifications, previous damage sustained to the vehicle, and all routine and preventative maintenance that is performed on the vehicle. In addition, the department should ensure that all service and repairs are performed by qualified technicians in accordance with manufacturer's recommendations. All repairs and inspections should be documented, and the record must include the date and description of what was done. These records must be maintained by a responsible person, such as the chief engineer or safety officer.

In addition to routine maintenance, fire departments should ensure that their vehicles are serviced and repaired in accordance with NFPA standards. One of the most important crash-related NFPA standards deals with annual weight certifications. Overweight vehicles are extremely dangerous and can lead to a number of serious issues which may result in a crash. It is important that every fire department weigh each vehicle on an annual basis in accordance with NFPA 1911. Once weighed, these certification records should be kept on file to prevent later accusations that a vehicle was overweight at the time of a crash.

If a vehicle is involved in a serious crash, it is beneficial to have the vehicle inspected before removing it from the crash scene. Important evidence such as the condition of the brakes and steering components are best examined before moving the vehicle. Also take note of whether or not the auxiliary braking device was turned on or off. If the auxiliary braking device was turned on, note what power setting it was set at. Most large police departments and state police agencies have trained truck inspectors. Having the vehicle properly inspected and examined by a trained truck inspector before moving the vehicle may prove invaluable. It is not unusual for a vehicle to be damaged as it is being towed or removed from the crash scene. By ensuring the vehicle damage can be properly tied to recovery efforts instead of the crash, fire departments may prevent a frivolous lawsuit by someone making false allegations that there was something wrong with the vehicle before the crash.

Special precautions should be taken by an investigator before allowing the tow company to move the vehicle. This is especially true when the air brake system is damaged. Remember that in the event of an air leak, the spring brake is designed to engage as a safety measure. In the event that the air brake system sustained damage, it is likely that the spring brake has engaged and it will be difficult to tow the vehicle. Many tow truck operators are inclined to readjust the slack adjuster, or back the brakes off, in order to release the spring brakes. If the tow truck operator backs the brakes off, an investigator will not be able to determine the pushrod stroke that was present at the time of the crash. Instead, investigators should ensure that the tow truck operator uses a caging bolt to *cage the brakes*. Caging the brakes involves the use of a caging bolt to disengage the spring brakes, allowing the truck to move. Caging the brakes will help to preserve valuable evidence, such as the pushrod stroke for each chamber. Do not allow the towing company to adjust the slack adjusters or recharge the system with air until the necessary measurements have been made by a trained investigator.

Traffic Pre-emption Devices

If the emergency vehicle is equipped with a traffic pre-emption device, investigators must confirm that the device was in use and working properly. Most pre-emption systems can be downloaded by local signal service personnel. The traffic control box will maintain a log of when the pre-emption system was activated. This log will help the investigator to determine the status of a traffic signal during an intersection crash.

Investigators must also examine if there were any issues that would have caused the pre-emption system to fail. A burned-out strobe light, malfunctioning detector head, or sight obstruction which blocked the signal to the pre-emption device may all factor into the crash investigation. Confirm whether the system was working, and if not, determine why.

NFPA Crash Reporting Requirements

The issue of crash investigation and reporting is addressed in NFPA 1451—Chapter 9. This chapter states that all crashes, injuries, fatalities, near misses, and violations of rules or laws must be investigated to determine the root cause. Once the root cause of a crash or violation is identified, the fire department must take whatever corrective action is necessary to prevent the incident from happening again.

The fire department safety officer is ultimately responsible for overseeing this program. The safety officer must manage the collection and analysis of any crash investigation that involves an on-duty crash involving a fire department vehicle or any personal vehicle used to transport a member. The safety officer must issue a report to the fire chief on a regular basis that summarizes the status, disposition, and corrective actions taken as a result of each crash. These reports and records must be maintained by the safety officer in accordance with NFPA 1521 "Standard for Fire Department Safety Officer".

The fire department must establish a data collection system that keeps permanent records of all on-duty crashes involving fire service vehicles and individual employee records related to on-duty crashes. Employee records must include the following:

1. On-duty crash history
2. Preventable versus nonpreventable crashes
3. Remedial training recommended or received as a result of a crash
4. Safety or crash review committee recommendations

5. All investigative or review committee reports related to the employee's crash

6. Transcripts of the employee's state driver's license record

Conclusion

A fire apparatus crash can be a trying time for a fire department, regardless of the fire department's size. It is important that all aspects of a proper crash investigation be discussed and agreed upon before a crash occurring. Once a crash occurs, the fire department investigator will be invaluable in providing assistance and information that will help identify the root cause of a crash.

NIOSH Firefighter Fatality Reports

The National Institute of Occupational Safety and Health (NIOSH) oversees the "Firefighter Fatality Investigation and Prevention Program." This program sends teams of investigators to the scene of firefighter fatalities to investigate and determine the root causes of the incidents. The investigative team will then make recommendations to the fire service so that the tragedy will not be repeated.

Over the past several years, NIOSH has developed an extensive database that contains all of the investigative reports authored by the program. The database contains scores of incidents that involve firefighter fatalities that were the result of a fire apparatus crash and provides an excellent source of crash-related scenarios. As NFPA 1451 Section 9.1.4 states that a fire department driver training program must include a review and critique of local and national crash scenarios, these NIOSH reports provide a useful tool for a driver training program.

This chapter will discuss each of the incidents and give my thoughts regarding how the crash occurred and what can be done to prevent a recurrence. The summary at the beginning of each case study is taken directly from the NIOSH firefighter fatality reports. The thoughts portion of each study are my own thoughts on how the material in this book relates to the crash.

F2013-26: Alabama

On October 17, 2013, a 28-year-old male volunteer probationary member lost his life after the pumper/tanker he was operating left the road and overturned. After calling 9-1-1 to report a structure fire in a house behind his residence, the victim donned his structural firefighting gear and waited for his department to arrive at the fire. After making an initial fire attack with other fire fighters, the victim left the hoseline and went to change the bottle of his SCBA. When he arrived at the engine, the pump operator stated she didn't know where the spare bottles were located on the apparatus. While still wearing his structural firefighting gear, the victim left the incident scene in his personal vehicle and responded to his department's nearby substation. He left the substation driving Engine 2, a 2,500-gallon pumper/tanker. While returning to the incident scene, Engine 2 left the road while traversing a curve and overturned. The probationary member received fatal injuries during the rollover and was pronounced dead on the scene. He was not wearing a seat belt.

Thoughts

In this case, the victim firefighter had only been a member of the department for 90 days. Before the crash, the victim had only operated a fire apparatus five times and had never driven under emergency conditions. This lack of experience contributed to the crash.

The fire department did not have a formal driver training program, nor did it have formal written policies. Instead, members would drive the apparatus with the fire chief until he felt comfortable with their abilities. While the gut-feel of an instructor is certainly a worthy endorsement for a new driver, a fire department must still have written policies and procedures for driver training.

The crash occurred as the apparatus was exiting a left-hand curve. As the apparatus was exiting the curve, the right-front tire left the roadway and dropped onto the soft shoulder. The apparatus struck a culvert on the roadside and overturned. While the police investigator did not believe excess speed was a problem, the investigator did believe that weight and water shift was a contributing factor to the crash.

Fire apparatus operators must understand how g-force, inertia, weight shift, and kinetic energy affect vehicle dynamics. This is especially true when a large fire apparatus navigates a curve. As the vehicle enters the curve, lateral g-force will cause the weight of the vehicle to shift to the front, outside corner of the vehicle. G-force and inertia will also cause the water in the tank to slosh. If the vehicle's water tank is not completely filled, or if the water tank is not equipped with baffles, the water may slosh inside the tank, shifting the vehicle's center of gravity. The shifting suspension and sloshing water may shift the center of gravity and artificially reduce the rollover threshold of the vehicle. If the vehicle is traveling too fast through the curve, the lateral g-force acting on the vehicle and the reduced rollover threshold will cause the vehicle to rollover.

In addition to a possible rollover, entering a curve too fast could lead to an understeer situation. In an understeer situation, lateral g-force will overwhelm the friction of the front tires, causing a loss of steering control. Instead of traveling through the curve, the vehicle will push straight ahead. If the vehicle pushes far enough, the front-outside tire will leave the roadway and ride onto the shoulder. If the shoulder is soft, the vehicle may begin to sink into the ground, resulting in a tripped rollover.

Fire apparatus operators must know where the curves in their response districts are located and slow down well in advance of a curve. If the apparatus operator enters the curve too fast, there will be no time or opportunity to slow down to a safe speed. It will be too late to slow down, and the fire apparatus will leave the road or rollover.

During this investigation, NIOSH mentions that the fire apparatus operator was wearing fire boots while driving the vehicle. While it is unknown if this was a factor in the crash, it is an interesting topic to discuss. When driving a vehicle, it is important to make sure that the driver is wearing comfortable footwear that allows him to safely move his feet between the pedals and feel the pedal through the sole of his shoe. This will allow for better control of the vehicle. If the driver's shoes or boots are large and cumbersome, his footwear may get caught between the pedals and cause a crash. Fire departments should consider lifting any requirement that a fire apparatus operator wear turnout gear while driving.

F2012-31: Illinois

On December 2, 2012, a 45-year-old male volunteer fire fighter died when he was struck by a backing fire apparatus at the scene of a rural brush fire that had extended into a vacant structure. The victim was one of two fire fighters on the first arriving fire apparatus. The driver was attempting to back the brush truck up a steep incline in the gravel roadway to get closer to the burning structure. The victim got out of the apparatus and positioned himself behind the apparatus on the driver's side in order to direct the driver. The victim tripped or fell and was struck by the backing brush truck. The incident occurred at night in a poorly lit rural area.

Thoughts

It is believed that the victim firefighter may have fallen down while acting as the backup person for the fire apparatus operator. After the victim fell to the ground, the apparatus operator continued to back up and ran over the victim. Fire apparatus operators must stop the vehicle anytime they lose sight of the backup person. If the backup person has fallen to the ground, or is standing directly behind the vehicle, they may be run over. If the driver loses sight of the backup person, they must stop the vehicle immediately until they regain sight of the backup person in their mirrors.

Fire apparatus have large blind spots directly behind the vehicle. These blind spots exist because the apparatus does not have a rear window like a passenger vehicle. Instead, the fire apparatus operator is limited to the small area that can only be seen in the side mirrors. It is important for anyone acting as a backup person to remember this fact and stay within sight of the driver.

The backup person should not stand directly behind the vehicle, as they will be out of sight and in the direct path of the vehicle should they trip and fall. Anyone acting as a backup person at night should be equipped with a flashlight, portable radio, and reflective vest.

Fire apparatus should be equipped with rearview cameras to help alleviate blind spots. These cameras are inexpensive and can be easily installed on older apparatus as a retrofit. While backup cameras are useful tools, fire apparatus operators must still be sure to use a backup person and not rely too much on the cameras.

It is believed that the victim firefighter may have become distracted by bystanders who were yelling at the fire apparatus operator to cut the wheel. When the firefighter became distracted, he may have tripped and fallen due to the uneven road surface. It is also believed that the driver may not have been able to fully concentrate on his rearview mirrors due to the complex task of reversing a manual transmission vehicle uphill. This situation lends itself to compromised situational awareness and distracted driving. Fire apparatus operators must remember that humans cannot multitask. They can only focus on one thing at a time. When more than one task is at hand, it is important to slow down and keep a close eye on the surroundings.

F2012-30: Indiana

On November 11, 2012, a 26-year-old male volunteer firefighter was killed when the tanker he was driving crashed en route to a grass fire. The victim drove his personal vehicle to the firehouse upon hearing his department dispatched to the scene of a grass fire. Upon arriving at the firehouse, he readied a water tanker to respond. Before responding, the victim asked other members if they wanted to go with him but they declined, saying enough resources were already en route. He boarded the tanker and left the station with lights and siren activated. The apparatus crashed approximately five miles from the firehouse and the victim was ejected and killed.

Thoughts

The water tender in this incident rolled over on a straight road. It is believed that the driver was attempting to turn to the right or change lanes to avoid traffic channeling devices that had recently been installed in the area of the crash. Regardless of the driver's intentions, his steering of the vehicle induced a rollover crash.

Rollover prevention training often highlights the danger of navigating a curve in the road. It is important to remember that turning from one road to another, switching lanes, or making an evasive maneuver is no different from rounding a curve. If the turn, lane change, or evasive maneuver is made too sharply, the vehicle may roll over depending on its speed. During these types of maneuvers the driver is creating his own curve in the road, even if the vehicle is driving on a straight road.

Fire apparatus operators must remember to slow down to a safe speed before turning the steering wheel. Many crashes are caused when the driver turns from one road to another at a speed that is too fast for conditions. Crashes can also result if the driver gives the steering wheel a hard turn to change lanes or avoid a hazard. These issues are especially important when the vehicle is traveling at a high speed. Drivers must keep the vehicle at a safe speed in case an evasive maneuver is needed.

F2012-23: Virginia

On July 16, 2012, a 30-year-old male volunteer firefighter died after being ejected from a fire engine. The victim, riding in the right front seat, was responding with one other firefighter to a reported motor vehicle crash. The fire engine traveled approximately 1.3 miles from the station when the driver lost control of the engine while rounding a curve. The engine left the paved road and crashed into trees on the right side of the roadway. The victim was ejected from the engine and landed in a wooded area. The victim was pronounced dead at the scene.

Thoughts

This is yet another curve related crash. In this case, the Virginia State Police estimated that the engine was traveling 55 mph in a posted 45 mph zone. During the investigation, the driver stated that it felt like the rear end was "picked up and moved over to the right." While it is difficult to speculate without seeing the full police report, it is likely that the vehicle entered the curve too fast and began to roll towards the passenger side. As the vehicle began to roll, the tires began to lose traction causing the rear end to yaw in a counterclockwise direction. During the yaw, the rear tires slipped off the road and onto the soft shoulder. Once the tires left the road, the driver lost complete control of the vehicle.

Fire apparatus operators must understand how g-forces affect a vehicle. If a fire apparatus rounds a curve too fast, lateral g-force will cause the vehicle's weight to shift. Weight shift will lead to a loss of steering control and the possibility of a rollover.

Drivers must understand that their vehicle may rollover even if it is traveling less than the posted speed limit. When driving through a curve, drivers must slow down well in advance of the curve and take the curve at a safe speed. Midway through the curve is not the time to realize the vehicle is traveling too fast. By then it will be too late.

F2012-06: West Virginia

On February 13, 2012, a 21-year-old male volunteer firefighter died after falling from the tailboard of a fire department tanker. The victim, acting as a spotter, had successfully guided the driver in backing the tanker to a dump tank. The driver of the tanker stayed in the driver's seat and watched the water gauge indicator lights on the pump panel through his side mirror. The victim, located on the tailboard, operated the dump valve to fill the folding tank. When the driver saw the tank-empty light flash, he left the fire scene to go to the water source to refill the tanker. The driver did not realize that the victim firefighter was still standing on the tailboard and could not get off.

Unknown to the driver, another tanker had accidently dumped 1500 gallons of water on the road while responding to the same fire and the water had turned to black ice. As the driver drove back to the water source to refill his tanker, he drove over the area of black ice and lost control. When the vehicle lost control, the victim was thrown from the tailboard and sustained fatal injuries.

Thoughts

This was a tragic case of Murphy's Law (what can go wrong will go wrong) and lack of communication. The driver of the water tender failed to communicate with his partner and confirm that his partner was safely clear of the apparatus. The driver also failed to conduct a walk around of the apparatus before moving the rig. This combination resulted in the victim firefighter being trapped on the tailboard as the apparatus drove away from the dump site.

This oversight could have ended with no injuries if the victim firefighter had been able to hold onto the apparatus until the vehicle made it to the fill site.

Unfortunately, the apparatus traveled over an area of black ice which was caused by water that had leaked from the faulty dump valve of a different water tanker. Unfortunately, the tanker that dumped the water was unable to communicate the hazard to other fire apparatus because they were operating on a different radio frequency. The second water tanker had also failed to mark the roadway to warn other apparatus of the unsafe condition. As a result, the victim's tanker lost control on the black ice, throwing the victim off the tailboard.

This case highlights the need for communication between the fire apparatus operator and the crew members, communication between different fire apparatus involved in fireground operations, and communication of hazards to others at the scene. Communication could be face-to-face, via radio, or by leaving an identifying mark such as a traffic cone or road flare to communicate a hazard.

F2011-21: Louisiana

On September 3, 2011, a 22-year-old male volunteer firefighter was fatally injured when the fire truck he was driving to a medical call crashed. The victim was traveling alone on a narrow, unmarked asphalt road. While attempting to negotiate a right-hand bend at the bottom of a slight grade, the victim lost control of the truck and ran off the left side of the roadway. The truck landed against a tree. Once extricated and transported to a local hospital, the victim succumbed to his injuries.

Thoughts

Investigators estimated the speed of the fire apparatus as it rounded the curve to be 50 mph. Investigators believe this speed was too fast to safely negotiate the curve on a wet roadway. This incident is another example of a fire apparatus crash that occurred as the vehicle rounded a curve. How a vehicle reacts in a curve will depend on the friction of the roadway and the height of the vehicle's center of gravity. Vehicles with high centers of gravity will most likely rollover, while vehicles with lower centers of gravity will most likely break traction with the roadway and lose steering control.

In this incident, the victim was operating a brush truck. Due to the relatively low center of gravity, the vehicle experienced a loss of traction rather than a rollover. Due to the wet roadway, the drag factor was reduced, making it easier for the vehicle to break traction and lose steering control.

It is easier to overwhelm the traction of the tires on a wet road. Fire apparatus operators must slow down more as they round a curve in wet weather. Drivers can't take a curve the same on a wet day as they normally would on a dry day. Slow down and account for weather conditions.

F2011-19: Massachusetts

On July 29, 2011, a 53-year-old career lieutenant died from injuries he received after a vehicle he was performing maintenance on crushed him. The victim had lifted the vehicle with a portable floor jack. While on a creeper, the victim positioned himself under the vehicle and was preparing to remove the oil drain plug when the vehicle came down on him. Firefighters came to check on the victim and discovered that the jack was no longer supporting the vehicle and the victim was motionless under the vehicle. The firefighters quickly repositioned the jack and attempted to raise the vehicle but the jack failed to support the vehicle's weight. The firefighters had to use hydraulic spreaders to raise the vehicle off of the victim. The victim later died at the hospital from injuries sustained during the incident.

Thoughts

Due to financial constraints, many fire departments are forced to conduct their own vehicle maintenance. While these financial issues are understandable, it is important for fire department administrators to understand the need to use qualified personnel and emergency vehicle technicians to perform routine maintenance and repairs.

For those departments that must do their own vehicle maintenance, ensure that the department has the proper equipment to do the work. Using old, worn-out, or improper equipment can result in a work-related injury or fatality. In this case, the floor jack may have been worn out to the point where it leaked hydraulic fluid. Furthermore, jack stands were not used as a safety measure. If a member is going to conduct vehicle maintenance or repairs that involve jacking a vehicle off the ground, proper safety measures must be taken. The member should also have a partner who can respond and assist should something happen such as a jack failure.

F2010-37: Indiana

On November 5, 2010, a 53-year-old male volunteer captain was fatally injured when he was struck and pinned against a wall by a fire department brush truck. The brush truck was having maintenance performed on the steering column by a firefighter/mechanic. As the victim and a firefighter passed in front of the brush truck, the firefighter/mechanic turned the ignition key to unlock the steering wheel, and the brush truck suddenly started and lunged forward, pinning the victim against construction materials stored along the wall. The victim succumbed to his injuries on scene.

Thoughts

In this case, the fire apparatus was equipped with a manual transmission. When the firefighter was conducting maintenance on the apparatus, he placed the manual transmission in low gear to keep it from moving as the truck sat parked and the ignition was turned off. Unfortunately, in order to help complete the repairs, the firefighter reached into the apparatus to turn the key and unlock the steering wheel. When this happened, the truck attempted to start and lunged forward, striking the victim.

Fire departments should have a policy that prohibits the apparatus from being started if the driver is not sitting in the driver's seat with his foot on the brake. It is not uncommon for a firefighter to reach in and start the truck so it can warm up while the fire crew dons their turnout gear. As evidenced in this tragedy, the truck may lunge forward with no one sitting in the driver's seat to apply the brakes.

This case also highlights the need to block the wheels anytime a fire apparatus is being worked on. While most fire departments don't require the use of chocks when the fire apparatus is sitting in station, it is a good idea to deploy the chocks when maintenance or repairs are being conducted on the apparatus. This will provide an additional level of safety should the apparatus be accidentally started with the truck in gear.

F2010-19: Virginia

On July 26, 2010, a 59-year-old male volunteer fire chief and a 67-year-old male volunteer firefighter died from injuries sustained after they were ejected during a rollover crash. The engine, with its lights and siren activated, was responding to a mutual aid residential structure fire. The crash occurred when the engine entered an intersection against a red light and was struck by a sport utility vehicle. The engine rolled over and both victims were ejected. Victim 1 was transported to a local hospital and pronounced dead. Victim 2 was pronounced dead at the scene. Both victims were reported to not be wearing their seat belts.

Thoughts

Unfortunately, many fire apparatus operators overestimate the effective range of their sirens. When approaching a 90° intersection, emergency vehicle sirens may only provide warning 40–80 ft in either direction. This limited effective range does not give civilian drivers enough time to hear and react to the approaching emergency vehicle. Because of this, emergency vehicle operators must come to a complete stop at negative right-of-way intersections and give plenty of time for civilian traffic to yield the right-of-way. Once civilian traffic has come to a stop, the emergency vehicle can then cross the intersection slowly and carefully.

F2010-15: Washington

On June 23, 2010, a 46-year-old male volunteer fire chief was killed when the rubber-tracked vehicle he was operating overturned at a brush fire. The incident occurred on a steep hillside covered with sage brush and short dry grass. The victim was accompanied on the rubber-tracked vehicle by another firefighter who was using the vehicle's 1-inch hand line to knock down hot spots along the flame front. Traveling uphill, the crew encountered rocky terrain and lost traction on a steep slope. The fire chief maneuvered the vehicle to the east to get through the rock outcropping. The vehicle turned uphill, lost traction again, quickly turned back to the west, and then overturned sideways and rolled downhill. The firefighter was ejected following the first roll and was not seriously injured. The vehicle rolled at least 3 times before coming to rest on the driver's side, facing west, pinning the fire chief beneath the vehicle's canopy. The fire chief died on the scene.

Thoughts

Specialized vehicles, such as the rubber tracked vehicle involved in this crash, can pose unique hazards to the firefighters who operate them. Firefighters often aren't accustomed to driving these vehicles because the vehicles are used infrequently. Fire departments that operate specialized vehicles must ensure that there are adequate training procedures in place for anyone designated to drive the vehicle. This training should include initial certification, as well as ongoing refresher training throughout the year.

Training programs for specialized vehicles should include classroom sessions, as well as hands-on training sessions in the terrain where they will be used. Preplanning the areas where the vehicle will be used will allow fire apparatus operators the opportunity to experience the capabilities and limitations of the apparatus under controlled conditions. The lessons learned during these training sessions can be later used in a real emergency.

F2010-07: Kansas

On February 10, 2010, a 69-year-old male fire chief died after being crushed between two fire apparatus. As a firefighter was backing a pumper into the station, he reported that he observed the fire chief outside of the station. After realizing that he had backed the pumper into the station too close to a parked tanker, the firefighter repositioned the pumper and parked it. The firefighter exited the pumper and yelled for the chief. The chief did not respond. When the firefighter searched for the chief, he found him unconscious and unresponsive on the station floor next to the pump panel of the tanker. Emergency medical services responded, and the chief was pronounced dead at the scene.

Thoughts

This tragedy involved a fire apparatus that was reversing into the fire station. Fire apparatus operators must have a thorough understanding of the dangers associated with driving a vehicle in reverse. Due to their design, fire apparatus have large blind spots directly behind the vehicle which make it impossible to see without using the side mirrors.

In this case, there were several factors that may have contributed to the fatality. The fact that the fire apparatus was older and had poorly designed side mirrors may have contributed to the blind spots around the apparatus. Fire departments that operate older apparatus should consider retrofitting the apparatus with better mirrors.

Another issue may have been the fact that the fire apparatus operator was backing the vehicle into a dark station from a sunlit front apron. Fire apparatus operators must remember how glare and lighting conditions can affect their ability to see. In situations where the driver is moving the vehicle from a well-lit area to a dark area, it may be helpful to take measures to increase the lighting in the darkened area. If the darkened area is a fire station, turn on the interior lights of the station. If the darkened area is outside, consider using auxiliary lighting on the apparatus, such as spot and flood lights, to assist with illuminating the darkened area.

Fire apparatus operators must also remember to stop the vehicle any time they lose sight of the person who is guiding the apparatus. If the driver loses sight of the backup person in his mirrors, they must stop the truck immediately until they regain sight of the person and know where they are.

F2009-30: Georgia

On December 26, 2009, a 37-year-old male career firefighter/paramedic was ejected from the front passenger seat of a fire department ambulance during a rollover. The ambulance was responding with lights and siren to a car fire with injuries. A motorist in a privately owned vehicle (POV) turned into the path of the ambulance as it was passing the POV. The ambulance struck the POV, lost control, and left the right side of the road. The ambulance overturned three times, and the victim was ejected. The driver sustained minor injuries. The victim's use of a seat belt could not be determined with certainty. The victim was airlifted by helicopter to a local trauma center where blunt force chest injuries and underlying medical conditions contributed to his death on December 28, 2009.

Thoughts

It was never determined if the victim in this crash was wearing his seat belt. Regardless, this incident is an important reminder of the need to wear a seat belt at all times. If a victim is ejected outside the protective envelope of the vehicle, the odds of sustaining a fatal injury increase substantially. Seat belts will help keep the occupants inside the vehicle where the occupant protection systems can do their job.

This incident is also an important reminder of the need to properly secure the loose equipment that is found in the passenger compartment of an emergency vehicle. If the vehicle is involved in a crash, a piece of loose equipment may become a dangerous projectile that can seriously injure or kill anyone in its path.

Lastly, fire apparatus operators must understand the need to drive defensively and never assume the actions of a civilian driver. Fire apparatus operators must learn to read the cues of other drivers that will allow them to anticipate their actions. Fire apparatus operators must also maintain a speed that will allow for an evasive maneuver should it be needed. Maintaining eye contact, watching the direction of another vehicle's steering tires, and keeping a close eye on brake lights and turn signals will help an emergency vehicle operator predict the actions of a civilian driver.

F2009-12: Maryland

On April 15, 2009, a 41-year-old male volunteer fire chief was fatally injured when the vehicle he was driving skid off the highway and struck a tree. The fire chief was driving a fire department pickup truck to a fire alarm when a car pulled out of an intersecting roadway to his left and into his path of travel. The victim tried to maneuver his vehicle around the car, but the vehicle turned sideways and skidded down the road. The vehicle went off the highway and struck a tree. An eyewitness to the incident called 9-1-1 and reported the crash. Emergency medical services arrived shortly thereafter and evaluated the victim's condition. He was pronounced dead at the scene.

Thoughts

Emergency vehicle drivers must be familiar with their response areas. Drivers must know the location of sight distance obstructions and road hazards. This will allow the emergency vehicle driver to anticipate an approaching danger and slow down.

Driver training programs should encourage members to drive the district on a regular basis to identify hazards, such as sight obstructions, roadway defects, or missing road signs. If a fire apparatus operator identifies a significant hazard, the fire department should have procedures in place to allow the issue to be quickly reported to the proper agency. Fire departments should also provide a method of communicating the hazard to other emergency vehicle drivers.

In this case, a sight distance obstruction may have contributed to the crash. Had this obstruction been identified and reported ahead of time, perhaps the civilian driver would have seen the approaching emergency vehicle and not pulled out in front of it.

Fire apparatus operators must also remember that evasive maneuvers are much more dangerous when a vehicle is traveling at a high speed. High speed results in more energy and a longer stopping distance should the vehicle need to stop. High speeds also increase the lateral g-force acting on a vehicle as it rounds a curve or attempts to make an evasive lane change or sudden steering maneuver.

F2009-10: New Jersey

On January 2, 2009, a 57-year-old male career firefighter was fatally injured when he was backed over while spotting an apparatus on the fire scene. The victim was the acting captain and responded in an engine with a crew of three to a reported working structure fire. While en route, the engine received a radio message to forward lay a supply line for an elevated master stream. Due to the location of the hydrant, the crew had to lay the supply line beneath a highway overpass. Upon arrival, the engine chauffeur had to drive around a police cruiser and tow truck in order to position the engine at an available hydrant. The engine then dropped off a firefighter at the hydrant to prepare for a forward lay when the incident commander advised the crew to do a reverse lay. The victim exited the engine to guide the chauffeur while he backed the engine around the police cruiser and tow truck. The victim walked down the officer's side of the engine and positioned himself at the rear on the officer's side. The firefighter positioned himself at the driver's side front bumper. The chauffeur was able to negotiate the engine around the police cruiser and tow truck before straightening up to lay a supply line into the scene. The victim walked backwards keeping eye contact with the chauffeur via the officer's side mirror. While backing, the chauffeur noticed the tow truck drive past him heading toward the scene. The chauffeur focused his attention on the tow truck momentarily when he felt the truck run over something. A police officer yelled for the chauffeur to stop the engine because something or someone was just run over. The victim was found underneath the engine just in front of the officer's side rear wheels. He was transported to a local metropolitan hospital where he was pronounced dead.

Thoughts

This incident highlights the issue of distracted driving. The fire apparatus operator was faced with a complex set of circumstances as he attempted to maneuver the apparatus in reverse. The fact that he was distracted by a nearby vehicle may have caused him to take his eyes off the road at the wrong time. The timing of this distraction was tragic, as it appears to have occurred at the same time the victim fell or was knocked to the ground.

Humans cannot multitask, they can only do one thing at a time. It is important to reduce distractions when driving a fire apparatus, especially when the vehicle is driving in reverse. Allowing the apparatus operator to focus on the driving task will lessen the chance of him not seeing a hazard.

It is also important to make sure that fire apparatus are inspected on a regular basis and all safety equipment is checked to make sure it is working. In this case, there is the possibility that the reverse safety system was inoperable. If the system had been working, perhaps it would have prevented this tragedy.

F2009-08: Georgia

On February 23, 2009, a 34-year-old male firefighter was killed and another firefighter was seriously injured in a single vehicle rollover crash. The fire apparatus was responding to a brush fire and lost its brakes as it approached an intersection. As the driver approached the intersection, he swerved to avoid traffic and overturned into a utility pole. The driver and passenger were not wearing seat belts and were ejected from the vehicle. The victim was pronounced dead at the scene, and the driver was transported to the hospital with serious injuries.

Thoughts

The fact that the fire service is forced to utilize outdated equipment is usually beyond our control. However, it is important to recognize the dangers posed by an older apparatus and keep a close eye on its mechanical condition. In this case, a 45-year old fire apparatus crashed when the brakes failed.

The vehicle in this crash rarely responded on runs and appeared to have pre-existing mechanical issues. When a vehicle reaches a point in its life that it has its own personality, it may be time to get rid of the vehicle. It is not the responsibility of a firefighter to risk his life riding in an unsafe vehicle if the local government doesn't care enough to replace it.

This incident is also a good example of training and experience. New drivers typically don't have the experience to recognize the symptoms of an unsafe mechanical condition. If a new driver is behind the wheel, it is important for senior members on the rig to pay close attention to any possible issues and point them out to the driver. It is also important for fire departments to implement graduated driver training programs and not allow drivers to operate the vehicle without proper supervision until they have been driving the apparatus for quite some time.

F2009-05: Massachusetts

On January 9, 2009, a 52-year-old male career lieutenant died and three male career firefighters were injured when the ladder truck they were riding in failed to stop while traveling down a hill. The ladder crew had just cleared a medical assist call before the incident. The chauffeur turned left out of a parking lot and immediately had to stop and back up to give the aerial ladder enough clearance to pass a utility pole. After clearing the pole, the chauffeur negotiated a right-hand downhill curve. As the apparatus descended the hill, the driver applied the brakes with no response. The chauffeur told the victim he had "no brakes" as the ladder truck gained speed while descending the hill. The chauffeur tried placing the ladder truck in neutral and applying the parking brake to slow the apparatus down, but there was still no response. The chauffeur was able to navigate the ladder truck through a busy intersection, crashing through two parked cars and a brick wall before coming to rest within a multistory residential complex. The victim was pronounced dead at the scene, and the chauffeur was extricated and transported to a local hospital where he was treated for serious injuries. The two firefighters riding in the crew compartment were also transported to the hospital with minor injuries.

Thoughts

This is another example of a fire apparatus crashing due to brake failure. This case highlights the need for proper inspection, maintenance, and repair procedures. It also highlights the need for trained technicians to properly make the repairs. While many fire departments are forced to conduct their own vehicle maintenance due to budget constraints, in-house repairs by uncertified personnel could result in danger to the vehicle and its crew.

The case also highlights the need to train drivers in the proper procedure for handling an air brake failure. Unlike a hydraulic brake system, pumping the air brakes will lead to a loss of air pressure, and a subsequent loss of braking efficiency. Putting the transmission in neutral will negate any braking by the drivetrain. Instead, the vehicle should be downshifted, or left in gear, to allow the drivetrain to help slow the vehicle. Fire department driver training programs should instruct drivers on how to handle a brake failure. Drivers must understand how the air brake system works, as well as how to stop the truck in the event of a brake failure.

F2009-04: Georgia

On December 31, 2008, a 24-year-old male career firefighter was killed and another firefighter was seriously injured after the victim lost control of the fire truck, struck a utility pole, and overturned in a ditch. The fire truck, with a crew of three firefighters, was responding to a reported chimney fire when the crash occurred. The fire truck was traveling through an intersection with a downhill grade. Georgia State Police investigators determined that the fire truck was traveling too fast for the road conditions. The victim was pinned between the driver's seat and the roof of the apparatus, upside down with his seat belt fastened. The officer, who was seriously injured in the crash, was not wearing a seat belt and was briefly trapped. The officer and third crew member (firefighter riding in the jump seat) initially tried to assist the victim before the arrival of other rescuers. The victim and the officer were transported to a local hospital where the victim was pronounced dead.

Thoughts

This crash occurred when the apparatus rolled over while turning from one road onto another. Remember that a fire apparatus turning from one road to another is no different from an apparatus that is rounding a curve. As the vehicle rounds the corner, lateral g-force will act on the vehicle. The amount of lateral g-force will depend on the speed of the vehicle and the radius of the turn. If the lateral g-force exceeds the coefficient of friction of the road, the vehicle will break traction and spinout. If the lateral g-force exceeds the rollover threshold of the vehicle, it will rollover.

Keep in mind that when a vehicle turns from one roadway to another, it usually turns with a very small radius. A curve with a small radius will result in a higher lateral g-force relative to the vehicle's speed. In this incident, the vehicle was estimated to only be traveling between 20–30 mph when it rolled over. The fact that the roadway was banked in the direction of the turn further lowered the rollover threshold of the vehicle, which meant less lateral g-force was required to roll the vehicle over.

This crash occurred on the first day the driver was assigned to the apparatus. While the driver had previous experience driving other fire apparatus, he was not familiar with the apparatus involved in the crash. Just because a driver is qualified to drive one piece of fire apparatus does not make him automatically qualified to drive all of the apparatus. Fire apparatus operators should complete a separate driver training program for each apparatus in the station.

F2008-27: Wisconsin

On July 23, 2008, a 76-year-old male volunteer fire chief was fatally injured when a pickup truck crashed through a bay door and pinned him against a parked tanker. The victim had just returned from a structure fire and had gone inside the fire station to write the fire report. Meanwhile, several firefighters noticed the assistant chief's pickup truck blocking the bay door. A firefighter wearing his bunker pants and wet rubber fire boots was pulling the pickup truck forward to clear the bay door when the truck accelerated suddenly. The pickup truck struck the bay door causing it to cave inward and pin the victim against a parked tanker. The victim was quickly removed from between the tanker and bay door and taken to a local hospital where he was pronounced dead.

Thoughts

Rubber fire boots are not conducive to driving a vehicle. The steel soles and bulky design make it difficult for a driver to feel the pedals beneath their feet. Furthermore, the size and shape of rubber fire boots may cause them to become trapped between the pedals or depress both pedals at the same time. Leather fire boots, or duty shoes, are more conducive to operating a vehicle. This is especially true if a member is driving a personal vehicle which has a much smaller foot well and pedals. In this incident, it is believed that the driver's wet, rubber-soled boot may have slipped off the pedal or depressed both pedals at the same time.

It should be noted that the crash investigators in this case used data from the striking vehicle's black box. As discussed in Chapter 26, these devices are useful investigative and training tools. Fire apparatus operators should understand the design, use, and limitations of factory installed and aftermarket event data recorders.

F2008-25: New York

On July 8, 2008, a 25-year-old male volunteer firefighter was fatally injured after being ejected during a fire truck rollover. The crash occurred as the fire truck was returning to the station after a propane gas fire. The fire truck was traveling down a paved road with a winding, steep grade. The road had a posted speed limit of 45 mph with a curve warning sign and recommended safe speed of 20 mph through an S-curve. The driver lost control of the fire truck, swerved off the left side of the road, returned to the pavement, and overturned on the right side of the road. The victim was not wearing a seat belt and was ejected out of the driver's side window. The pumper's 725-gallon water tank detached from the truck body and landed on top of the victim. The victim was pinned underneath the water tank and died.

Thoughts

Fire departments must understand the unique training issues posed by an older vehicle. The vehicle involved in this crash was 30 years old, and the victim had just commented to another firefighter how difficult it was to shift gears. The driver may have been so focused on dealing with the manual transmission that he allowed the vehicle to drift off the road and crash.

If a fire department operates an outdated vehicle, the department must have an intense driver training program which focuses on the unique issues related to driving the vehicle. This is especially true in fire departments that do not handle many calls. In slower departments, the issue of driver training is made worse by the fact that many apparatus operators do not receive much time behind the wheel. In these types of departments, it is important to give each driver plenty of opportunity to take the rig out on the road and practice.

Another issue that may have contributed to this crash was the fact that the water tank separated from the chassis. Regardless of whether this separation occurred before or during the crash, it is important to understand the hazards of modified or retrofitted vehicles. Any modifications or repairs must be performed by certified emergency vehicle technicians as per the manufacturer's specifications.

F2008-22: Ohio

On July 7, 2008, a 58-year-old male volunteer firefighter was fatally injured after the engine he was driving left the roadway and overturned several times. The victim was en route to a vehicle fire and was approximately 1.4 miles from the fire station when the incident occurred. The engine left the road after the driver misjudged the clearance around a large farm tractor stopped in the oncoming lane. The victim was not wearing a seat belt and was ejected from the vehicle. The victim was airlifted to a medical center and was later pronounced dead. It was later determined that the victim was driving with a blood alcohol level in excess of the legal limit.

Thoughts

Every fire department should provide alcohol-awareness training to all members. This training should teach fire apparatus operators, line officers, and firefighters the symptoms and negative effects of alcohol ingestion. If a member suspects that the driver is under the influence of drugs or alcohol, they must speak up. Turning a blind eye and ignoring the situation may lead to a crash.

Drivers must also understand how to handle the fire apparatus if it should drift off the roadway. As this NIOSH report discusses, it is not uncommon for a fire apparatus to leave the road because of the driver's actions. When this happens, the vehicle should be brought to a safe stop and then brought slowly back onto the roadway. Fighting to get the vehicle back on the roadway, especially at higher speeds, may cause the vehicle to go into the oncoming lane. If the driver attempts to correct and get back into their own lane, the vehicle may trip itself and rollover.

F2008-14: Colorado

On April 15, 2008, at approximately 1535 hours, a 30-year-old volunteer fire chief and a 38-year-old volunteer firefighter died when their vehicle drove onto a bridge that had collapsed from wildfire damage. The victims were driving through thick black smoke which obscured their forward vision.

Thoughts

Forward sight distance defines the distance a driver can see in front of the vehicle. At no time should the stopping distance of the vehicle exceed the forward sight distance. If the forward sight distance is reduced by fog, snow, rain, or smoke, the driver should reduce speed accordingly. Failure to reduce speed will result in the driver overdriving the environmental conditions. The driver will not have enough distance to perceive, react, and skid to a stop when a hazard emerges into view.

In this case, smoke obscured the forward sight distance of the driver. One can only assume that once the driver realized the bridge was burned out, he did not have enough time to stop the truck before driving onto the bridge. As a result, the bridge failed and the vehicle plunged below.

F2008-10: Louisiana

On March 28, 2008, a 33-year-old male volunteer firefighter was fatally injured after the tanker truck he was driving left the roadway and overturned. The victim was en route to a structure fire, took an unfamiliar route, and failed to negotiate a 90° curve to the right. The tanker left the roadway rolling onto the driver's side then slid through a ditch into a row of pine trees, crushing the cab. The victim was extricated by emergency personnel, air-lifted to a local medical center, and later pronounced dead.

Thoughts

The vehicle involved in this crash was a 3000-gallon tanker which overturned while rounding a curve. At this point in the book, readers should be very familiar with the danger of rounding a curve. This fact is especially true for the driver of a water tanker. As the vehicle rounds the curve, lateral g-force will act on the vehicle. The amount of lateral g-force is directly related to the vehicle's speed and the radius of the curve. If the lateral g-force exceeds the rollover threshold of the vehicle, it will rollover.

Drivers must know the location of the curves in their first-due and mutual aid districts. When driving in an unfamiliar area, fire apparatus operators should keep a close eye out for warning signs which will indicate an upcoming curve or other hazard in the roadway. Drivers must slow down well in advance of the curve and take the curve at a safe speed. Midway through a curve is not the time to realize the vehicle is going too fast; by then it will be too late.

The NIOSH report also addressed the issue of staffing fire apparatus with more than one member. The driver of the vehicle should concentrate on driving. A fire officer or firefighter should be operating sirens, air horns, and mobile radios. Anytime fire apparatus operator take their hands off the wheel or eyes off the road to complete a task, the chance of a crash increases dramatically.

F2008-05: Colorado

On February 23, 2008, a 33-year-old male volunteer captain was fatally injured when the engine he was driving to a medical emergency left the roadway and rolled one-half time onto its roof. The victim was ejected from the apparatus and was pronounced dead at the scene.

Thoughts

It is believed that the victim in this crash may have attempted to reach over the engine box to reach a mobile radio. In so doing, the driver would have had to unbuckle the seat belt and reach over with only one hand on the wheel. This type of manual and visual distraction could easily lead to a loss of control. The loss of control and subsequent crash are made worse when one considers that the driver had removed his seat belt to reach for the radio.

This is another example of the need for more than one firefighter to staff a responding apparatus. When the driver is forced to use the radio, sound the siren, and complete other tasks while driving, he is taking his mind off his primary job of safely operating the vehicle. Whenever possible, fire apparatus should be staffed with a minimum of two people.

In situations where a vehicle will be operated with only a driver, it is important to design the cab so that everything is within easy reach. If drivers will be forced to use a radio and sound the siren, these controls must be located in a location that prevents drivers from having to reach or take their eyes off the road.

F2007-30: North Carolina

On August 11, 2007, a 43-year-old male volunteer firefighter died while he was responding to an automatic alarm. Less than a half-mile from the station, the victim rounded a curve in an engine. The driver was forced off the right side of the road and onto the shoulder when he swerved to miss an oncoming vehicle that had crossed into his lane of travel. The victim steered the engine back onto the road and crossed the centerline. The victim overcorrected and again ran off the right side of the road. The engine collided with a fenced embankment, skidded sideways, struck a tree, and overturned.

Thoughts

If the apparatus tires should drop off the road, bring the vehicle to a safe stop and evaluate the problem. It is not worth the risk of fighting to get the vehicle back onto the road, only to have the vehicle cross into the oncoming lane. Once the vehicle goes into the oncoming lane, the apparatus operator will be forced to re-correct to get back into the proper lane. When this occurs, it is not uncommon for the driver to turn the steering wheel too hard, which will place excessive g-force on the vehicle. If the g-force exceeds the rollover threshold of the vehicle, the apparatus will trip itself and rollover.

F2007-25: North Carolina

On March 24, 2007, a 45-year-old male volunteer firefighter and a 19-year old male volunteer firefighter died while they were responding in a tanker to a structure fire. The tanker crossed the double-yellow line of a two-lane state roadway, and as the driver steered to the right, the tanker began skidding sideways. The tanker ran off the road, then skidded back onto the roadway and overturned several times before coming to rest upside down. One firefighter was completely ejected, and the other firefighter was partially ejected from the cab of the tanker. Both of the victims were found unresponsive and were pronounced dead at the scene by emergency medical services.

Thoughts

This incident is straightforward: the driver turned the wheel too hard, thus over-correcting, and caused the vehicle to rollover. Both occupants were not wearing seat belts and were either ejected, or partially ejected, from the vehicle. This incident highlights the need to wear seat belts.

Once a vehicle occupant is ejected from the vehicle, the odds of sustaining fatal injuries increase substantially. While many firefighters believe the size of the apparatus will protect them in a crash, the size of the apparatus will actually make the vehicle more prone to rollover. Once a vehicle enters a rollover, an unre-strained occupant will be flung around within the cab and possibly ejected. There is no excuse for not wearing a seat belt. As firefighters, we must lead by example and demonstrate to the public that we practice what we preach.

F2007-17: Connecticut

On May 19, 2007, a 37-year-old male career captain was fatally injured when the engine he was riding in collided with a ladder truck at an intersection. Both apparatus were from the same department and were responding to a reported kitchen fire. A total of eight firefighters were involved in the collision. According to the police report, the captain and the driver of the engine were not wearing seat belts. They were both ejected from the apparatus and suffered multiple injuries. The captain succumbed to his injuries three days after the collision and the driver remained hospitalized for several weeks. The six injured firefighters were transported to local hospitals with nonlife-threatening injuries.

Thoughts

When firefighters think of an intersection crash, it is usually a fire apparatus crashing into a civilian vehicle. As this incident highlights, crashes with other emergency vehicles must also be considered. If multiple units are responding to an emergency, fire apparatus operators must constantly assess where they may encounter other responding vehicles.

Use the mobile radio when approaching an intersection and advise other responding apparatus of potential conflicts. However, keep in mind that other agencies, such as police and medical units, may not be able to monitor the same radio channel. Knowing the usual routes that other apparatus will take to a location will help the fire apparatus operator anticipate where a conflict may occur.

It is imperative to come to a complete stop at red lights and stop signs. While state vehicle codes may allow an emergency vehicle to slow down as necessary, NFPA and national best practices require a complete stop at a negative right-of-way intersection. As discussed in Chapter 6, most drivers will not see or hear the approach of the apparatus until it is too late. This is especially true of other responding emergency vehicles, as they will not be able to hear other sirens over the sound of their own.

This incident also brings up the issue of brake inspections. As discussed in other sections, fire departments must ensure that the pushrod strokes on all of the airbrake chambers are inspected on a monthly basis. If a brake is out of adjustment, it will reduce the braking efficiency of the apparatus and lead to longer stopping distances.

F2006-25: Alabama

On July 26, 2006, a 17-year-old female volunteer junior firefighter died after the tanker truck she was riding in went off a narrow one-lane bridge. The tanker was en route to a structure fire and failed to negotiate a sharp curve at the approach to the bridge. The tanker crashed through the guardrail and landed upside down below the bridge. The driver and two other firefighters riding in the tanker were injured in the incident. The victim was extricated by emergency personnel and pronounced dead at the scene.

Thoughts

Four firefighters were riding in an apparatus that only had two seats. The victim firefighter was actually sitting on the floor between the driver and passenger seats. When the tanker crashed, she was trapped in the cab and sustained fatal injuries.

The number of firefighters riding in the apparatus cannot exceed the number of seats that are equipped with seat belts. If one of the riding positions is equipped with a malfunctioning seat belt, that riding position must be taken out of service. If the apparatus only has two seats that are equipped with seat belts, only two firefighters should be riding the apparatus.

This incident is another example of a water tanker that rolled over while rounding a curve. As stated in previous case studies, fire apparatus operators must slow down well in advance of a curve and take the curve at a safe speed. Midway through the curve is not the time to realize the vehicle is going too fast and hit the brakes. By then it will be too late.

F2006-06: Texas

On November 22, 2005, a 28-year-old male volunteer firefighter died while responding to a mutual aid call for a major grass fire. The victim was driving the department's tractor trailer tanker on a two-lane state highway. The tanker went off the road, overturned, and came to rest in a field. The victim was ejected and found lying unresponsive on the roadway. He was pronounced dead at the scene.

Thoughts

The vehicle involved in this crash was a surplus military fuel tanker that had been converted to carry water. The vehicle was not modified or upgraded in any way after it was acquired by the fire department. As the truck was originally designed to carry fuel, loading it with water would make the vehicle overweight because water weighs considerably more than fuel.

Fire departments should proceed with caution anytime a used vehicle is acquired from another industry and modified for the fire service. Vehicles that were not originally designed for the fire service may not meet the stringent design standards or have the necessary safety equipment to serve as a frontline fire apparatus. If a fire department must take possession of a former civilian vehicle, the department should consult with a professional fire apparatus manufacturer to make sure the vehicle will be able to handle the rigors of the fire service.

F2006-05: Alabama

On November 28, 2005, a 25-year-old male career firefighter died from injuries he sustained after the engine he was driving collided with a tractor trailer. The victim was driving the engine behind a rescue truck that was dispatched from the same station. The victim was approaching an intersection when the tractor trailer pulled out in front of him. The victim struck the tractor trailer, and the force of the impact pinned him in the driver's seat. Rescue crews extricated the victim and flew him to a local trauma center where he died as a result of his injuries.

Thoughts

Apparatus operators cannot assume that a civilian driver will see or hear an approaching emergency vehicle. Due to the limited effective range of an emergency vehicle siren, civilians are left with little time to perceive and react to the approaching emergency vehicle. This is why fire apparatus operators must come to a complete stop at red lights, stop signs, and negative right-of-way intersections.

Fire apparatus operators must be especially cautious when responding in a caravan with other emergency vehicles. Fire apparatus operators who are driving in the second or third caravan position must pay close attention to the behavior of civilian drivers. It is common for a civilian driver to see the first emergency vehicle pass and assume that there are no other emergency vehicles approaching. If the civilian assumes that no other emergency vehicles are approaching, he may suddenly pull in front of the other emergency vehicles driving in the caravan.

There are allegations that the brakes failed in this incident. Fire apparatus operators must conduct routine checks on the condition of the air brake system. Apparatus operators should complete the suggested NFPA air brake test on a routine basis to ensure that the brakes are working properly. If there are any issues, the vehicle should be placed out of service immediately.

F2005-35: Louisiana

On December 2, 2005, a male career captain died and two firefighters suffered severe injuries when their apparatus struck a passenger van at a four-way intersection. After striking the van, the apparatus left the road and overturned 1¼ times. The crew was responding Code 3 to a reported gas leak when the incident occurred. The permanent stop lights had been replaced by temporary stop signs as a result of an extended power outage caused by Hurricane Katrina.

Thoughts

During severe weather, fire apparatus operators may face unique hazards that are not routinely encountered. Fire department SOPs should address such hazards, including malfunctioning traffic signals and roads blocked by downed trees or wires. In this case, the fire apparatus proceeded through two temporary stop signs at highway speed without coming to a complete stop.

It is imperative for fire apparatus operators to come to a complete stop at any negative right-of-way intersection. This is especially true when an intersection is controlled by a temporary traffic control device. If a traffic light is not functioning, the intersection has been rendered even more hazardous than normal. Fire apparatus operators must proceed with extreme caution and drive defensively when approaching an intersection controlled by a temporary traffic control device or when the traffic signal is completely out of service.

F2005-28: California

On August 6, 2005, a 23-year-old male career firefighter died after he was ejected from the open cab of an engine during a crash. The crew was responding nonemergency in a reserve engine. The engine was traveling at approximately 45 mph in a heavy rainstorm when the driver lost control. The vehicle left the road, traveled down an embankment, and struck two trees before coming to rest on the roadway below. The engine's auxiliary braking system was engaged at the time of the incident. The victim was found lying unresponsive under the running board on the passenger's side of the vehicle. The victim was removed and transported to an area hospital where he was later pronounced dead.

Thoughts

Fire apparatus operators must understand the need to turn off auxiliary braking devices when driving on slick roads. Using an auxiliary braking device on a slick road could cause the rear tires to lock up and induce a loss of control. This is especially true when a fire apparatus is rounding a curve or turning a corner.

Driver training programs should include extensive drive time with the auxiliary braking system deactivated. Many fire apparatus operators complete their entire driver training time in good weather conditions. Shortly after being certified to drive the apparatus, they are called upon to drive in bad weather. The driver shuts off the auxiliary braking device as instructed and is suddenly shocked by the change in the fire apparatus's braking power. Drivers must have experience driving with the auxiliary braking device deactivated so they are able to anticipate the change in stopping distance.

It is also important for fire apparatus operators to complete a training program that is specific to each piece of apparatus. This is especially true in departments that operate multiple apparatus out of one station. Just because a firefighter is qualified to drive one truck does not automatically make them qualified to drive them all. Each apparatus should have its own training program that addresses the issues unique to each piece.

F2005-27: Utah

On June 21, 2005, a 52-year-old male volunteer chief died from injuries sustained during a tanker rollover. The chief was driving a military surplus truck that had been converted into a tanker. He was traveling to a nearby town to obtain a state inspection. While driving on the gravel road at approximately 40 mph, the left-front tire ruptured causing him to lose control, leave the road, and roll over several times. The chief was ejected and died on the scene from his injuries.

Thoughts

The vehicle involved in this crash was a surplus military vehicle that was originally designed to carry fuel. When the vehicle was modified to carry water as an off-road fire apparatus, it became significantly overweight. This excess weight would have contributed to the tire failure which ultimately caused this crash.

In addition to the vehicle being overweight, it was routinely parked outside in a desert environment. The desert environment would have further degraded the tire. While many fire apparatus tires never wear down the tread enough to need replacement, they are still subject to the negative effects of ozone and ultraviolet light which dry out and degrade the rubber. When a degrading tire sits for days or weeks on end without moving, tire failure is inevitable. NFPA recommends that tires be replaced every 7 years, provided they don't need to be replaced sooner due to damage or tread wear.

NFPA 1451 requires fire departments to provide training in how to handle a tire blowout. Unfortunately, many fire departments fail to do this. Fire department driver training programs should address the proper method for handling a tire blowout and find ways to simulate a catastrophic tire failure (see Chapter 25).

F2005-15: Texas

On April 23, 2005, a 27-year-old male career firefighter sustained a fatal head injury when he fell from an enclosed cab. The incident occurred shortly after leaving the station as the truck was en route to a reported structure fire. It is believed the victim was reaching to close a rear passenger door that had opened during a right turn when he fell out of the quint and landed on the pavement. The victim was treated at the scene and transported to a local hospital by ambulance. He died two days later from his injuries.

Thoughts

Nearly every case study discussed so far involves a lack of seat belt use. This case is no different. The victim firefighter reached over to secure a door that had opened when the apparatus rounded a corner and he fell out of the truck and sustained a lethal head wound.

Wearing a seat belt must be mandatory for anyone riding a fire department vehicle. If a member fails to wear their seat belt, they should face immediate discipline. This incident would probably have never happened if the firefighter was wearing his seat belt. The same can be said for countless other firefighter fatalities experienced in recent years.

F2005-12: Florida

On August 23, 2004, a 22-year-old male career firefighter/emergency medical technician died after the ambulance in which he was riding crashed into a tree. The ambulance, with a patient on board, had proceeded through an intersection with lights and siren activated when the driver lost control on a wet road. The victim, who was belted into the captain's chair in the patient compartment, died on impact and was pronounced dead at the scene. A paramedic who was also sitting in the rear of the ambulance was ejected and sustained serious injuries. He was transported to a local hospital via ambulance.

Thoughts

The tires on the ambulance involved in this crash had worn down to the tread-wear indicators. The lack of sufficient tire tread may have caused the tires to hydroplane and induce a loss of control. This loss of control caused the vehicle to leave the roadway, rollover, and strike a tree.

Fire departments must keep a close eye on the condition of the apparatus tires. Underinflation and excess tread wear will reduce the tire's ability to move water out of the way so it can grip the road. Drivers must remember to slow down in wet weather and give the tire tread time to push the water out of the way so the rubber tire face can contact the road surface.

If a tire is spinning too fast because the vehicle is moving too quickly, the tread will not have sufficient time to move the water out of the way. Instead, a wedge of water builds up in front of the tire. Eventually, the tire will climb atop the wedge of water and lose contact with the road. The tire will hydroplane and the vehicle will lose control.

F2005-1: California

On August 14, 2004, a 25-year-old female career firefighter died when she apparently fell from the tailboard and was backed over by an engine. The victim and her crew had been released from the scene of a residential fire. The road was blocked by other apparatus, so the victim's crew began backing to an intersection approximately 300 ft away in order to proceed forward. The victim took her position on the tailboard as the tailboard safety member and signaled the driver to begin backing. A captain acting as the traffic control officer guided the backing operation from the road on the driver's side, behind the apparatus. When the captain turned and walked into the intersection to stop cross-traffic, the victim apparently fell from the tailboard and was run over by the engine.

Thoughts

The victim in this incident was riding on the tailboard and acting as the tailboard safety member. The purpose of this position is to help guide the driver while the vehicle is being driven in reverse. During the backing maneuver, the member fell off the tailboard and was backed over.

No members should be riding on the outside of a fire apparatus at any time. This includes assisting with backing maneuvers, repositioning apparatus, or for special events such as parades. The days of firefighters holding on for dear life are long over.

When assisting with backing operations, no member should be standing or walking directly behind the apparatus. If the member trips or falls, they will end up directly in the path of the backing apparatus. Anytime a member is assisting with backing maneuvers, they should stand to the side of the apparatus, in view of the driver via the rearview mirrors. In the event the member trips or falls, they will be clear of the vehicle's path and within view of the driver.

F2004-43: Illinois

On April 27, 2004 a 34-year-old male part-time firefighter died after the engine in which he was riding crashed into an engine from another department as they passed through an intersection. Both engines were responding to the same structure fire. The force of the impact caused the front of one engine to collapse inward, crushing the unrestrained driver whose legs were pinned between the seat and the dashboard. He was extricated and transported to the hospital for treatment. The rear passenger of the engine received minor injuries and was transported to a local hospital where he was treated and released. The victim, who was riding unrestrained in the officer's seat, was ejected from the vehicle. He was transported to a local hospital where he was pronounced dead on arrival.

Thoughts

This is another example of two fire apparatus colliding at an intersection. In fact, one of the apparatus was equipped with a traffic preemption device that allowed it to take control of the traffic signal. Despite being equipped with such technology, the apparatus still collided with another emergency vehicle that failed to stop for a red light.

Fire departments must be wary of responding to incidents when other agencies are also responding. Fire apparatus operators must think ahead and anticipate where potential intersection conflicts could occur. Even if the apparatus is equipped with a traffic preemption device, the apparatus operator must still be wary of vehicles failing to stop for the red light on the cross street. Approach green lights cautiously and make sure traffic on the cross street is stopped.

F2004-19: Massachusetts

On April 30, 2004, a 58-year-old male career firefighter sustained a fatal head injury when he fell from a moving cab-forward engine. The engine was responding to a reported gas odor with a fire apparatus operator and an officer in the cab, and two firefighters, including the victim, seated in the open jump seats. Upon departure from the station, the engine made a right turn from the apron onto the street. During this turn, the victim fell out of the driver side jump seat door, landing on the street and striking his head. He was treated at the scene for head trauma and transported to a local hospital. He died from his injuries three days after the incident.

Thoughts

This is another example of a firefighter falling out of a moving fire apparatus and sustaining fatal injuries. Fire apparatus should not be moved until all members are properly positioned in their seats and wearing their seat belts. This is especially true in an older apparatus that does not have an enclosed cab. While most fire departments now have fully enclosed cabs, there are still a number of open-cab apparatus in use across the country.

Even if the apparatus has a fully enclosed cab, it is still necessary to wear a seat belt. Should the apparatus be involved in a crash, the seat belt will keep the occupant properly positioned in the seat, allowing the vehicle's occupant-safety systems to do their job. In the event of a door latch failure, a belted member will be less likely to fall out of the apparatus if a door should suddenly open, especially if the vehicle is rounding a curve or making a turn.

F2004-15: Florida

On March 3, 2004, a 40-year-old male forest ranger died after he lost control of the brush truck he was driving to a controlled burn. The truck experienced a catastrophic blowout of the right-front tire, left the road, struck a culvert, and overturned. It came to rest on its roof in a ditch filled with approximately 2 ft of water. The victim was pronounced dead at the scene.

Thoughts

This crash occurred after a tire blowout. Immediately following the blowout, the driver lost control and the vehicle crashed into a culvert. The tire was 6 years old at the time of the failure.

Driver training programs must address the proper way to handle a tire blowout. As discussed in Chapter 11, the proper method for handling a tire blowout is not to forcefully apply the brakes. Instead, the driver should give the vehicle enough acceleration to increase forward momentum and regain control of the vehicle. Once the driver has regained control, he can gently decrease speed and come to a stop.

Fire departments must also keep a close eye on the condition of their vehicle tires. NFPA provides recommendations on when to replace the tires, including an age limit of 7 years. Due to infrequent use, most fire apparatus tires don't wear the tread down below NFPA standards. However, tires are still subject to rubber degradation caused by environmental conditions. An old and degraded tire is more prone to a blowout.

F2004-03: Kansas

On November 17, 2003, a 53-year-old male career captain died when the department car he was driving was involved in a single-vehicle crash. The victim was responding to a reported commercial structure fire when his car left the roadway, hit a tree stump, overturned, and slid into a utility pole, pinning him in the vehicle. He was extricated from the vehicle and transported to a local medical center where he was pronounced dead upon arrival.

Thoughts

At the time of this crash, there were wet leaves on the road and a slight mist in the air. Fire apparatus operators must be keenly aware of how wet roads and slick conditions will affect the dynamics of a moving vehicle. Slick roads will result in increased total stopping distances and reduced critical curve speeds. Apparatus operators must remember to slow down when the road conditions deteriorate and account for the change in vehicle dynamics caused by slick roadways.

F2003-33: Nebraska

On August 6, 2003, a 43-year-old career firefighter/emergency medical technician died after the ambulance he was driving was struck from behind and pushed into a straight truck. The victim and a paramedic were conducting a nonemergency transport between two hospitals. The ambulance was traveling through a highway work zone, and as the ambulance driver slowed down to move around a line painting crew, a tractor trailer struck the rear of the ambulance and pushed it into the straight truck. Although the victim was wearing a seat belt, the cab sustained such extensive damage that he was fatally injured.

Thoughts

This crash occurred when an ambulance slowed down in a work zone and was struck from behind by a large tractor trailer. Emergency vehicle operators should be especially wary when they have to slow down for any reason on a high-speed roadway. Drivers should keep a close eye on their rearview mirrors and always leave enough room around the vehicle to move out of the way should a vehicle encroach from the rear. If drivers spot a vehicle closing from behind, they should pull out of the lane of travel and quickly drive the vehicle to a point of safety.

F2003-30: North Carolina

On July 28, 2003, at approximately 1730 hours, a 23-year-old male volunteer lieutenant was killed and a 19-year-old male volunteer firefighter was seriously injured during a single-vehicle rollover crash while responding in a POV to a confirmed trailer fire. The POV was traveling southbound at approximately 80 mph on a two-lane state road when it drifted off the right side of the roadway. The lieutenant lost control after he attempted to bring the POV back onto the roadway and overcorrected. The POV overturned several times, struck a wood utility pole, and ejected both firefighters. Emergency medical services responded within minutes of the incident, and the victim was flown via life-flight helicopter to the hospital, where he was pronounced dead.

Thoughts

The victim in this crash was driving his personal pickup truck at 80 mph on a straight road. The vehicle drifted to the right, and the driver overcorrected, causing the vehicle to lose control and roll over multiple times. The victim was ejected and sustained fatal injuries.

Any time a vehicle is driven at a high speed, the margin for error is significantly reduced. It is much easier to regain control of a vehicle at 35 mph than at 80 mph. This is because at a higher speed, a vehicle will travel a much greater distance in a much shorter time. Slower speeds give the driver more time and distance to correct the problem before the vehicle shoots off the side of the road.

This crash also involved a volunteer firefighter's personal vehicle. Fire departments must have policies and procedures in place to ensure that their firefighters drive their personal vehicles in a safe and prudent manner. Most states do not grant vehicle code exemptions to a firefighter responding in a personal vehicle. Instead, members must adhere to all of the vehicle code regulations, such as driving the speed limit and stopping at negative right-of-way intersections. Fire departments must ensure that volunteers are not violating the law as they drive to the firehouse or to the scene of an emergency.

Volunteer firefighters must also check with their insurance companies to make sure they have coverage should they be involved in a crash while responding to the firehouse or to an incident scene. Under these circumstances, some insurance companies will not cover a vehicle involved in a crash. If this is the case, fire departments must provide supplemental insurance policies for the members who respond to calls in their personal vehicles.

F2003-23: New Mexico

On June 26, 2003, a 46-year-old male volunteer assistant chief was fatally injured after being ejected from a water tanker as a result of a rollover crash. The victim was traveling to a wildland fire on an unpaved road within a national forest. The tanker failed to negotiate a curve, rolled over, left the road, and rolled several more times into a canyon. The victim was ejected from the cab during the rollover and was found lying unresponsive on the ground. He was pronounced dead at the scene.

Thoughts

The vehicle involved in this crash was nearly 50 years old and only responded to two or three calls per year. A postcrash inspection revealed that the vehicle was significantly overweight and had 3 tires with less than 1/32 of an inch of tire tread. Furthermore, the state police believe that the crash occurred when the braking system failed as the vehicle rounded a curve.

For some reason, fire departments have no problem driving a 50-year-old vehicle. This is especially troublesome when one considers the rigors that a fire apparatus will undergo as it serves as an emergency vehicle. Imagine the fit an insurance company would have if a nationally recognized gasoline company decided to transport gasoline in a 50-year-old tanker. This type of scenario would not be allowed to happen in any industry other than the fire service.

It is understandable that financial and budget issues often affect a fire department's ability to purchase a new vehicle. However, the fire service must do a better job of advertising why so many of our fire apparatus are outdated and obsolete. Using resources such as the NFPA 1901 standard and Annex D, fire departments must educate the public on the need for new fire apparatus. The days of driving a 50-year-old fire apparatus to an emergency call must end.

F2003-20: Wyoming

On May 22, 2003, a 16-year-old female junior firefighter died after the tanker truck she was riding in overturned while responding to a brush fire. The tanker truck drifted off the roadway causing the driver to lose control of the truck and overturn. The victim was ejected and trapped beneath the front passenger door. She was extricated by emergency personnel and transported to a county hospital where she was pronounced dead upon arrival.

Thoughts

The driver involved in this crash was under the influence of alcohol. The driver, who was suspended from driving any apparatus at the time of the crash, arrived at the station and was seen by several other members. At no time did any of the members report the driver to an officer, even though he was suspended from driving the apparatus.

Fire departments must have policies and procedures in place to address the possibility of a member arriving for duty under the influence of drugs or alcohol. This type of awareness training should be commonplace in the fire service. What is the procedure if the crew suddenly becomes aware of the fact that the driver is under the influence of alcohol or drugs? The vehicle must be brought to a stop immediately and the driver relieved of duty.

Planning these scenarios ahead of time will prevent confusion should it actually happen in real life. Driving down the road in an emergency vehicle with an intoxicated driver is not the time to wonder how to handle the situation. Hopefully, a thorough driver training program would have addressed the hazards of intoxicated driving and prevented this situation in the first place. But should it occur, members must know what to do.

On another note, the vehicle involved in this crash was yet another modified water tanker that was overweight. The vehicle had previously served as a jet fuel tanker. The vehicle was purchased by the fire department and used to carry water. As several other case studies have shown, converting a fuel delivery vehicle to a water tanker often results in an overweight vehicle.

F2003-19: Kentucky

On June 16, 2003, a 30-year-old male volunteer firefighter was fatally injured after his POV hydroplaned and struck a billboard post. At 1830 hours, during a heavy rainstorm, the fire department was dispatched to a high-water emergency. The victim was responding in his POV to the fire department to pick up a fire apparatus. The victim was heading east on a two-lane paved road when he drove over a large pool of water which caused him to hydroplane and lose control. He traveled off the westbound shoulder of the road and struck a billboard signpost. Within approximately 6 minutes, the victim was extricated. He was pronounced dead at the hospital at 1940 hours.

Thoughts

This crash occurred when the victim's personal vehicle hydroplaned on wet roads. Heavy rain can cause pools of water on the road which may overwhelm the tread of the tires and cause the vehicle to hydroplane. Emergency vehicle drivers must remember to slow down during wet weather. By slowing down, the driver will give the tire tread time to push the water out of the way and allow the tire to grip the road.

F2003-17: Tennessee

On May 18, 2003, a 28-year-old male volunteer firefighter was seriously injured when he fell from a moving pickup truck. He had completed a three-day training course and at the time of the incident was being transported within the training grounds. The victim was riding on the lowered tailgate of a pickup truck when he fell onto the road. He suffered severe head trauma and was treated at the scene by fellow firefighters and on-site emergency medical services. The victim was transported by medical helicopter to a local trauma center where he died from his injuries on May 24, 2003.

Thoughts

It is tempting for firefighters to jump in the back of a department pickup truck or sit on the tailboard of an apparatus to go for a short ride. While this may save effort and energy, especially after a long firefight, riding on the outside of a vehicle is a dangerous practice. Should the vehicle hit a bump in the road or begin to slide on rough terrain, the firefighter may be thrown off the vehicle where they may strike their head on the ground or be run over by the truck. Fire departments must have policies and procedures in place that forbid any member from riding on the outside of the apparatus for any reason.

F2003-15: Ohio

On April 3, 2003, a 56-year-old male volunteer firefighter was fatally injured after the pumper/tanker he was driving to a brush fire overturned. The vehicle was heading east on a two-lane county road and had just rounded a left-hand curve when it drifted off the right side of the pavement. The vehicle re-entered the road and crossed to the opposite lane where it went off the road and overturned, trapping the victim. The victim was initially conscious and able to call in the crash but died approximately 6 minutes later.

Thoughts

This is another crash involving a water tender that left the roadway after rounding a curve. It has been stated numerous times in this section: fire apparatus operators must know where the curves in their district are located and drive them at a safe speed. If the vehicle is traveling too fast for the radius of the curve, lateral g-force will cause the vehicle to drive off the road or tip over. Curves are dangerous—respect them.

F2003-14: Oregon

On March 19, 2003, a 53-year-old male volunteer captain was killed while en route to a hazardous materials training exercise. The victim was riding on the front passenger side of an engine when it left the roadway and careened into a ditch. After traveling over a hundred feet in the ditch, the engine struck a tree on the front passenger side of the vehicle. The victim was pinned in his seat by the wreckage, and although extrication was attempted shortly thereafter, the victim died of his injuries while inside the apparatus.

Thoughts

The fire apparatus involved in this crash encountered an oncoming vehicle on a narrow roadway. As the driver attempted to avoid the oncoming vehicle, the rear tires left the roadway and entered a ditch that was 40 inches deep. The apparatus traveled approximately 165 ft before striking a tree on the front, passenger side.

Anytime a fire apparatus leaves the roadway, there are numerous issues that must be accounted for. Attempting to bring the vehicle back onto the road at highway speed may cause the driver to overcorrect and enter the oncoming lane. If the driver steers sharply to bring the apparatus back into the proper lane, the vehicle may trip itself and rollover.

If the vehicle leaves the roadway, the driver must also account for the fact that there may be large obstacles such as trees or other heavy objects on the side of the road. If the driver strikes one of these objects, the energy associated with the heavy fire apparatus will result in severe crush damage to the vehicle. If the object strikes a portion of the apparatus where someone is sitting, the occupant will suffer significant injuries. For this reason, the apparatus operator must take measures to slow or stop the vehicle as soon as possible. The longer the vehicle travels off the roadway, the greater the risk of striking a roadside object.

Fire departments should also take the time to preplan the roadways in their district and determine if the road is safe for fire apparatus. Some roads may need to be restricted due to height, width, or weight restrictions, unsafe road conditions, bridges, or tunnels. If there is a road-related hazard, fire departments will have to preplan alternate routes for larger apparatus.

F2003-07: California

On January 13, 2003, a 46-year-old female career firefighter died from injuries she received after falling from a moving open-cab engine. The engine was responding to a reported airport emergency with an officer and a fire apparatus operator in the cab and two firefighters seated in the open-cab jump seats. As the engine rounded a bend and accelerated up a slight grade to enter a highway, the victim lost her balance and fell from the apparatus. The victim was treated at the scene for multiple traumatic injuries and transported to a local hospital. She died from her injuries five days after the incident.

Thoughts

This is yet another example of an unrestrained firefighter falling out of an apparatus. The apparatus was an open-cab reserve piece that was being used while the full-time apparatus was out of service. Due to the open cab, firefighters were required to wear earmuff style hearing protection while riding the apparatus. It is believed that the earmuffs slid away from the victim firefighter, causing her to unlatch her seat belt and reach for them.

Once a firefighter is seated and belted into the apparatus, under no circumstances should the seat belt be released. In this case, the firefighter lost her life attempting to retrieve a piece of personal protective equipment. As the s report indicates, fire departments should find ways to mount or secure required protective equipment so that it does not become lost or slide away during a response. Firefighters may be tempted to unbuckle their seat belts to retrieve the equipment, which makes them susceptible to a fatal incident such as this one.

F2003-05: Texas

On January 19, 2003, a 28-year-old male career firefighter died when the ambulance he was riding in crested a hill and was struck head-on by an oncoming vehicle. Following the collision, both vehicles caught fire and the victim was entrapped. The driver was transported by ambulance to a local hospital where he was treated for numerous injuries and discharged four days later. The victim was pronounced dead at the scene.

Thoughts

The ambulance involved in this crash was struck head-on by a vehicle which encroached into the ambulance's lane of travel. Unfortunately, the collision occurred at the crest of a hill, which would have made it extremely difficult for the ambulance driver to see the approach of the oncoming vehicle. As a result, little time was left for an evasive maneuver.

Drivers must remember to account for sight distance issues caused by hills, curves, or sight obstructions in the road. If the vehicle is approaching a hill crest or curve, use caution and anticipate the possibility of a vehicle suddenly encroaching in the emergency vehicle's lane of travel. Always look for an out which will provide an escape route for the emergency vehicle should the driver need to take evasive action.

F2002-42: North Carolina

On June 13, 2002, an 18-year-old female volunteer emergency medical technician was fatally injured after being involved in a single-vehicle crash while driving the fire department medic unit to a confirmed structure fire. The victim was en route to the fire on a rural state highway when the medic unit veered across the road, overturned, and entered a ditch. After entering the ditch, the medic unit overturned again, struck a barn, and overturned two more times in a field. The victim was ejected and found unresponsive. She was transported to a local hospital where she was later pronounced dead.

Thoughts

The medic unit involved in this crash is believed to have been traveling 75 mph in a posted 55 mph zone. Considering the narrow width of the roadway, this speed left little room for error should the vehicle drift outside its lane of travel or need to make an evasive maneuver.

As a vehicle's speed increases, the margin for error decreases. A vehicle that drifts slightly at 35 mph will not travel nearly as far as a vehicle that drifts at 75 mph. At 75 mph, the driver will have little time and distance to correct a potentially fatal mistake.

The NIOSH report also makes mention of the layout and ergonomic design of fire apparatus, specifically the controls located inside the cockpit of the vehicle. All vehicles, especially those designed to be operated by one person, must be carefully laid out so the driver does not have to reach and search for control switches. If the driver needs to search for a control switch, he may become distracted, taking his eyes off the road and potentially causing the vehicle to drift outside its lane of travel. If the vehicle drifts at high-speed, it could lead to a disastrous loss of control.

F2002-42: North Carolina

On September 23, 2002, a 32-year-old female career firefighter was fatally injured when the tanker truck she was driving overturned while returning to the fire station. The tanker truck drifted off the roadway onto the shoulder, causing the driver to overcorrect and lose control. The truck skidded across the roadway, entered a ditch, and overturned. The victim was entrapped in the truck and was pronounced dead on the scene.

Thoughts

For an unknown reason, the water tanker involved in this crash drifted off the right side of the road on an open straightaway. The driver overcorrected, crossing into the oncoming lane, and subsequently traveled off the left side of the roadway where the vehicle overturned. As discussed in previous case studies, if the vehicle drops off the road edge, the vehicle should be brought to a safe stop instead of fighting to get the vehicle back onto the road. If the vehicle should suddenly mount the road edge as the driver fights to get it back onto the road, the vehicle will have a tendency to shoot across into the oncoming lane. Examination of the scene shows that if the driver had simply stopped the vehicle, it would have been much safer to slowly bring it back onto the road under controlled conditions.

F2002-39: Tennessee

On September 5, 2002, a 17-year-old male volunteer firefighter was fatally injured after the tanker truck he was driving overturned while responding to a brush fire. The tanker was traveling south on a two-lane county road when it drifted off the right side of the road, causing the driver to lose control. The driver tried to correct his direction of travel but was unable to recover. The tanker went down a slight embankment and overturned, trapping the victim. The victim was extricated and then transported to a local hospital by ambulance where he was pronounced dead.

Thoughts

The victim in this crash was a 17-year-old junior firefighter who was driving a water tanker. Motor vehicle crashes are the leading cause of death for young people because they often lack the experience and good judgment to safely operate a motor vehicle. Under no circumstances should someone so young be allowed to operate a fire apparatus, especially a water tanker—especially not under emergency conditions.

Most insurance companies require a driver to be at least 21 years of age before they are allowed to operate an emergency vehicle. This is because older members tend to have more maturity and life experience, which makes them better suited to drive a large vehicle under emergency circumstances. Several of the case studies discussed in this section involve drivers who had only driven the fire apparatus a few times before the crash. Driver training programs should account for this fact and use a graduated system, where the driver is not allowed to drive the vehicle alone until they have driven to multiple emergencies with an instructor.

This incident also addresses the issue of control switch positioning and radio mounting. As the water tanker rolled over on a straightaway, it is surmised that the driver may have been reaching for the radio when the vehicle lost control. Because distracted driving is a common cause of motor vehicle crashes, emergency vehicles should be designed to reduce distracted driving. Control switches, radios, and microphones should be easily reachable by drivers so they do not have to reach over and take their eyes off the road or their hands off the wheel.

Fire departments should also implement policies that require a minimum of two firefighters on every call. A second firefighter will be able to work the radio, sound the siren, and determine a safe route of travel while the driver simply drives. If the vehicle must be driven by the "driver only," a great deal of thought should be given on how to properly lay out the cockpit so drivers do not have to reach over or take their eyes off the road to work the controls.

F2002-36: Texas

On August 8, 2002, a 28-year-old male volunteer firefighter was fatally injured when he was run over by the left-front tire of a brush truck. The victim was attacking a grass fire with a charged hoseline from a work platform on the front of a moving brush truck. The brush truck was making a U-turn through heavy smoke when another vehicle skidded into it. The victim was ejected from the left side of the work platform and run over by the brush truck. The victim was pronounced dead at the scene.

Thoughts

A firefighter was riding on the outside of a fire truck when the truck was struck by another vehicle, ejecting the victim off the rig. At no time should a firefighter be riding on the outside of a fire apparatus (with the exception of certain brush vehicles that have a designated and secured seat with a restraint system). Anyone riding on the outside of a vehicle is susceptible to being thrown off if the vehicle is involved in a collision or encounters rough terrain. If the driver does not have time to perceive and react after the firefighter is ejected, the vehicle may run over the firefighter.

This incident also deals with limited sight distance caused by smoke from the brush fire. Fire apparatus operators must account for limited sight distance caused by rain, fog, snow, or smoke. If the driver cannot see far ahead due to a sight obstruction, they must slow down and drive accordingly. In this case, it would appear that the driver of the brush truck did not see an approaching vehicle due to the smoke. As a result, the driver made a U-Turn in front of the oncoming vehicle, which resulted in a collision. Fire apparatus should attempt to drive out of the smoke before making any turns or maneuvers.

F2002-21: Pennsylvania

On May 4, 2002, a 14-year-old male junior volunteer firefighter was fatally injured while responding to a fire alarm on his bicycle. He was on his way to the fire station when he crossed a T-intersection without stopping and was struck by an automobile. The victim was treated at the scene and then transported to a local hospital. He was later transported by helicopter to a nearby children's hospital where he was pronounced dead the following day.

Thoughts

This incident involved a junior firefighter who was riding a bicycle to the firehouse in response to a call. While the incident does not deal directly with a fire apparatus, it does bring up the issue of unconventional vehicles. Fire departments that allow members to respond to the firehouse in an unconventional vehicle must have policies and procedures in place to address the issue.

If the fire department allows members to ride bicycles to the firehouse, the members operating the bicycles should be required to wear a helmet and reflective vest. The bicycle should also be equipped with a state required headlamp and marker lights. Training should be provided to the members operating the bicycles so they understand the issues associated with riding a bicycle on a public highway. Many bicyclists overestimate their visibility and assume motorists will see them from a long way off. This is often not the case.

It is also believed that the victim in this incident drove through a stop sign and into the path of an oncoming car. Fire apparatus operators must be aware of tunnel vision and the associated effects that tunnel vision will have on a person's ability to operate a vehicle. If the fire apparatus operator is focused solely on arriving at the scene, they may tune out everything else around them as they drive the vehicle to the call.

The phenomenon of tunnel vision explains the need for the NFPA recommended practice of "challenge and response." Challenge and response protocols call for the officer of the apparatus to continually ask the driver what his intentions are while driving to the scene. While some may view the practice as back-seat driving, it provides a second set of eyes in the front passenger seat while also ensuring that the driver sees approaching hazards such as traffic signals and stop signs.

F2002-18: Kansas

On April 11, 2002, a 61-year-old male career fire chief providing mutual aid at a scene of a motor vehicle incident died after being struck by a fire truck. A 26-year-old male volunteer fire chief from another department lost control of the fire truck after his brakes failed as he was arriving on the scene. The driver received injuries and was transported by ambulance to a regional hospital where he was hospitalized and then discharged the following day. The victim was transported by ambulance to a regional hospital where he was pronounced dead.

Thoughts

This crash occurred as a result of a fire apparatus losing its brakes while responding to a call. The police investigation determined that the cause of the brake loss was oil leaking from one of the brake lines contaminating the friction surface in the metal brake drum. The brake was also inoperable due to a piece of plastic that was wedged in the drum after an earlier repair.

Losing one brake due to drum contamination should have left the vehicle with some braking ability. While it is difficult to speculate so long after the crash, the other brakes may have started to fade as they attempted to burn off the excess energy that the broken brake was not burning off. This is known as a "cascade failure" and would have resulted in complete loss of the braking system.

The malfunctioning brake was also affected by a piece of plastic wedged in the drum assembly. The plastic came from a dust cover that had been glued back together with adhesive and caulk. While it is understandable that some fire departments must make their own vehicle repairs due to budget constraints, this practice should be discouraged. Repairs should only be made by trained and certified emergency vehicle technicians. All repairs and preventative maintenance should be documented and kept for the entire life of the apparatus.

The driver put the apparatus in neutral when he lost the brakes. Putting a vehicle in neutral will take away engine braking, which may have helped to slow the apparatus. If the brakes are lost, drivers should attempt to downshift and use the engine and drivetrain to slow the vehicle. Driver training programs should reinforce this technique as part of the training program.

F2002-16: Alabama

On April 7, 2002, a 29-year-old female volunteer firefighter was fatally injured when the engine in which she was riding overturned on the way to a brush fire. The victim was one of three volunteer firefighters responding in the vehicle. The engine had just rounded a curve when it went off the road onto the shoulder, causing the driver to lose control and the engine to overturn. The driver and the front-seat passenger were assisted out of the engine through the windshield area and transported by ambulance to a local hospital. The victim was entrapped and had to be extricated by firefighters. The victim was transported to a regional medical center where she was pronounced dead upon arrival.

Thoughts

The engine involved in this crash was traveling 74 mph in a posted 45 mph zone. That's ridiculous. To drive a large fire apparatus nearly twice the speed limit is completely unacceptable. The associated increase in perception and reaction distance and total stopping distance is substantial. Furthermore, there are few apparatus that could safely negotiate a curve in the road at 74 mph due the large g-force acting on the vehicle's center of gravity. Having read this far into this book, the fact that the apparatus rolled over while rounding a curve should not be surprising to anyone.

F2002-10: Kentucky

On March 2, 2002, a 48-year-old male volunteer firefighter/engineer was fatally injured when the tanker truck he was driving was struck by a freight train as he attempted to traverse a private, unguarded railroad crossing. The victim was returning to the station from a live-burn training exercise. The train struck the tanker truck on the passenger side just in front of the door. The victim was ejected and found lying approximately 75 ft from the track. He was transported to a hospital where he was pronounced dead.

Thoughts

NFPA 1500 (6.2.10) requires all fire apparatus to come to a complete stop at an unguarded railroad crossing. While trains are required to sound their horns as they approach a railroad crossing, it is nearly impossible for a fire apparatus operator to hear the horn over the sound of the engine and other sound producing devices in the cab of the apparatus. Drivers should pull up to within a safe distance of the crossing, shut down any noise producing equipment, roll down the window, and listen. Once it has been determined safe to do so, the apparatus can cross the tracks. See Chapter 9 for details on railroad crossing safety.

F2002-04: Ohio

On January 21, 2002, a 26-year-old male volunteer firefighter was fatally injured after being ejected from his POV as a result of a motor-vehicle crash while responding to a confirmed house fire. The victim was traveling toward the fire station on a county road when he crested a hill and drove over an icy section of road. His vehicle left the road, struck two mailboxes, traveled up an embankment, and overturned several times. The victim was ejected through the sunroof and was found lying unresponsive on the road. He was transported via ambulance to a local hospital and was life-flighted shortly thereafter to a regional trauma center where he was pronounced dead the next morning.

Thoughts

The victim in this crash crested a hill and drove over a patch of ice, which resulted in a loss of vehicle control. Fire apparatus operators must anticipate icy conditions based on the existing ambient conditions. If there is the possibility of black ice or ice patches on the road, the apparatus operator should slow down accordingly.

This incident also highlights the issue of forward sight distance. It would appear that the victim did not see the ice on the road ahead due to the limited forward sight distance created by the hill crest. Whenever a driver approaches an area that can't be seen due to limited sight distance, it is important to slow down during the approach. Slowing down will shorten the total stopping distance should a hazard suddenly emerge in front of the vehicle. Traveling at high speeds will not give the driver enough time to perceive, react, and bring the vehicle to a stop once the driver sees the hazard.

F2001-36: Oregon

On August 19, 2001, a 52-year-old male volunteer firefighter died after the right front tire of his tanker truck blew out and he lost control. The truck struck a large boulder and tree, entrapping the victim in the cab. He was extricated from the truck, and the medical examiner pronounced him dead at the scene.

Thoughts

This crash occurred when the fire apparatus's tire blew out and the driver lost control. The tire that blew out was 23 years old. According to the police report, the tire appeared brittle, and the steel cords that help hold the tire together were rusted from exposure. It should be no surprise that a tire in this condition blew out on the highway.

While it is true that the tire tread on most fire apparatus tires never wears down past the minimum state or NFPA requirements because the tires don't get that much use, the tires are still susceptible to damage caused by environmental conditions. Tires are admittedly expensive to replace; however, it is much cheaper to replace a set of tires than have to replace the entire vehicle if it crashes after a tire blowout. Remember that NFPA recommends that tires be replaced every 7 years.

Driver training programs must also address how to handle a tire blowout should it occur. Using simulated training, drivers should practice how to respond in the event of a tire blowout. See Chapter 11 for further information.

F2001-17: Alaska

On March 6, 2001, a 41-year-old male volunteer assistant chief died after the engine he was driving left the road, overturned, and struck a tree. The victim and a firefighter had responded at 1445 hours to a structure fire. At 1503 hours, the victim lost control of the engine on an icy, snow-packed gravel road. The engine left the road, overturned onto the passenger's side, and struck a tree. The victim was thrown into the passenger's side, landing on top of the firefighter. The victim was entrapped and had to be extricated from the vehicle. Emergency medical technicians performed CPR on the victim for approximately 20 minutes before pronouncing him dead at 1530 hours at the scene.

Thoughts

While it is unknown what caused the apparatus to initially lose control, it should be noted that the engine had a 1000-gallon, unbaffled water tank. The lack of baffles may have led to a liquid surge which initiated the loss of control. At the very least, it compounded the loss of control. It should also be noted that the roadway was covered with packed snow and a thin layer of ice. The reduced coefficient of friction of the roadway may have contributed to the initial loss of vehicle control. If the driver applied the brakes too hard, or if an auxiliary braking device activated,[1] the rear tires may have locked up, causing the apparatus to fishtail as described by witnesses.

When driving on low friction surfaces, fire apparatus operators must remember to anticipate the need to stop and slow down well ahead of the required stopping point. Drivers must use a light application of the brake pedal and remember to turn off auxiliary braking devices in inclement weather (as required by manufacturer recommendations) to prevent the rear tires from locking up and potentially causing the vehicle to lose control.

F2001-06: Kentucky

On January 12, 2001, a 21-year-old volunteer firefighter responding to a grass fire died from injuries he received when the tanker he was driving struck a utility pole and overturned. The tanker was traveling eastbound on a two-lane state road, and after traversing a slight curve, the tanker drifted off the road. The victim overcorrected and lost control. The tanker continued to travel across the left side of the roadway, entered a ditch, hit an embankment, and struck a utility pole. The tanker overturned and then continued to slide before coming to a rest on the passenger side with the victim entrapped under the tanker. The victim was extricated and pronounced dead on the scene.

Thoughts

When the water tanker involved in this crash exited a left-hand curve, the passenger side tires drifted off the roadway. The driver attempted to bring the vehicle back onto the roadway, at which time he overcorrected, crossed into the oncoming lane, and rolled over on the right side of the roadway. By now, the scenario of a water tanker drifting off the road, overcorrecting, and rolling over should seem quite familiar to the reader. Fire apparatus operators must remember to slow down for a curve and traverse it at a safe speed. Midway through the curve is not the time to realize the vehicle is traveling too fast and attempt to correct. By then it will be too late.

This is also another example of an overweight vehicle that had poorly maintained tires. Due to their size and handling characteristics, water tankers must have a thorough preventative maintenance schedule to make sure the vehicle is only operated in tip-top shape. Any mechanical defect, such as excess weight, worn-out tires, or out-of-adjustment brakes, could lead to a potential disaster.

F2001-01: Kentucky

On November 16, 2000, a 19-year-old male volunteer firefighter died and a 17-year-old junior fighter was injured when the victim lost control of the tanker truck he was driving. At 1900 hours, the fire department was performing a water shuttle training exercise. At approximately 2000 hours, the victim was en route to the station with the junior firefighter as a passenger. The tanker was traveling eastbound on a two-lane state road when it drifted off the edge of the road. The victim steered the vehicle back onto the roadway and traveled a short distance on the center line. The victim then overcorrected in an attempt to steer the tanker back into the right lane. The tanker left the right side of the roadway and traveled down an embankment and through a ditch line. The tank separated from the chassis, became airborne, and rotated approximately 180° before landing on the ground. The truck overturned and rolled upside down before coming to rest on its roof. The junior firefighter was ejected from the tanker and sustained serious injuries. The victim was trapped inside the cab of the truck. He was removed and taken to a local hospital where he was pronounced dead.

Thoughts

This incident represents another water tanker that drifted off the road, overcorrected, and crashed. The vehicle was being operated by a 19-year old driver, once again demonstrating the issues associated with young drivers. Most insurance companies require fire apparatus operators to be at least 21-years of age because younger drivers typically lack the experience and maturity to safely operate a large emergency vehicle.

This vehicle drifted off the road on a straightaway. While the report does not indicate what may have caused the vehicle to drift off the road, fire apparatus operators must be especially conscious of staying within their lane of travel. Distractions, reaching for equipment, or taking the driver's hands off the wheel may all cause the vehicle to drift out of its lane. In a large vehicle at high speeds, once the vehicle drifts off the road it will be very difficult to get it back onto the road again. This is especially true for a water tanker, as the liquid load may surge and cause the vehicle to rollover.

F2000-41: Alabama

On September 27, 2000, a 33-year-old male career firefighter died after attempting to board a ladder truck. At 1808 hours, the fire department was dispatched to a motor-vehicle incident involving two vehicles, with injuries and a possible entrapment. The captain on duty was outside the station when the driver, along with a firefighter and the victim, proceeded to the bay area of the station. The driver stopped and looked quickly at the map board and then made his way to the truck. The firefighter and the victim walked towards the rear of the truck. The firefighter told the victim that he would run to the back of the station to alert the captain of the call. The driver got in and started the truck, then activated the lights and sounded the air horns to let the captain know that the truck was ready to leave. At 1809 hours, the driver radioed that the truck was responding to the call. The firefighter donned some of his gear and boarded the jump seat, facing forward on the captain's side. The captain boarded the truck, and the driver stated that he had heard both jump-seat doors close. He checked his rear view mirrors and both of the jump-seat doors were in the closed position. At approximately 1810 hours, the truck exited the bay area of the station with lights and sirens activated. The truck moved down the station's driveway and paused briefly to wait for traffic heading westbound to yield for the truck. The truck crossed over the three westbound traffic lanes and made a left turn. Note: According to several civilian witnesses' statements, during this time the victim was running beside the driver's side jump-seat door while trying to catch up to the truck. The truck continued to travel in an inside turning lane, heading eastbound. The truck began to accelerate as it was straightening out from making the left turn. With the turning lane ending at an intersection, the driver looked in the rear view mirror on the captain's side and made a right turn to merge into the center lane of traffic at the intersection. As the truck approached the intersection all three personnel on the truck felt a "thump." Upon checking both mirrors the driver saw the victim lying on the ground. The driver told the captain that the victim had fallen out of the truck. At 1811 hours, the captain radioed to central dispatch to request assistance for an injured firefighter. The victim had been run over by wheels of the truck and sustained fatal injuries as a result.

Thoughts

This was a tragic case of a firefighter who missed the responding apparatus and was running alongside it to try and jump on the truck. Fire apparatus operators must have a method of confirming that all of the assigned crew members are on the apparatus and properly seated and belted. In apparatus that have a common cab, this is as easy as turning around and looking. In some apparatus, the procedure is more complicated because there may be a partition in between the driver's area and the jump seats.

In a career department, the driver will almost always know who is supposed to be on the apparatus when it responds. If the cab is separated by a partition, the crew members should yell up to the driver to confirm that they are ready to go. Some apparatus may even have intercom systems in place that will make this procedure easier.

In volunteer departments, it is not uncommon for members to run into the firehouse as the apparatus is pulling out the door, grab their gear, and attempt to get on. In these situations, the fire department must have a strict policy that prohibits members from attempting to board a moving fire apparatus.

F2000-39: Illinois

On April 29, 2000, a 43-year-old male career Lieutenant died as the result of injuries he received when the truck he was responding in was struck by a pickup truck. At approximately 1150 hours, Central Dispatch notified a career fire department of an automatic alarm at a residential structure. An engine and ladder truck responded to the call. As the ladder truck approached an intersection, the driver made a rolling stop. As the driver proceeded through the intersection, a civilian pickup truck ran the stop sign and collided with the apparatus. The victim was ejected from the passenger-side door of the truck and received massive head injuries. He was transported to a local hospital where he was pronounced dead. The other three firefighters and the driver of the truck received medical attention for their injuries and were released.

Thoughts

It is believed that at the time of the crash, the victim firefighter was not wearing his seat belt. Had the victim been wearing his seat belt, he would have probably stayed within the protective envelope of the apparatus and not sustained fatal injuries. The front passenger door may have also had a malfunctioning door latch, which allowed the door to come open during the crash. When the malfunctioning door latch was combined with a firefighter who was not wearing a seat belt, it resulted in a firefighter being ejected and killed.

There is no excuse for not wearing a seat belt. It is required under NFPA standards, and many states do not provide any exemption for emergency responders. Fire apparatus operators should not put the rig in drive until they are sure everyone is seat belted, and no firefighter should undo their seat belt while the rig is in motion.

If a seat belt or door latch is not functioning correctly, the riding position should immediately be taken out of service. Firefighters would not fight a fire with a malfunctioning hose or air pack, so why should the apparatus be any different? If there is a problem with the vehicle's safety equipment, put the apparatus out of service and notify repair personnel.

F2000-35: Pennsylvania

On July 2, 2000, a 17-year-old male volunteer junior firefighter died after his POV collided with a farm tractor. At 1535 hours, the volunteer fire department was dispatched to a structure fire, and the victim responded in his POV to the fire station. En route to the station, the victim was traveling in the southbound (right) lane on a two-lane township road. The victim negotiated a curve and drifted into the northbound (left) lane of the road. Traveling northbound was a farm tractor with an attached front-end loader. The victim's POV struck the left side of the front-end loader, which entered the windshield and pushed through the A-post. The victim was found unconscious and was transported by a helicopter to an area hospital where he was pronounced dead approximately one hour after the incident.

Thoughts

The 17-year old victim of this crash crossed into the oncoming lane while rounding a curve. When the victim exited the curve, he drove into a farm tractor that was on the other side of the curve. As a result of the collision, the firefighter received fatal injuries.

This crash highlights the issue of forward sight distance. Because corn was growing along the roadside, the victim was unable to see into the curve. As a result, the victim did not see the farm tractor in the oncoming lane. Had the victim seen the farm tractor, he may have attempted to maintain his lane of travel rather than cross into the oncoming lane while rounding the curve. Anytime the driver's forward sight distance is obscured, the driver should slow down accordingly and maintain his lane of travel.

F2000-33: Missouri

On May 27, 2000, a 27-year-old male volunteer firefighter died after losing control of the pumper truck he was driving. At 1522 hours, the fire department was dispatched to a motor-vehicle incident with unknown injuries. At 1523 hours, the victim responded in an engine. Also responding to the scene was a captain in his POV and an assistant chief in a department vehicle. At 1539 hours, the victim radioed to central dispatch that the call was unfounded and he was returning to the station. En route to the station, the engine was traveling northbound on a two-lane state road with the captain following behind. The engine drifted off the right side of the roadway and passed over a large hole created by erosion at the end of a drainage culvert. The victim applied his brakes in an attempt to steer the engine back onto the road, overcorrected, and lost control. The engine rotated counterclockwise, then came back onto the roadway and overturned. The engine rolled approximately one and three-quarter revolutions as it continued to skid, traveling northbound. The victim was ejected from the engine and sustained fatal injuries as a result of the collision.

Thoughts

This crash is another example of an apparatus drifting off the right side of a straight road, the driver overcorrecting, and the vehicle rolling over. Fire apparatus operators must strive to avoid distractions and do everything they can to prevent the tires from leaving the road surface. Once the tires leave the road surface, the vehicle is prone to any number of disastrous consequences.

This incident may have been exacerbated by the fact that it had recently rained and the soil on the side of the roadway was soft. Soft soil would have caused the tires to sink into the ground more than usual. If the tires sink into soft ground, it will be more difficult to get the vehicle back onto the road once it drifts off. The driver will be more prone to overcorrect, as he fights to bring the vehicle up over the road edge and back onto the road surface. If the vehicle drives off the road, the driver should slow down and come to a stop. The crew should evaluate the situation and then bring the vehicle back onto the road at a safe speed.

F2000-19: North Carolina

On March 17, 2000, a 31-year-old male career firefighter/engineer died after the apparatus that he was driving collided with an Amtrak train at a railroad crossing. The victim was returning to the station after his apparatus was cancelled when the alarm was determined to be false. Truck 1, followed by Squad 1, stopped behind a civilian vehicle on the west side of a railroad crossing consisting of three sets of tracks. The safety gates at the crossing were down, the warning lights were activated, and a freight train was moving slowly on the tracks. The freight train stopped after it cleared the crossing to wait for a signal ahead. Truck 1 started to go around the first safety gate and over the track. Witness 1, in a vehicle behind Truck 1, saw a tanker car at the end of the freight train that obstructed the northbound view of the tracks, and he heard a train whistle. Witness 2, waiting in a vehicle on the east side of the crossing, saw a southbound Amtrak train approaching. He also saw Truck 1 driving around the first safety gate, so he honked his horn and flashed his headlights to warn the driver of the Amtrak train. Truck 1 continued around the safety gate and traveled into the path of the train, which struck the truck's left front corner and the bucket of the aerial ladder. The victim was ejected, landing behind the truck's left rear dual wheel. He was killed instantly.

Thoughts

The fire apparatus operator drove around railroad gates that were blocking the road. As the apparatus drove around the gates, it was struck by an oncoming train. Under no circumstances should a motorist drive around railroad gates that are deployed to block the roadway. In this case, the driver did not wait for the gates to go up and we can only assume that he thought the initial train had passed. Little did the driver know that the gates were still down because another train was approaching. When the driver went around the gate, tragedy struck.

In the event that a malfunction is suspected and it is believed that no train is coming, patience is the key. Contact the railroad company and ascertain if the track is clear to cross. The phone number of the railroad company should be located somewhere on the gate crossing equipment. If it is not feasible to contact the rail company, send members down the tracks to determine if it is clear for the apparatus to cross (see Chapter 9).

If a fire department encounters a guarded train crossing that deploys itself when a train is not approaching, it is important to notify the railroad company so the issue can be addressed. If motorists routinely encounter a guarded rail crossing that is defective, they will begin to believe it is a false deployment and drive around it.

F2000-18: Missouri

On January 17, 2000, a 47-year-old male volunteer fire chief died after he lost control of the truck he was driving. At 1512 hours, the volunteer fire department was dispatched for mutual aid to a working structure fire. At 1523 hours, the victim notified dispatch that Truck 960 was en route to the scene. As the victim approached a slight curve in the roadway, the truck began to skid, and the right tires traveled onto the shoulder. Continuing on the shoulder, the truck entered a ditch, became airborne, crossed over a side street, and struck a center median. The truck crossed another side street, struck a guardrail, and flipped end over end until it landed in a concrete culvert. The victim was killed instantly.

Thoughts

The engine involved in this crash was traveling 58 mph when it rounded a curve and lost control. More than likely, this speed was too fast for the radius of the curve. If a vehicle travels too fast through a curve, the g-force acting on the vehicle will cause it to lose traction, or in the case of a vehicle with a high center of gravity, to rollover.

The seat belts had also been removed from the driver's area of the apparatus. While it is one thing to not wear a seat belt, it is another thing for the fire department to completely remove the seat belt from the apparatus. If there is a riding position that does not have a functioning seat belt, that riding position must be taken out of service. If the driver's seat does not have a functioning seat belt, the entire apparatus must be taken out of service.

The tires on this apparatus were also worn down, and the water tank was not equipped with baffles. If the water load had sloshed inside the tank as the apparatus rounded the curve, it would have caused the vehicle's center of gravity to shift. If the center of gravity shifts, it may reduce the rollover threshold of the vehicle, and it will take less lateral g-force to roll the vehicle over. When the shifting center of gravity was combined with bald tires that could not properly grip the wet road, this vehicle was destined to lose control.

F2000-17: California

On February 11, 2000, a 25-year-old male volunteer firefighter died as the result of injuries he received when the engine he was driving crashed into a tree. A 35-year-old female volunteer firefighter who was a passenger in the engine was also injured. At 1530 hours, the volunteer component of this combination fire department was notified by central dispatch of a motor-vehicle incident involving injuries. Engine 1 responded to the scene with two firefighters. The assistant chief responded in his POV. En route to the scene of the motor-vehicle incident, the driver of the engine swerved to the right side of the road to avoid an oncoming vehicle. The engine traveled onto the soft shoulder and continued for approximately 230 ft. To avoid striking a utility pole, the driver steered the engine sharply back onto the road. He overcompensated, and the engine traveled across both lanes, left the road, and struck a large tree. The passenger was ejected from the cab of the engine and landed approximately 30 ft away. The victim was trapped in the engine and was removed approximately 1½ hours later. The victim and the passenger were taken by ambulance to an area hospital. The passenger was hospitalized for 5 days, and the victim died 3 days after the incident.

Thoughts

The fire apparatus operator moved over to the right because of an oncoming vehicle. When the apparatus operator moved to the right, the right side of the apparatus dropped off the road and onto a soft shoulder. The vehicle traveled over 230 ft along the shoulder until the driver swerved to avoid a utility pole.

If the vehicle drops off the roadway, slow down and bring the vehicle to a safe stop. Fighting to get the vehicle back on the road could lead to an overcorrection and subsequent rollover crash. Continuing to drive with the wheels off the roadway will expose the vehicle to countless hazards and obstacles on the roadside which the apparatus may drive into.

The fire apparatus operator involved in this crash did not have a valid license. Fire departments should check each fire apparatus operator's driver's license at least twice a year. Fire departments must be sure that all apparatus operators are driving with a valid license and that each operator has the appropriate license endorsements required by the respective state.

F2000-10: Indiana

On October 28, 1999, a 57-year-old captain and a 23-year-old fire apparatus operator died after the tanker they were responding in veered off the road and rolled several times in a corn field. The incident occurred while they were responding to a mutual aid call which had been dispatched as a grass fire threatening nearby structures. The tanker was traveling west on a two-lane state road. As the tanker approached a curve, it drifted towards the shoulder of the road. Just past the curve, it veered off the road and into a corn field. The tanker rolled on the passenger side and continued to roll several times. One victim was ejected out the driver's side window. The other victim was entrapped in the crushed upside-down cab and had to be extricated. He was transported to a local hospital and remained conscious and able to communicate, but died 7 days after the incident. Victim 2 was transported by ambulance to a local hospital then flown by life flight helicopter to a trauma center. He remained hospitalized and was conscious, but was unable to communicate and died 86 days after the incident.

Thoughts

The police investigator believes that the fire apparatus was traveling in excess of 55 mph when it lost control and rolled over. Considering the size of the vehicle and the fact that it was carrying 1800 gallons of water, this speed would have been excessive. Fire apparatus have low rollover thresholds (see Chapter 4). If the lateral g-force placed on this vehicle as it rounded the curve exceeded the vehicle's rollover threshold, it would cause the vehicle to rollover. This is why so many fire apparatus roll over while rounding curves.

F2000-06: Texas

On November, 14, 1999, a 46-year-old male volunteer firefighter died while responding to a mutual aid call. The career department was battling fires in approximately 200 hay bales and requested an additional supply of water. Upon receiving the call for assistance, the victim responded in the department's tanker truck. The tanker truck rounded a curve near the center line and drove off the right shoulder of the northbound lane. In an attempt to steer the top-heavy truck back onto the road surface, the victim overcorrected and lost control of the vehicle. He locked the truck's air brakes but could not bring the truck under control. The truck crossed over the center line and reached the shoulder of the southbound lane. The truck tires dug into the shoulder, causing the truck to flip and roll. The victim was partially ejected through the rear window of the cab. The truck came to rest upside down with the victim partially trapped within the truck, between the top of the cab and the ground. The victim was extricated from the truck and transported to a local hospital where he was pronounced dead.

Thoughts

The apparatus involved in this crash was a modified vehicle that had previously served as a utility company boom truck. The fire department acquired the vehicle, removed the boom, and installed a 1180-gallon water tank. Due to the short wheel base and the aftermarket water tank, the vehicle's stability was significantly reduced.

The driver was rounding a curve when he lost control and eventually rolled over. Once again, if a vehicle rounds a curve too fast, the tires will lose traction, or the vehicle will roll over. Either way, the vehicle will lose control. Drivers must appreciate the hazards posed by a curve in the road and slow down well in advance of the curve. Midway through the curve is not the time to realize you're driving too fast and try to slow down. By then it will be too late.

F2000-01: Virginia

On December 18, 1999, a 22-year-old male volunteer firefighter died and a 30-year-old volunteer district chief was injured after the rescue truck he was driving veered off the road and struck an oncoming car and then a tree. The incident occurred while they were responding to a reported gas leak at a private residence. The driver and the victim responded in Rescue 49. While en route, the driver looked down at the dashboard to lower his response priority by shutting off his lights and siren. While doing so, the truck's right side tires dropped off the road and into a ditch. When the driver tried to bring the truck back onto the roadway, he overcompensated, causing the truck to cross the oncoming lane of traffic and strike an oncoming car. Just before striking the car, the driver started steering the truck back to the right to avoid a second collision. The truck struck a tree and flipped onto the driver's side. Seconds after the wreck, the driver crawled out of his loose seat belt and out of the truck. The driver was taken by an ambulance to a local hospital where he was treated and released. The victim, who was trapped inside the vehicle, was removed, and taken to a local hospital where he was pronounced dead.

Thoughts

The fire apparatus operator took his eyes off the road to find a switch for the emergency lights. The switch was located 23 inches from the steering wheel, requiring the driver to reach over with one hand to manipulate the switch. If a driver takes his eyes off the road, he will be prone to have a crash because he will not see a hazard ahead. Furthermore, if a driver has to reach for something, there is a tendency to pull the wheel in the direction of the reach, which could cause the vehicle to leave the road.

The police calculated the speed of the apparatus at 51 mph in a posted 35 mph zone. When a fire apparatus is exceeding the speed limit, there is less room for error. The vehicle will travel considerably farther if the vehicle drifts off the road, making it more difficult to regain control of the apparatus before it travels into the oncoming lane or strikes an object. When distracted driving is combined with excess speed, the chance of a crash will greatly increase.

99-F45: Pennsylvania

On November 2, 1999, a volunteer firefighter sustained a traumatic head injury after falling off a responding open-cab ladder truck. This injury led to his death the following day. The victim was part of a volunteer crew which also included a driver, an officer, and three other firefighters. The crew was responding to provide mutual aid assistance to an adjoining community. As the ladder truck was leaving the station, the victim was reported to be standing behind the officer and firefighter 1 in the open crew compartment of the vehicle. None of the responding personnel reported wearing seat belts. Shortly after the ladder truck left the fire station and completed the second turn, firefighter 1 realized the victim was missing and signaled to have the ladder truck stopped. The crew dismounted the ladder truck and ran back to the victim, who was lying in the roadway. They gave him emergency medical care, and he was transported to a local hospital where he died the following day.

Thoughts

This is another example of a firefighter falling from a moving vehicle, resulting in fatal injuries. Fire apparatus should not be moved unless all firefighters are seated and securely belted. The same lateral g-force which causes fire apparatus to rollover will cause unrestrained firefighters to be flung from the rig.

99-F44: Ohio

On September 13, 1999, a 29-year-old male volunteer firefighter was killed as a result of injuries sustained in a motor-vehicle incident that occurred while he was responding to a kitchen fire. Central dispatch notified the fire department at 0656 hours of a kitchen fire. The victim left his residence in his POV. It is believed that he was heading directly to the fireground, and as he approached the fireground, he saw that no apparatus had arrived on scene. He was proceeding to the fire station to get the apparatus when his POV collided with a tandem dump truck that was turning onto the road he was traveling. The driver of the dump truck had stopped at a stop sign, and when he looked for oncoming traffic, he saw the victim's vehicle approximately 450 ft away. As he started to turn left, the driver of the dump truck realized the victim's vehicle was quickly coming toward him. In an attempt to avoid a collision, the driver stopped the dump truck. Due to a tall corn field near the roadway, the victim may not have seen the truck until the collision. The victim's vehicle struck the dump truck's front axle, and the victim was killed instantly.

Thoughts

When an emergency vehicle is approaching an intersection, the driver must be wary of potential hazards. These hazards include vehicles that are turning left in front of the emergency vehicle or vehicles that are entering the road from a side street. When approaching an intersection where the emergency vehicle has the right of way (green light or thru-street), emergency vehicle operators should slow down, cover the brake, and prepare for a potential hazard. Slowing down will reduce the total stopping distance of the emergency vehicle, making it easier to bring the vehicle to a stop if another vehicle encroaches, or at least lessen the severity of a crash should it occur. This is known as defensive driving.

99-F42: California

On October 16, 1999, four volunteer firefighters responded to a request for assistance in fighting wildland brush fires that were endangering several structures. The brush fires were burning approximately 60 miles east of the department's home location. The firefighters traveled in two different vehicles. Two firefighters brought the engine, and one male and one female firefighter drove a POV. After meeting up with the two firefighters on the engine, the crew of four proceeded down the highway to look for structures endangered by fire. A determination was made to relocate the POV, so the firefighter stopped the engine across the highway from the POV. The victim and one of the firefighters intended to ride together in the POV; however, unknown to the firefighter in the engine, the victim picked up some soft drinks from the POV and proceeded back across the highway to the engine. The victim apparently stepped up on the sideboard of the engine and was going to pass the soft drinks through the pass-through window when the chief started to move the engine forward. The victim lost her balance, and she turned and jumped toward the center of the highway, landing in a squatting position and off balance. She then fell backward toward the slow-moving engine, and her head was caught by the rear dual wheels.

Thoughts

The victim firefighter in this crash jumped onto the sideboard of the engine just as the driver began to move the vehicle. Firefighters must realize the danger of a moving fire apparatus and stay clear. In this case, there was a lack of communication between the victim and the driver. Before approaching an apparatus with a driver sitting in the driver's seat, a firefighter on the outside of the vehicle should make eye contact with the driver. Making eye contact uses nonverbal communication to notify the driver that a member is approaching. This should provide the driver enough notice to keep the vehicle from moving as the firefighter gets closer.

99-F36: Texas

On October 5, 1999, a captain, the driver, and a firefighter from Engine 33 responded to a medical call that had been dispatched as a patient with shortness of breath. Traveling north, the engine approached a four-way intersection that was crossed on the north side by an overpass supported by concrete columns and controlled by electronic traffic lights. The traffic signal was red for the engine's direction of travel, so the driver initially reduced the engine's speed, and after checking that traffic had cleared, increased the engine's speed and began traveling through the intersection. At the same time, a civilian vehicle traveled through the intersection into the engine's path. The driver of the engine was unable to avoid the automobile, and the two vehicles collided. The driver lost control of the engine which then struck one of the concrete columns supporting the overpass. The engine struck the column on the driver's side, and the victim was ejected through the windshield, landing in a lane for oncoming traffic. The engine continued past the column and came to a stop in the same lane next to the victim. The driver was knocked unconscious, and the firefighter riding in the rear crew compartment received minor injuries. The victim was flown by life-flight helicopter to a nearby hospital where he was pronounced dead upon arrival. The two injured firefighters were transported by ambulance to the hospital, where the driver was admitted in critical condition, and the firefighter was treated for his injuries and released. The civilian driver of the automobile was not injured.

Thoughts

In this case, the fire apparatus slowed at a red light and then proceeded into the intersection against the red light. As the fire apparatus crossed the intersection, it collided with a civilian vehicle that had traveled into the intersection. The fire apparatus subsequently struck a bridge support, resulting in the victim firefighter being ejected from the apparatus.

Fire apparatus must come to a complete stop at a negative right-of-way intersection. The apparatus must give civilians time to see and hear the fire apparatus and then give the apparatus the right-of-way. Once the fire apparatus begins to cross the intersection, it should do so at a slow and cautious speed. If the apparatus is traveling slowly, it will be able to come to a stop much quicker should a civilian vehicle enter the intersection and encroach into the apparatus's path of travel. While a collision may not be avoided under these circumstances, the severity of the crash will be reduced. The faster a vehicle is traveling, the more energy it will have. The more energy a vehicle has, the more damage and injury to those involved.

99-F33: South Carolina

A 34-year-old male volunteer firefighter died after the engine he was driving veered off the road and rolled two times before coming to rest. The incident occurred while the victim and another firefighter were responding in an engine to a motor-vehicle crash involving injuries. While en route, the tires on the engine's right side dropped off the road surface. As the victim attempted to bring the engine back onto the roadway, he overcompensated, causing the engine to cross the oncoming lane of traffic. The engine crossed a small ditch, rolled across another roadway, crossed another ditch, and rolled again before coming to rest in a resident's yard. Both firefighters were thrown from the engine, and the victim was killed instantly.

Thoughts

The engine involved in this crash was rounding a curve at 55 mph in a posted 35 mph zone. As the engine rounded the curve, the right tires dropped off the roadway. The driver overcorrected to get the vehicle back on the road, which induced a rollover crash. During the crash, the unrestrained victim was ejected from the vehicle and killed.

At 55 mph, the lateral g-force on the apparatus would have been significant as the vehicle rounded the curve. The lateral g-force would have acted to push the vehicle off the road, as the driver struggled for control. Once the driver overcorrected and the vehicle began to rollover, there was little the driver could do to stop the crash from happening. Drivers must slow down when approaching a curve and traverse the curve at a safe speed. Failure to do so will result in a crash.

F99-16: North Carolina

A 28-year-old male volunteer firefighter died after the engine he was driving veered off the road, overturned onto the passenger side, and struck a tree. While returning to the station, the victim apparently steered the engine toward the shoulder of the road to provide more room for oncoming traffic. As he approached the shoulder, the right-rear dual tires went off the road. Due to insufficient hard shoulder and a steep incline beside the road, the victim was unable to get the engine back onto the road. Continuing off the road and into the ditch for approximately 280 ft, the engine overturned onto the passenger side, striking a large tree. The victim was killed instantly.

Thoughts

The victim in this crash was driving the apparatus and moved to the side of his lane to accommodate oncoming traffic. When the driver moved over in his lane, the passenger side tires dropped off the roadway. The vehicle traveled 280 ft until it overturned and struck a tree.

If the vehicle leaves the roadway, it should be brought to a safe and controlled stop. Fighting to get the apparatus back onto the roadway may cause the vehicle to shoot back across the oncoming lane. Once in the oncoming lane, the driver may strike an oncoming vehicle or attempt to steer the vehicle back into the original lane of travel. If the driver steers too sharply, the g-forces acting on the vehicle will cause the vehicle to trip itself and rollover. An unrestrained occupant may be ejected onto the roadway and sustain fatal injuries.

Notes

1. The NIOSH report does not mention if the vehicle was equipped with an auxiliary braking device.

Understanding the Math

Why Understand the Math

This section will explain the math behind some of the basic driving principles. While I would not hold it against anyone for skipping this section, I can't emphasize enough the importance of its contents. Understanding the math behind the driving is what separates the true professional driver from someone who simply holds and turns a steering wheel. As the public expects emergency responders to be highly trained professionals, it is imperative that every driver have a grasp of these basic concepts. I promise that while the math may look daunting, it's no harder than pump school.

Before discussing the math associated with a moving vehicle, we must first understand the concept of energy. Understanding the amount of kinetic energy that is generated by a moving fire apparatus, and more importantly, how this energy must be controlled, should be an essential element of any fire apparatus driver training program.

Calculating the Kinetic Energy of a Moving Vehicle

It is important for a fire apparatus operator to understand how speed affects the kinetic energy of a moving vehicle. The faster the fire apparatus is traveling, the more kinetic energy it will possess. More kinetic energy will result in a longer stopping distance and a more severe crash.

To calculate the energy of a moving vehicle, we must know only two things: how much it weighs and how fast it is going. Once we have determined these two factors, we can use the following equation to calculate how much kinetic energy the vehicle will have as it drives on the highway.

$$KE = \frac{W \times S^2}{30}$$

Where

KE = Kinetic Energy in foot pounds

W = Weight of the vehicle in pounds

S = Speed of the vehicle in miles per hour (mph)

As an example, let's compare the kinetic energy of a 40,000-lbs. fire truck versus a 4000-lbs. civilian vehicle if both are traveling 35 mph.

Fire Truck (40,000 lbs.)	Civilian Vehicle (4,000 lbs.)
$KE = \dfrac{W \times S^2}{30}$	$KE = \dfrac{W \times S^2}{30}$
$KE = \dfrac{40,000 \times 35^2}{30}$	$KE = \dfrac{4,000 \times 35^2}{30}$
$KE = \dfrac{40,000 \times 1225}{30}$	$KE = \dfrac{4,000 \times 1225}{30}$
$KE = \dfrac{49,000,000}{30}$	$KE = \dfrac{4,900,000}{30}$
KE = 1,633,333 foot pounds of energy	KE = 163,333 foot pounds of energy

A 40,000-lbs. fire truck traveling at 35 mph will have **ten times** the energy of a civilian vehicle traveling at the **exact same** speed. This is an important teaching point for anyone who operates a large fire apparatus. If the fire apparatus operator drives to the fire station in a small sedan and jumps behind the wheel of a large apparatus, he must understand the increased amount of energy he must now keep under control.

Now let's examine how an increase in speed will affect the amount of kinetic energy each vehicle will possess. We will assume that the two vehicles increase their speed from 35 mph to 55 mph.

Fire Truck (40,000 lbs.)	Civilian Vehicle (4,000 lbs.)
$KE = \dfrac{W \times S^2}{30}$	$KE = \dfrac{W \times S^2}{30}$
$KE = \dfrac{40,000 \times 55^2}{30}$	$KE = \dfrac{4,000 \times 55^2}{30}$
$KE = \dfrac{40,000 \times 3025}{30}$	$KE = \dfrac{4,000 \times 3025}{30}$
$KE = \dfrac{121,000,000}{30}$	$KE = \dfrac{12,100,000}{30}$
KE = 4,033,333 foot pounds of energy	KE = 404,333 foot pounds of energy

When the fire apparatus increases speed from 35 mph to 55 mph, the amount of energy increases from 1,633,333 foot pounds of energy to 4,033,333 foot pounds of energy. Simply increasing the speed of the apparatus 20 mph results in the vehicle's kinetic energy increasing over 246%. This is why it is so important to control the speed of the vehicle especially when driving a large fire apparatus.

"Feet per Second" Instead of "Miles per Hour"

Fire apparatus operator must understand the definition of speed. Speed is a distance traveled over a unit of time. The faster an object is moving, the more distance it will cover in a shorter amount of time. If we say a vehicle is traveling "60 miles per hour," this means it will travel 60 miles in one hour.

When discussing vehicle safety, it will be better for the fire apparatus operator to understand speed in terms of feet per second instead of miles per hour. Thinking of speed in terms of feet per second will give the driver a better appreciation for the amount of distance the vehicle will travel in a short period of time. As an example, a vehicle that is traveling 60 mph will cover 87 ft in just one second.

To convert miles per hour into feet per second, simply multiply the speed in miles per hour by 1.466.

To further reinforce this concept, use a hands-on training demonstration to show fire apparatus operators just how far the vehicle will travel in one second.

Have each driver measure 87 ft in a parking lot. Discuss the fact that at 60 mph, this distance will be covered in the blink of an eye. When this concept is combined with a hands-on demonstration, it is a very effective way to show drivers how speed affects safety.

MPH	Multiply by 1.466	FPS
5	x 1.466	7.3
10	x 1.466	14.6
15	x 1.466	21.9
20	x 1.466	29.3
25	x 1.466	36.6
30	x 1.466	43.9
35	x 1.466	51.3
40	x 1.466	58.6
45	x 1.466	65.9
50	x 1.466	73.3
55	x 1.466	80.6
60	x 1.466	87.9
65	x 1.466	95.2
70	x 1.466	102.6
75	x 1.466	109.9
80	x 1.466	117.2
85	x 1.466	124.6
90	x 1.466	131.9
95	x 1.466	139.2
100	x 1.466	146.6

How Weather Affects the Skid to Stop Distance

As discussed in Chapter 1, one of the most important concepts a fire apparatus operator must come to understand is how weather affects the skid to stop distance of a vehicle. A wet or snow covered road will significantly increase the skid to stop distance due to the reduced coefficient of friction. Table B–1 provides examples of the stopping distance of a passenger vehicle with a braking efficiency of 100% traveling 50 mph in various weather conditions. Table B–2 provides examples of the stopping distance of a fire apparatus with a braking efficiency of 65% traveling 50 mph in various weather conditions.

TABLE B–1.

Skid to Stop Distance of a Passenger Vehicle Traveling 50 mph			
	Dry Asphalt	Wet Asphalt	Snow/Ice
Vehicle Speed	50 mph	50 mph	50 mph
Coefficient of Friction	0.80	0.60	0.20
Slope of the Road	0% (Level Road)	0% (Level Road)	0% (Level Road)
Braking Efficiency	100%	100%	100%
Adjusted Drag Factor	0.80	0.60	0.20
Calculate	$SD=\dfrac{S^2}{(30)(f)}$ $SD=\dfrac{50^2}{(30)(0.8)}$ $SD=\dfrac{2500}{24}$	$SD=\dfrac{S^2}{(30)(f)}$ $SD=\dfrac{50^2}{(30)(0.6)}$ $SD=\dfrac{2500}{18}$	$SD=\dfrac{S^2}{(30)(f)}$ $SD=\dfrac{50^2}{(30)(0.2)}$ $SD=\dfrac{2500}{6}$
Skid Distance	104.1 ft	138.8 ft	416.6 ft

TABLE B–2.

Skid to Stop Distance of a Fire Apparatus Traveling 50 mph			
	Dry Asphalt	Wet Asphalt	Snow/Ice
Vehicle Speed	50 mph	50 mph	50 mph
Coefficient of Friction	0.80	0.60	0.20
Slope of the Road	0% (Level Road)	0% (Level Road)	0% (Level Road)
Braking Efficiency	65%	65%	65%
Adjusted Drag Factor	0.52	0.39	0.13
Calculate	$SD=\dfrac{S^2}{(30)(f)}$ $SD=\dfrac{50^2}{(30)(0.52)}$ $SD=\dfrac{2500}{15.6}$	$SD=\dfrac{S^2}{(30)(f)}$ $SD=\dfrac{50^2}{(30)(0.39)}$ $SD=\dfrac{2500}{11.7}$	$SD=\dfrac{S^2}{(30)(f)}$ $SD=\dfrac{50^2}{(30)(0.13)}$ $SD=\dfrac{2500}{3.9}$
Skid Distance	160.2 ft	213.6 ft	641.0 ft

Note that the adjusted drag factor is the coefficient of friction of the road surface adjusted for the slope of the road and the braking efficiency of the vehicle. The adjusted drag factor is calculated using this formula:

$$\textit{Drag Factor } (f) = (\mu)(n) \pm m$$

Where

μ = the coefficient of friction of the roadway

n = the braking efficiency of the vehicle

m = the slope of the roadway

How Road Slope Affects the Skid to Stop Distance

Table B–3 provides examples of the stopping distance of a passenger vehicle with a braking efficiency of 100% traveling 50 mph on two roads, each with a different slope. Table B–4 provides examples of the stopping distance of a fire apparatus with a braking efficiency of 65% traveling 50 mph on two roads, each with a different slope.

How Braking Efficiency Affects the Skid to Stop Distance

As long as the braking system is working properly, a personal vehicle with hydraulic brakes will have 100% braking efficiency. However, a fire apparatus will have a reduced braking efficiency due to the rubber composition of the truck tires and the lag time associated with the air brake system. Studies have shown that the braking efficiency of a fire apparatus could be as low as 65% when compared to a personal car. Fire apparatus operators must understand how reduced braking efficiency will affect the skid to stop distance of the fire apparatus. Table B–5 examines the stopping distance of a fire apparatus and a personal vehicle which are both traveling 50 mph.

TABLE B–3.

Skid to Stop Distance of a Passenger Vehicle Traveling 50 mph		
	10% Uphill Grade	**10% Downhill Grade**
Vehicle Speed	50 mph	50 mph
Coefficient of Friction	0.80	0.80
Slope of the Road	(+0.10)	(−0.10)
Braking Efficiency	100%	100%
Adjusted Drag Factor	0.90	0.70
Calculate	$$SD=\frac{S^2}{(30)(f)}$$ $$SD=\frac{50^2}{(30)(0.9)}$$ $$SD=\frac{2500}{27}$$	$$SD=\frac{S^2}{(30)(f)}$$ $$SD=\frac{50^2}{(30)(0.7)}$$ $$SD=\frac{2500}{21}$$
Skid Distance	92.5 ft	119.0 ft

TABLE B–4.

Skid to Stop Distance of a Fire Apparatus Traveling 50 mph		
	10% Uphill Grade	**10% Downhill Grade**
Vehicle Speed	50 mph	50 mph
Coefficient of Friction	0.80	0.80
Slope of the Road	(+0.10)	(−0.10)
Braking Efficiency	65%	65%
Adjusted Drag Factor	0.62	0.42
Calculate	$$SD=\frac{S^2}{(30)(f)}$$ $$SD=\frac{50^2}{(30)(0.62)}$$ $$SD=\frac{2500}{18.6}$$	$$SD=\frac{S^2}{(30)(f)}$$ $$SD=\frac{50^2}{(30)(0.42)}$$ $$SD=\frac{2500}{12.6}$$
Skid Distance	134.4 ft	198.4 ft

TABLE B–5.

	Car—100% Braking Efficiency	Fire Truck—65% Braking Efficiency
Speed	50 mph	50 mph
Coefficient of Friction	0.80	0.80
Slope of the Road	0%	0%
Braking Efficiency	100%	65%
Adjusted Drag Factor	0.80	0.52
Calculate	$SD=\dfrac{S^2}{(30)(f)}$ $SD=\dfrac{50^2}{(30)(0.8)}$ $SD=\dfrac{2500}{24}$	$SD=\dfrac{S^2}{(30)(f)}$ $SD=\dfrac{50^2}{(30)(0.52)}$ $SD=\dfrac{2500}{15.6}$
Skid Distance	104.1 ft	160.2 ft

At 50 mph, the skid to stop distance of the fire apparatus is 56 ft longer than the passenger car. This is an important teaching point. Fire apparatus operators must remember to leave extra room between the fire apparatus and the vehicle in front of them. If a the fire apparatus operator is following too closely and the civilian vehicle should suddenly attempt to stop, the fire apparatus will not be able to stop before slamming into the back of the civilian vehicle.

Calculating Perception and Reaction Distance

The distance a vehicle travels as the driver perceives and reacts to a hazard will depend on the speed of the vehicle and how long it takes the driver to perceive and react. To determine the perception and reaction distance of a vehicle, we must first convert the speed from miles per hour to feet per second. Once the speed has been converted from miles per hour to feet per second, we simply multiply the *speed* of the vehicle (in feet per second) by the *perception and reaction time* of the driver.

STEP 1—Convert speed from miles per hour (mph) to feet per second (fps) by multiplying miles per hour by 1.466.

Miles per Hour (mph)	Multiply by 1.466	Feet per Second (fps)
5 mph	x 1.466	7.3 fps
10 mph	x 1.466	14.6 fps
15 mph	x 1.466	21.9 fps
20 mph	x 1.466	29.3 fps
25 mph	x 1.466	36.6 fps
30 mph	x 1.466	43.9 fps
35 mph	x 1.466	51.3 fps
40 mph	x 1.466	58.6 fps
45 mph	x 1.466	65.9 fps
50 mph	x 1.466	73.3 fps
55 mph	x 1.466	80.6 fps
60 mph	x 1.466	87.9 fps

STEP 2—Multiply the speed in feet per second by the perception time of the driver. This will calculate the how far the vehicle will travel as the driver perceives and reacts to a hazard. This is known as the perception reaction distance. Table B–6 assumes that the driver reacts with an average perception reaction time of 1.6 seconds.

TABLE B–6.

Speed in Miles per Hour	Speed in Feet per Second	Perception Reaction Time of 1.6 Seconds	Perception Reaction Distance
5 mph	7.3 fps	1.6 sec	11.7 ft
10 mph	14.6 fps	1.6 sec	23.4 ft
15 mph	21.9 fps	1.6 sec	35.1 ft
20 mph	29.3 fps	1.6 sec	46.9 ft
25 mph	36.6 fps	1.6 sec	58.6 ft
30 mph	43.9 fps	1.6 sec	70.3 ft
35 mph	51.3 fps	1.6 sec	82.0 ft
40 mph	58.6 fps	1.6 sec	93.8 ft
45 mph	65.9 fps	1.6 sec	105.5 ft
50 mph	73.3 fps	1.6 sec	117.2 ft
55 mph	80.6 fps	1.6 sec	129.0 ft
60 mph	87.9 fps	1.6 sec	140.7 ft

Perception-Reaction Distance Changes as the Speed of the Vehicle Increases

As the speed of a vehicle increases, it will cover more distance as the driver perceives and reacts to a hazard. Therefore, the vehicle will come upon the hazard more quickly. To better understand this concept, examine table B–7. In table B–7, we will assume that a driver perceives and reacts to a hazard in 1.6 seconds. Notice how the perception-reaction distance increases as the speed of the vehicle increases. If the vehicle is traveling 30 mph, it will cover 70.3 ft as the driver perceives and reacts to a hazard. If the vehicle is traveling 60 mph, it will cover 140.7 ft as the driver perceives and reacts to a hazard. A longer perception and reaction distance will reduce the margin of error for the driver. If a hazard should suddenly appear on the road ahead, a driver traveling at a faster speed may not have enough time and distance to take evasive action.

TABLE B–7.

Speed in Miles per Hour	Speed in Feet per Second	Perception Reaction Time of 1.6 Seconds	Perception Reaction Distance
5 mph	7.3 fps	1.6 sec	11.7 ft
10 mph	14.6 fps	1.6 sec	23.4 ft
15 mph	21.9 fps	1.6 sec	35.1 ft
20 mph	29.3 fps	1.6 sec	46.9 ft
25 mph	36.6 fps	1.6 sec	58.6 ft
30 mph	43.9 fps	1.6 sec	70.3 ft
35 mph	51.3 fps	1.6 sec	82.0 ft
40 mph	58.6 fps	1.6 sec	93.8 ft
45 mph	65.9 fps	1.6 sec	105.5 ft
50 mph	73.3 fps	1.6 sec	117.2 ft
55 mph	80.6 fps	1.6 sec	129.0 ft
60 mph	87.9 fps	1.6 sec	140.7 ft
65 mph	95.2 fps	1.6 sec	152.4 ft
70 mph	102.6 fps	1.6 sec	164.1 ft
75 mph	109.9 fps	1.6 sec	175.9 ft
80 mph	117.2 fps	1.6 sec	187.6 ft
85 mph	124.6 fps	1.6 sec	199.3 ft
90 mph	131.9 fps	1.6 sec	211.1 ft
95 mph	139.2 fps	1.6 sec	222.8 ft
100 mph	146.6 fps	1.6 sec	234.5 ft

Perception-Reaction Distance Changes as the Perception Time of the Driver Increases

Now let's examine what will happen if the speed of the vehicle remains the same but the driver's perception and reaction time changes. Consider a driver who is distracted by something inside the vehicle. If it takes the driver a longer period of time to perceive and react, the vehicle will cover a greater distance during the process. Table B–8 examines the perception and reaction distance for a vehicle traveling 50 mph (73.3 fps) based on different perception-reaction times for the driver. Note that the longer it takes the driver to perceive and react to a hazard, the further the vehicle will travel in that time.

TABLE B–8.

Perception Reaction Distance at 50 mph		
Perception Reaction Time	Speed (50 mph)	Perception Reaction Distance
0.5 sec	x 73.3 fps	36.6 ft
1.0 sec	x 73.3 fps	73.3 ft
1.5 sec	x 73.3 fps	109.9 ft
2.0 sec	x 73.3 fps	146.6 ft
2.5 sec	x 73.3 fps	183.2 ft
3.0 sec	x 73.3 fps	219.9 ft
3.5 sec	x 73.3 fps	256.5 ft
4.0 sec	x 73.3 fps	293.2 ft

Calculating Lateral G-Force

Fire apparatus operators must have a thorough understanding of lateral g-force, as it is one of the primary reasons why fire apparatus lose control or rollover. The amount of lateral g-force acting on a vehicle as it rounds a curve is directly related to the speed of the vehicle and the radius of the curve.

Keep in mind that the term "curve radius" does not always mean the vehicle is actually driving through a curve in the road. Anytime the driver turns the wheel, he creates a curve. The radius of the curve is directly related to how sharp the driver turns the steering wheel. If the driver turns the wheel sharply, the vehicle will traverse a curve with a much smaller radius than if the driver had only turned the wheel slightly. This may include turning a corner or making an evasive maneuver on a straight road.

As the radius of the curve decreases (gets sharper) or the speed of the vehicle increases, the lateral g-force acting on the vehicle will increase. This is why speed control and gentle use of the steering wheel are such important skills for a fire apparatus operator. If the driver turns the wheel too sharply, or turns the wheel at a very high speed, he may create enough lateral g-force to cause the apparatus to lose control or rollover. To better understand this concept, let's examine the lateral g-force acting on a vehicle if it traversed a flat 60-ft curve at both 15 mph and at 30 mph.

$$Lateral\ G\ Force = \frac{Speed^2}{(Curve\ Radius)(15)}$$

15 mph	30 mph
$GForce = \dfrac{Speed^2}{(CurveRadius)(15)}$	$GForce = \dfrac{Speed^2}{(CurveRadius)(15)}$
$GForce = \dfrac{15^2}{(60)(15)}$	$GForce = \dfrac{30^2}{(60)(15)}$
$GForce = \dfrac{225}{900}$	$GForce = \dfrac{900}{900}$
$GForce = 0.25$	$GForce = 1.00$

If a vehicle rounds the flat curve at 15 mph, the vehicle will experience 0.25 gs. If the vehicle rounds the same curve at 30 mph, it will experience 1.00 gs. By doubling the speed at which the vehicle rounds the corner, the lateral g-force on the fire apparatus increases 4 times.

Consider a pumper with a rollover threshold of 0.55. If the driver took this corner at 15 mph, the vehicle would have experienced 0.25 gs, which is below the rollover threshold of the vehicle. Because the amount of lateral g-force is below the rollover threshold of the vehicle, it would have safely driven around the corner. However, if the driver had taken the corner at 30 mph, the vehicle would have experienced 1.00 g which is greater than the rollover threshold of 0.55. At 30 mph, the vehicle would have rolled over.

Calculating Critical Curve Speed

Every curve in the road has a critical speed. If the vehicle exceeds the critical curve speed, the tires will break traction with the road and the vehicle will lose control. The critical curve speed is calculated using the following formula:

$$Critical\ Speed\ of\ a\ Curve\ (MPH) = 3.86\sqrt{(R)[(f) + or - (e)]}$$

Where

R = radius of the curve (in ft)

f = the drag factor of the roadway

e = superelevation (bank) of the road

As the radius of the curve gets smaller (gets sharper) or the road gets more slippery, the critical speed will go down. This is important for a fire apparatus operator to remember, especially when driving in wet weather or on a snow- or ice-covered road. Consider the following example. On a dry day, the drag factor of the road is 0.75. On a wet day, the drag factor of the road is 0.4. Look at what happens to the critical speed when the road gets wet and slippery. The critical speed drops from 40.9 mph to 29.8 mph. A driver who traversed the curve at 35 mph on a dry day would find himself spinning off the road if he drove through the same curve at the same speed on a wet day.

Dry Day	Wet Day
Critical Speed = $3.86\sqrt{(R)\ [(f)+(e)]}$	Critical Speed = $3.86\sqrt{(R)\ [(f)+(e)]}$
Critical Speed = $3.86\sqrt{(150)[(.75)+(0)]}$	Critical Speed = $3.86\sqrt{(150)[(.4)+(0)]}$
Critical Speed = $3.86\sqrt{112.5}$	Critical Speed = $3.86\sqrt{60}$
Critical Speed = (3.86)(10.60)	Critical Speed = (3.86)(7.74)
Critical Speed = 40.9 mph	Critical Speed = 29.8 mph

Now let's examine what happens to the critical speed as the curve gets sharper. If the curve has a radius of 300 ft, the critical speed would be 57.9 mph on a dry day. If the radius of the curve decreases to 150 ft, the critical speed would drop to 40.9 mph on a dry day. As the curve gets sharper, the critical speed will go down. Keep in mind that vehicles with a high center of gravity, such as a fire truck, will most likely rollover before they break traction with the road.

Curve Radius = 150 ft	Curve Radius = 300 ft
Critical Speed = $3.86\sqrt{(R)\ [(f)+(e)]}$	Critical Speed = $3.86\sqrt{(R)\ [(f)+(e)]}$
Critical Speed = $3.86\sqrt{(150)\ [(.75)+(0)]}$	Critical Speed = $3.86\sqrt{(300)\ [(.75)+(0)]}$
Critical Speed = $3.86\sqrt{112.5}$	Critical Speed = $3.86\sqrt{225}$
Critical Speed =(3.86)(10.60)	Critical Speed = (3.86)(15)
Critical Speed = 40.9 mph	Critical Speed = 57.9 mph

Calculating Rollover Speed

In Chapter 3, we discussed lateral g-force. If the lateral g-force acting on a vehicle's center of gravity is greater than the its rollover threshold, the vehicle will rollover. The speed at which a vehicle will rollover can be calculated using the following formula:

$$V=\sqrt{(R)(g)(RT)}$$

Where

V = Speed in feet per second

R = Radius of the curve in feet

G = gravity which is 32.2 fps/sec

RT = Rollover threshold

As an example, let's consider the rollover speed of a standard passenger vehicle versus a large fire apparatus. The rollover threshold of the passenger car is 1.5, while the rollover speed of the fire apparatus is 0.49.

Curve Radius of 150 ft	
Average Car	Fire Apparatus
Speed = $\sqrt{(R)(g)(RT)}$	Speed = $\sqrt{(R)(g)(RT)}$
Speed = $\sqrt{(150)(32.2)(1.5)}$	Speed = $\sqrt{(150)(32.2)(.49)}$
Speed = $\sqrt{7245}$	Speed = $\sqrt{2366}$
Speed = 85.1 fps	Speed = 48.6 fps
Speed = 58.0 mph	Speed = 33.1 mph

A standard passenger vehicle can round a curve with a 150-ft radius curve at 58 mph before it rolls over. A fire apparatus rounding the same curve will rollover at 33 mph. This is due to the high center of gravity, and resulting instability, of the fire apparatus. Fire apparatus operators must understand the need to slow down when rounding a curve or making an evasive maneuver in a fire apparatus.

Index

Symbols

5th wheel lockout 197

15-passenger vans 199

  distractions and 203

  fatigue and 203

  federal studies of 199

  implementing safety procedures
      for 201–204

  number of passengers and 202

  operator age of 201

  rollovers and 200–201

  seatbelts and 203

  speed of 202

  tire pressure and 204

90° intersection 345

A

ABS. *See* antilock braking systems (ABS)

acceleration 27, 215

  curves and 64–65

  lateral 27, 45

  trailers and 215

accelerometer 323

accidents and vehicle maintenance 344

accountability and driver training 293

active rollover protection (ARP) 54

active suspension system 54

active warning systems 111–112

adaption and visibility 255

adjusted drag factor 404–405. *See
     also* drag factor

aerial devices 243

age

  nighttime driving and 250

  perception and reaction time and 3

air bags 103–104. *See also* seatbelt

air brake systems 10, 121–141. *See
     also* brakes

  adjustment of 129

  brake fade of 126–127

  braking force of 131

  inspection of 124–126, 129, 133

  primary and secondary systems 122

  service brake side of 122

  spring brake side of 122–125

air imbalance 130

air pressure 145, 151. *See also* tires

alcohol 371. *See also* blood alcohol
     concentration

  driver training and 319

  metabolizing of 163–164

all-wheel drive 222. *See
     also* transmissions

ambient noise 78

  crash investigation and 333

  factors affecting 78

  sirens and 78

  speed and 79

  windows and 82

ambulances 107–108

  children and 108

  seatbelts and 107

anchor points 238–239

anchor straps 236

annual weight certifications 128
antilock braking systems (ABS) 7, 8,
        11, 70, 139–142. *See also* braking
  driver training and 139, 315
  mechanical lag time of 13
  warning light 140
apparatus. *See also* vehicles
  backing 243–246
  crashes between 94
  high cost of 183
  instability of 48
  obtaining funding for 185
  off-road driving and 221
  out of service criteria for 187–189
  restrictions of 178–179
applied stroke methods 132
ARP (active rollover protection) 54
articulated vehicles 191
attenuation 79
autoleveling headlight systems 252.
        See also headlights
auxiliary braking systems 135. *See
        also* engine retarder
  curves and 65
  types of 67
available forward sight distance 19.
        See also sight distance

B
BAC. *See* blood alcohol concentration
        (BAC)
backing. *See also* case studies, backing
  accident avoidance and 243–246
  back-up alarms 246
  best practices 243–246
  cameras 244, 340
  crashes 243
  distractions during 247
  driver training and 322
  mirrors, use of during 244
  operations 196–197, 243–248
  safety 243–246
  signal use during 245
  spotters for 244–246
  tiller ladders 196–197
  trailers and 217
  visibility and 246

baffles 41–43, 390
  case studies 43, 338–339
  lack of 383
  modified 175–176
  modified vehicles and 175
bicycles 378
black boxes 52, 287–289
  crash investigation and 332
  driver training and 320–321
  resolving complaints 288–289
black ice 263, 263–264
  case studies 267–268
  crashes involving 267–268
blind spots 244
  case studies on 340
  driver training and 310
blood alcohol concentration
        (BAC) 161–164
  eight minimum rule and 165–168
  legal limits laws and 162
  metabolizing alcohol and 163–164
brake chambers
  bolt-type 134
  clamp type 134
brake fade 126–127, 206
  cascade failure 128
  fluid fade 128
  friction fade 127
  mechanical fade of drum
        brakes 126–127
  modified vehicles and 174, 175
  prevention of 128–132
brakes. *See also* antilock braking
        systems; brake fade; braking
  adjustment of 129
  air 10
  auxiliary braking systems 135
  bolt type brake chambers 134
  capability of 75
  crash investigation and 334–335
  drag and 8–10
  efficiency of 9–10, 13–14, 215
  failure of 350
  force of 65, 67
  hydraulic 9
  imbalance of 130–131
  inspection of 359

braking
  caging the brakes 334–335
  curves and 64–65, 66–67
  curves and distance efficiency 265
    distance 1, 6–7
    drag and 8–10
    energy related to 6
    kinetic energy and 6–7
    normal maneuvers of 7
    skid to stop formula 10
  electronic stability control and 66–67
  energy and 6–8
  fishtailing and 67
  kinetic energy and 6–7
  skidding and 9
  trailers 217–220
Bridgestone/Firestone 152

C

caffeine nap 276–277. *See also* sleep
caging the brakes 334–335
calculating
  critical curve speed 411–412
  feet per second 401
  kinetic energy 399–401
  lateral g-force 33, 409–410
  perception and reaction
      distance 406–409
  rollover speed 412–413
  rollover threshold 47–49
  skid to stop distance 402–406
  stopping distance 1
  untripped rollovers and 47–49
cameras 244–245
  backup 244, 340
  driver training and 316
  inclement weather and 245
  resolving allegations with 289–290
  thermal imaging 264
caravaning 90–92
  case studies on 361–362
case studies
  backing 339–340, 346–347, 349–350,
      365
  black ice 267–268, 342–343
  blind spots 340

brake failure 350, 351–352
  caravaning 91–92, 361–362
  distracted driving 92, 350–351,
      356–357, 393
  driver's license, invalid 390–391
  drunk driving 355–356
  fatigued driving 277
  fire boots 339, 353–354
  improper equipment 344, 364–365,
      386–387, 389–390, 392–393
  inclement weather 67, 361–362,
      362–363, 368, 382–383
  intersection related 67, 92, 345–346,
      358–359, 361–362, 366–367,
      394–395, 396–397
  lack of communication 342–343,
      385–386, 395–396
  lack of experience 337–339, 387
  lack of training 305
  loss of traction 343–344, 346–347, 397
  maintenance related 344
  night driving 339–340
  overcorrecting 357, 358–359, 369,
      383–384, 384–385, 388, 390–391,
      392–393, 397
  railroad crossing 388–389
  rollover 345–348, 350–357, 369, 388,
      397
    involving a water tanker 43,
        337–339, 340–341, 359–360, 363,
        370–371, 382–385, 389–393
  seatbelt 337–339, 345–346, 350,
      352–353, 354–355, 355–356,
      359–360, 363–364, 366–367,
      386–387, 389–390, 394
  sight obstruction 349, 355, 387,
      394–395
  speed, unsafe 340–341, 343–344,
      389–390, 391–392, 393, 397
  tire blowout 367
  yaw 341–342
CDL (commercial driver's license) 121,
    295
cell phones 271–272
center of gravity 46–47, 52
  rollover threshold and 37
  weight shift and 37

centrifugal force 27
chains 159–160
challenge and response 379
child transport 108
chocks 228
choker chain 237
circadian rhythms 275. *See also* sleep
civilian response to sirens 83
clamp type brake chambers 134. *See also* brakes
clevis 236
coasting. *See* rolling
cocaine. *See* impaired driving
coefficient of friction 8, 404
  crash investigations and 329–330
  roadways and , 9
  skid to stop distance and 402–404
  wet roadways and 12
cold air pressure 151
commercial driver's license (CDL) 121, 295
compliance with NFPA requirements 184
compression release engine retarder 135. *See also* engine retarders
cone courses 319
conspicuity 249
contact patch 59, 143–144
  roadways and 59
  skidding and 59
  tires and 59
containment method 41
contrast and night driving 250
cornering 58
cornering power and corrugations 28, 181
counter-steering 69, 72
couplers 208
crack pressure 130
crashes. *See also* case studies, rollovers; crash investigation
  backing and 243
  black ice and 267–268
  causes of 143
  delta-V and 103
  helmets and 105

  impaired driving and 161
  intersections and 94
  kinematics of 105
  marijuana and 228
  medical surveillance and 298
  nonrollover 103
  prevention of 243–246
  reportable and nonreportable 326
  securing of equipment 106
  stress related 301
  trains and 111, 301
crash investigation 335
  black boxes and 332
  caging the brakes and 334–335
  collecting evidence for 327–332
  crash records and 335–336
  identifying witnesses 331
  identifying witnesses statements 330–331
  law enforcement and 326–327, 330–331
  NFPA standards for 334–336
  notification procedures 326
  photo-video evidence 330, 332
  procedures for 325–327
  pushrod stroke and 334–335
  sirens and 332–335
  traffic preemption devices and 335
  vehicle inspections and 333–335
crawl ratio 223
critical speed and rollovers 412
crumple zones 104
crushing 7
curves 373, 390
  braking in 66–67
  centrifugal force from 27
  critical curve speed
    calculation of 411–412
    driver training and 320
    weather and 267
  dangers of 27
  g-force and 409–410
  radius and g-force 409–410
  sight distance and 19
cushion when cornering, rollover threshold and 33

D

date of manufacture on tires 156
debris 329
defensive driving 4. *See also* driving
  case studies about 348–349
  at night 255
defensive parking 243
delta-V 103–104
deuce-and-a-half 174, 177
diesel engines 136
differentials 225–226
  limited slip 225
  locking 225
  open 225
disc brake system 123. *See also* brakes
disciplinary action 294
distractions. *See also* driving
  while backing 247
  case studies on 350–351, 356–357, 393
  cognitive 270–272
  drugs or alcohol 371
  manual 270
  perception and reaction time 4
  phones 271–272
  studies about 271
  visual 269
downshifting 135
drag factor 7–10
  adjusted 404–405
  braking and 8–10
  calculating 8
  crash investigations and 329–330
  formula for 404
  friction circles and 59
  roadways and 8–9
  rollovers and 40, 46
  slope and 10
draw bar 208
driveline 135–136
driveline retarder. *See* engine-retarder
driver's licenses 297–298
driver training
  alcohol and 319
  antilock braking systems and 139, 315
  backing and 322
  black boxes and 320–321
  blind spots and 310
  disciplinary action 294
  DOT laws and regulations for 299
  exercises 302–303
  fire officers 297
  for fire apparatus operators 296–297
  for hazard training and 303–304
  inclement weather and 311–312
  intersections and 313–314
  lesson plans 296
  negative right-of-way and 313
  NFPA and 292, 318–319
  nighttime driving and 250–251
  notice of approach and 313–314
  off-road driving and 321–322
  online 322
  point of no escape and 314–315
  policies and procedures for 292–294
  practical 300
  risk management and 304–305
  selecting instructors for 295–296
  sight stopping distance and 312–313
  sirens and 316–317
  speed and 308, 401–402
  stopping distance and 309
  tire blowouts and 314
  tire inflation and 318–319
  trainees 299–300
  training records 298–299
driving. *See also* driver training
  driving under the influence (DUI) 161, 170
  driving while impaired (DWI) 161
  driving while intoxicated (DWI) 161
  fatigue and 272–277
  key elements of safety 17
  mud and 229
  nighttime 249–256
    conspicuity 249
    contrast 250
    glare 250
  off-road 226
  rain and 258–262
  sand and 229
  stress and 300–301
  surfaces 226

drum brake system 123. *See
  also* brakes
drum contamination 379
drunk driving. *See* impaired driving
  blood alcohol concentration 161–
    164
  case studies on 355–356
  driver training and 319
D-shackle 236
dual tires 52, 152–153
dynamic method of baffles 43

E
eight hour minimum rule. *See* blood
    alcohol concentration (BAC)
electro-magnetic retarder 136–137.
    See also engine retarders
electronic stability control systems
    (ESC) 31–32, 53–54
  braking and 66–67
  definition of 53
  rollover prevention and 31–32
emergency lights
  for notice of approach 87–88
  for notice of right-of-way 87–88
emergency response 23–26, 84–85
emergency vehicles. *See* ambulances;
    apparatus; vehicles
Emergency Vehicle Safety
    Initiative 102
energy. *See* kinetic energy
engine retarders
  compression release engine 135
  electro-magnetic 136–137
  exhaust 136
  skidding and 58
  transmission 136
  wet roads and 58, 65, 67
equipment
  chocks 228
  choker chains 237
  driver training and 303
  eye protection 104
  helmets 104–105
  securing of 106
  spotters and 245
  thermometers 263

ESC system. *See* electronic stability
    control systems (ESC)
European Road Safety
    Observatory 273
event data recorders. *See* black boxes
evidence. *See also* crash investigation
  photographs 330
  statements 330–331
  video 332
exhaust retarder 136
expectancy 254
eye protection 104

F
fatigued driving 272–277
  15-passenger vans and 202
  case studies on 277
  prevention of 276–277
  studies of 273
Federal Emergency Management
    Agency (FEMA) 300
Federal Highway Administration 96
fifth-wheel hitches 208
fire apparatus. *See also* apparatus;
    vehicles
  headlights 250–253
  hose 279–281
  railroad crossings and 113
  safe speed for 48
  sirens 77
  steering-induced rollover of 37–38
  tire grip 59
fire apparatus operators
  auxiliary braking system training
    for 67
  controlling g-force while driving
    and 33–37
  correcting skids 72
  inability to hear other emergency
    vehicle sirens and 94
  medical surveillance of 298
  railroad crossings and 113
  safety and 55
  selection of 296–297
  sirens and 77
  understanding rollover
    thresholds 32

fire apparatus rollover. *See* rollovers
Fire Department Occupational Safety
    and Health Program 167–168
fire departments
  administrators 293
  discipline and 294
  operating policies and
     procedures 292–294
  training records 298–299
  unions and 294
fire helmets. *See* helmets
fire hose. *See* hose
fire officer training 297
fishtailing 67
flat tires 147
fog 262–263
formulas. *See* calculating
forward sight distance. *See* sight
    distance
four-wheel drive 222. *See*
    also transmissions
  high 224
  low 223
friction 59. *See also* coefficient of
    friction
  friction circle 59–67, 139
    dry road and 60–61
    radius of 60–62
    truck tires and 60
    wet roads and 62–65
  lateral 59
  longitudinal 59
  rain and 258
  skidding and 59
  uses of 59
front tire skids 70
front-wheel drive 221. *See*
    also transmissions
full tire skids 70. *See also* skidding
furrows 329

G
g-force 27–43, 397. *See*
    also longitudinal g-force;
    latitudinal g-force
  combined 28
  driver training and 307

g-meter 307–308
  spin out and 29
  vehicle handling and 28
  weight shift and 37–38
  while driving 33–37
glare. *See* nighttime driving
global positioning preemption
    system 97–98. *See also* traffic
    preemption devices
gooseneck hitch 208. *See also* hitches
gouges 328
governors 52
green lights
  stale 92
  steady 94
grip and tires 59–60, 143
gross combined vehicle weight
    rating 207
gross trailer weight rating 207
gross vehicle weight rating 176, 207
ground clearance at railroad
    crossings 113
guidelines for driver training 293
Guide to IAFC Model Policies and
    Procedures for Emergency
    Vehicle Safety 165

H
handling modified vehicles 175
hand signals 245
hazards
  conditions
    inclement weather 257–267
    nighttime 249–256
  driver training and 303–304
  materials 373
  perception of 3
headlights 250–253
  autoleveling systems 252
  driver training and 309–310
  halogen 251
  high beams 256
  high intensity discharge (HID) 251
  LED 251
  nighttime driving and 250
  range of 251–252
headspace 107

hearing protection 104
heat energy 6
helmets
  crashes and 105
  rollovers and 105
  storage of 105
  when to wear 104–105
hitches 207–209
  classes of 208–209
  class recommendations 208
  coupler 208
  draw bar 208
  fifth-wheel 208
  gooseneck 208
  hitch ball 208–209
  weight carrying 208
  weight distributing 208
hose 279–281
  driving over safely 280
  loading 280
  securing 281
hydraulic brakes 10, 404. *See also* brakes
hydroplaning 143–147, 259–260
  air pressure 145–146
  case studies on 364–365
  depth of water and 146
  handling 146
  rain and 259–260
  speed of 145–146
  tire tread depth and 145

I

IAFC. *See* International Association of Fire Chiefs (IAFC)
IAFF (International Association of Firefighters), policy and procedure manual writing 294
ice 263–264
  case studies on 381–382
  inclement weather and 263–264
idle speed 228
IIHS (Insurance Institute for Highway Safety) 53
impaired driving
  blood alcohol concentration and 162–165
  case studies on 169–171, 277
  definition of 161
  discipline for 168–169
  national recommendations for 165–168
  prescribed medication and 164
  symptoms of drug and alcohol abuse 168–169
inattention blindness 271
inclement weather 257. *See also* case studies, inclement weather
  driver training and 311–312
  fog 262–263
  ice 263–264
  nighttime driving and 253, 257–264
  rain 258–262
  skid to stop distance and 402–404
  sleet and freezing rain 264
  snow 260–262
  stopping distance and 265–266
  wind 265–266
  windshields and 253
induction 136
inertia 41
inflation 143–144. *See also* tires
  driver training and 318–319
  maximum permissible inflation pressure of tires 158
  underinflation 152–153
infrared-activated traffic preemption system 97. *See also* traffic preemption
insertion loss 333
inspecting and reading tires 155–157
instructors of driver training programs and selection 295–296
  NFPA and 295
Insurance Institute for Highway Safety (IIHS) 53
International Association of Fire Chiefs (IAFC) 165
  policy and procedure manual writing and 294
International Association of Firefighters (IAFF), policy and procedure manual writing 294

intersections. *See also* case studies, railroad crossings; railroad crossings
  apparatus vs. apparatus crashes at 94
  crossing 88–89
  driver training and 313–314
  negative right-of-way at 87
  notice of approach to 87–88
  rolling through 95
  safety and 87, 88–89
  sight obstructions at 90
  stale green lights and 92
  traffic preemption devices and 96–99
inverse square law 80

J
jackknifes 191–193, 215
  prevention of 194–196
  tiller ladders and 191–193

K
kinetic energy 206
  braking distance and 6–7
  calculation of 399–400
  crashes and 105
  crushing and 7
  heat energy and 6
  rollovers and 105
  skidding and 7
  speed and 12

L
lag time 10, 121
lap belts, dangers of 104
lateral acceleration 27, 45
lateral friction 59
lateral g-force 28, 33, 45, 47, 48, 50. *See also* g-force
  calculation of 33, 409–410
  friction circle on dry roads and 60
  rollover threshold and 30
  skidding and 58
  spin out and 29
  untripped rollovers and 47
  weight shift and 38

law enforcement and crash investigation 326–327, 330–333
left tire skids 70. *See also* skidding
legal limit laws. *See* blood alcohol concentration (BAC)
light-emitting traffic preemption system 96
limited sight distance 21–22. *See also* sight difference
limited-slip differential 225
liquid products and weight differences 174
liquid slosh 40–41
liquid surge. *See* surge and baffles
load index and tires 157
loading information and tires 149
localization and noise 82
location of hazard and perception and reaction time 3
locking differential 225
longitudinal friction 59
longitudinal g-force 28, 33. *See also* g-force
  skidding and 57
  weight shift and 38
low beams 255–257. *See also* headlights
lower light levels and two-lane highways 253
LT (light truck tire) 156

M
M+S 156
  tires and 157
maintenance
  tires and 155
  trailers and 218–219
malfunctions of seatbelts 109. *See also* seatbelts
maneuver-induced rollovers 47. *See also* rollovers
manual transmission 228. *See also* transmissions
marijuana and crashes 277
math, importance of 399
maximum load rating of tires 158

maximum permissible inflation
  pressure of tires 158
mechanical lag
  distance and 1
  time and 10, 13
medical surveillance programs 298
Michelin 153
microsleep 275. *See also* sleep
minimum rollover threshold 31–32.
  See also rollovers
Miranda Warnings 331
mirrors, adjustment of 244. *See
  also* rearview mirrors
modern vehicles and sirens 77
modified vehicles 173–182
  brake fade and 174
  handling of 175
  overweight issues and 173–174
  skidding and 175
  tank baffles and 175
  tire blowouts and 175
multitasking 271

N
National Fire Protection Association
  (NFPA). *See* NFPA
National Highway Traffic Safety
  Administration (NHTSA) 156,
  200, 252, 253
  studies by 252
National Institute of Occupational
  Safety and Health (NIOSH) 165
  firefighter fatality and 337, 374
National Transportation Safety Board
  (NTSB) 73, 118
negative right-of-way
  driver training and 313
  intersections crossing and 87
  rolling an intersection and 95
  stale green lights and 92
Nevada Bureau of Land Management
  Engine Company 43
NFPA 302, 382, 387
  crash investigation and 334–336
  driver training and 318–319
  hose loads and 281

NFPA 1002 292
NFPA 1041 295
NFPA 1451 99, 153, 183, 187, 280, 292,
  295, 302–304, 335, 337, 363
NFPA 1500 88, 90, 104, 106, 167–168,
  230, 288, 292, 381
NFPA 1521 335
NFPA 1901 30, 41, 105–106, 135, 153,
  175, 184, 287
NFPA 1911 128, 145, 147, 155, 188,
  334
  Annex D 155
  Chapter 6 189
  record checks and 297
  requirements
    compliance and 184
    seatbelts and 108–109
  selecting instructors and 295
NHTSA. *See* National Highway Traffic
  Safety Administration (NHTSA)
nighttime driving 249–256
  case studies on 339–340
  contrast of 250
  driver training and 310–311
  glare and 250
  with headlights 250
  inclement weather and 253, 257–264
  pedestrians and 249–250, 254
  studies on 252
  visibility and 250
  windshield wipers and 260–261
Nighttime Glare and Driving
  Performance 253
NIOSH. *See* National Institute of
  Occupational Safety and Health
  (NIOSH)
noise
  ambient 78
  attenuation and 79
  inverse square law and 80
  localization and 82
  sound frequencies and 79
  windows down and 82
nonemergency calls and sirens 84–85.
  See also sirens
nonreportable crashes 326. *See
  also* crashes

nonrollover crashes 103. *See also* crashes
notice of approach 24, 87–88
  driver training and 313–314
  intersections and 87–88
  safety and 24
notice of right-of-way and sirens 87–88. *See also* sirens

O

occupant protection systems 104. *See also* seatbelts
off-road driving 221–232, 226
  driver training and 321–322
  seatbelts and 230
  surface 226
  terrain 226
  vehicle operations 226–230
off-track 216
on-board video systems 289–290
open-cab ladder truck 394
open differential 225. *See also* differentials
Operation Lifesaver 113, 118–119
out of service procedures 189
over-acceleration 58
over-breaking and skidding 57–58
overcompensation 391
overcorrecting 357–359, 369
  case studies on 357
oversteering 54, 68–70, 72, 222
  correction and 72
  main factors of 68–69
  stability loss and 68
  yaw and 68
overtillering 198
overweight issues and modified vehicles 173–174

P

P-18 rollover testing 40
parking brakes and wheel chocks 286–287
parking, defensive 243
part-time four-wheel drive. *See* four-wheel drive
passenger vehicles and tires 156

passive warning systems at railroad crossings 111–112
pedestrians
  nighttime driving and 249–250, 254
  two-lane highways and 254
perception 380
  reaction distance and
    calculation of 406–409
      increase in perception time and 409
      speed and 408
  reaction distance changes, calculating 408
  reaction time and 2–4
    age of driver and 3
    average 2, 407
    distractions and 4
    increase in 409
    location of hazard and 3
    nature of hazard and 3
    surrounding circumstances and 3–4
  time increases 409
personal protective equipment 101
phases of stopping distance 1
point of no escape 314–315
policies
  driver training programs and 292–294
  procedure manuals and 294
policy and procedure manual writing at IAFC 294
potholes 180–181
power 136
p (passenger vehicles) 156
pressure 226
  15-passenger vans and 204
prevention of
  crashes 243–246
  rollovers 29, 51–52
procedures
  driver training programs and 292–294
  railroad crossings and 117–119
programming and risk management 305
push 68

pushrod stroke 130–134
  crash investigation and 334–335

R
(r) radial tires 156
radio-activated traffic preemption
     systems 97–98
radius 33
  of friction circle 60, 62–63
  g-force and 36
railroad crossings 111. *See also* case
     studies, railroad crossings;
     intersections
  active warning systems 111–112
  design of 111–113
  fire apparatus operators and 113
  ground clearance at 113
  passive warning systems 111–112
  procedures at 117–119
  safety and 113–117
  stalled or hung-up apparatus at 117
  unguarded 381
rain
  driving 258–262
  friction and 258
  hydroplaning and 259–260
  inclement weather and 258–262
  visibility and 258
  window fog and 260
  windshield wipers and 260–261
rating of tires
  speed and 157
reaction distance 1, 14, 380
  calculation of 406–409
reaction time 2–4
rearview cameras. *See* backing,
    cameras
rearview mirrors 206, 244–246, 248,
    251, 281, 340, 347, 365, 368, 385
rear-wheel drive 222. *See*
    also transmissions
record checks and NFPA 297
recovery operations 233
recovery points, properly rated 235
recovery straps 233, 237
  calculation of 241
  rating of 236

usage of 240–241
use of 240–241
response time
  SCBA and 101–102
  seatbelts and 101–102
rigging strap 236
right-of-way 87
right tire skids 71. *See also* skidding
risk management
  driver training and 304–305
  program and 305
road defects
  corrugations 181
  potholes and 180
  rutting 180
  soft spots and 181
  surface types and 181
roads. *See also* road defects
  conditions and skid distance 12,
    14–15
  contact patch and 59
  drag factor of 8–9. *See* coefficient of
    friction; drag factor
  geometry of 255
  slope and 10
  surface types 181
  wet and coefficient of friction 12,
    62–63
rolling an intersection 95
  negative right-of-way and 95
rollovers 216–217, 341, 388, 392, 397.
    See also case studies, rollovers;
    rollover threshold
  15-passenger vans and 200–201
  common scenarios of 45
  dangers of 102–103
  dynamics of 45–47
  electronic stability control and 31–
    32
  helmets and 105
  kinematics of 105
  lethal nature of 55
  maneuver-induced 47
  prevention of 29, 51–52
  roof crushes and 106, 106–107
  safety and 52–53
  seatbelts and 102–103

securing of equipment and 106
speed and 47–49, 51, 308–312
   calculating critical curve 412
stability control systems and 31
steering-induced 37–38
threshold 32
tiller ladders and 195–198
tripped 46–47, 54, 73, 339
untripped 47, 54
USAF P-18 rollover testing 40
warning systems and 52, 54
water tankers and 338–339
rollover threshold 29, 30–33, 46–47,
        226, 227. *See also* rollovers
   calculation of 47–49
   center of gravity and 37
   fire apparatus vs civilian vehicles 32
   improvement of 50–51
   lateral g-force and 30
   maintaining a cushion when
        cornering 33
   minimums of 31–32
   requirements to be NFPA
        compliant 31
   rollover speed and 47
   static stability factor for
        calculating 30–32
   tilt table test and 31
   weight shift and 37, 338
roof crushes 106–107
   injuries from 106
   rollovers and 106–107
rotochamber-type brake
        chambers 134. *See also* brakes
r (radial) tires 156
rutting 180

S

saccade 3
safety 111, 233
   air bags and 103–104
   and railroad crossings 113–117
   at intersections 87, 88–89
   backing apparatus and 243–246
   notice of approach and 24
   observers and hose loading 280

rollovers and 52–53
seatbelts and 101–102, 109–110
spotters and 244–246
sag loss 286
sand, driving in 229
SCBA. *See* self-contained breathing
        apparatus (SCBA)
scratches in crash investigation
        328
seatbelts 364, 386. *See also* case
        studies, seatbelts
   in ambulances 107–109
   importance of 103
   in 15-passenger vans 203
   lap belts and 104
   malfunctions 109
   malfunctions of 228
   NFPA requirements and 108–109
   off-road use of 230
   response time and 101–102
   rollovers and 102–103
   safety and 101–102, 109–110
   SCBA and 101–102
   sensors and 108
   special events and 109
   three-point harnesses 104
securing equipment
   crashes and 106
   rollovers and 106
self-contained breathing apparatus
        (SCBA)
   response time and 101–102
   seatbelts and 101–102
sight distance 312. *See also* stopping
        sight distance
   available forward 19
   curves and 19
   definition of 17
   emergency response and 23–26
   forward 18–19
   limited 21–22
   sight obstruction 19
   speed limits and 21–22
sight stopping distance and driver
        training 312–313
sirencide 309

sirens
  ambient noise and 78
  attenuation of 79
  civilian response to 83
  crash investigation and 332–335
  driver training and 316–317
  fire apparatus operators and 77
  inverse square law and 80
  localization and 82
  modern vehicles and 77
  nonemergency calls and 84–85
  notice of approach to intersections
      and 87–88
  range of 77–80
  sirencide 309
  sound frequencies of 79
  wail siren 79
  windows and 82
skidding 50, 215, 390. *See also* skid
      distance
  braking efficiency and 9
  cause of 57
  contact patch and 59
  cornering and 58
  correction of 72
  counter-steering and 69, 73
  distance 6–7, 12
  engine retarder and 58
  friction and 59
  front tire 70
  full tire 70
  kinetic energy and 7
  lateral g-force and 58
  left or right tire 70
  longitudinal g-force and 57
  modified vehicles and 175
  over-accelerating and 58
  over-breaking and 57–58
  passenger vehicle vs fire apparatus
      and 14
  preventing 75–76
  roadway conditions and 12, 14–15
  slope 10, 13
  tanking and 71
  tiller ladders and 195–196
  tires 7
  turning and 58

  types of 70–71
  wet/dry switch 73
skid distance 12, 14
  calculation of 403
  speed and 12
skid marks and 328
skid to stop distance
  braking efficiency and 404–406
  calculation of fire apparatus 405
  calculation of passenger vehicle 405
  skid to stop distance and slope
      and 13
  slope and 404
  weather and 402–404
sleep
  caffeine nap 276–277
  circadian rhythms and 275
  disorders 274–276
  lack of 274
  medications for 276
  microsleep 275
sleet, freezing rain, driving in 264
slide out. *See* spin out
slope 10
  roadways and 10
  skidding and 13
  stopping distance and 12–13
slosh and inertia 40–41
snatch block 238–239
snow
  deep 262
  driving in 260–262
snow tires 158–159
  identification 159
  M+S 159
snub braking process 137–138. *See
      also* braking
soft spots 181. *See also* road defects
sound-activated traffic preemption
      system 97
sound frequencies 79
speed 12, 341, 369, 375, 380, 390, 393
  15-passenger vans and 202
  ambient noise and 79
  driver training and 308
  in feet per second 401–402
  kinetic energy and 12, 399–401

limits, sight distance and 21–22
notification device 52
rollovers and 47–51
skid distance and 12
stopping distance and 12
stopping sight distance and 21–22
unsafe
  case studies 340–341
speed limits
  stopping sight distance and 21–22
spin out
  lateral g-force and 29
spotters 244–246
  equipment 245
  safety and 244–246
  visibility and 246
stability control system. *See* electronic
    stability control systems (ESC)
stability loss 68. *See also* electronic
    stability control systems (ESC);
    oversteering; understeering
stale green lights
  intersections and 92
  negative right-of-way and 92
standards. *See* NFPA
statements
  crash investigation and 330–331
  recording of 331
static stability factor for
    calculating 30–31
  rollover threshold 30–32
steady state curve 60
steering-induced
  rollovers and 37–38
stickiness. *See* coefficient of friction;
    drag factor
stomp, stay and steer 139, 315
stopping distance 215, 312. *See
    also* total stopping distance
  calculating total for 1
  definition of 1
  driver training and 309
  factors affecting 12–15
  phases of 1
  slope and 10, 12–13, 13
  speed and 12
  truck tires and 10

weather and 265–266
stopping sight distance 17–20. *See
    also* stopping distance
  factors affecting 18
  speed limits and 21–22
stress related crashes
    301
struts and weight shift 69
  weight shift and 40
supply line 279–280, 349
surface types and road defects 181
surge and baffles
    41–42
surge braking systems 211
surrounding circumstances,
    perception and reaction time
    3–4
suspensions 50
  active suspension system 54

T
tank baffles. *See* baffles
tanking 71
terrain and off-road driving 226
three-point harnesses 104. *See
    also* seatbelts
threshold braking 139. *See also* braking
tiller ladders
  backing operations 196–197
  jackknifes and 191–193
  rollovers and 195–196
  skid control 195–196
  turning of 197
tiller operator 191
  overtillering and 198
tilt table test 31
T-intersection 378
tires 144, 370. *See also* air pressure;
    inflation
  ages of 155
  aspect ratio 156
  blowouts and 147–148, 382
    proper response to 153–154
  Bridgestone/Firestone 152
  case studies on 367
  cold air pressure and 151
  contact patch and 59

tires (*continued*)
  date of manufacture 156
  degrading of 363
  driver training and 314
  dual assembly 152–153
  dual tires 52, 152–153
  feedback from 50
  flats of 147
  grip 59–60, 143
  inspection of 155–157
  load index 157
  loading information of 149
  LT (light truck) 156
  M+S 157
  maintenance of 155
  maximum load rating 158
  maximum permissible inflation
    pressure 158
  modified vehicles and 175
  passenger vehicles and 156
  ply composition and materials
    used 158
  pressure of 148–155, 153, 158
  reading of 155–157
  (r) radial 156
  speed rating 157
  traction letter 158
  tread and 390
    depth of 145
    truck checks 155–156
  truck 10, 60
  underinflation 152–153
  U.S. DOT Identification Number 157
  wear rate 155, 158
  wheel diameter 157
  width 156
Tire Safety: Everything Rides on It *156*
tongue weight 207
torque steering and imbalance 130–
    131, 222
total stopping distance 11–12, 14, 18.
    See also stopping distance
  calculation of 2
towing
  capacity 205–206
  packages 206
  recovery points and 234–235

  straps 237
  trailers 206
  vehicles 205–206
    205–206
track width 52
tractor trailers
  dynamics of 191
  trailer sway and 218
traffic gaps and safe maneuvers
    24–26
traffic preemption devices 96–99
  comparison of different types 98–99
  crash investigation and 335
  global positioning preemption
    system and 97–98
  infrared-activated 97
  intersections and 96–99
  light-emitting 96
  radio-activated preemption systems
    and 97–98
  sound-activated preemption system
    and 97
trailers 205–219
  acceleration and 215
  backing 217
  braking 210–215
    adjustments 212
    breakaway systems 213
    surge braking systems 211
    sway control systems 213
  coupler 208
  draw bar 208
  fifth-wheel hitches and 208
  gooseneck hitch and 208
  gross combined vehicle weight rating
    for 207
  hitch ball and 208
  hitch class recommendations for 208
  maintenance and inspections
    of 218–219
  operations of 213–214
  rollover and 216–217
  selection of 209–210
  sway of 217
    roads and 218
    tractor trailers and 218
    weight distribution and 217

wind and 218
  tongue weight and 207
  tow capacity and 205
  tow packages and 206
  tow vehicle and 205–206
  turning of 216
  weight carrying hitch and 208
  weight distributing hitch and 208
training exercise 373
trains, crashes and 301
transmissions
  all-wheel drive 222
  four-wheel drive 222
  front-wheel drive 221–224
    four-wheel high 224
    four-wheel low 223
  rear-wheel drive vehicle 222
  retarder 136
tread
  tires and 390
  wear indicators 364
tree trunk protector 236
tripped rollovers 46–47, 54, 73. *See also* rollovers
truck checks and tire tread 155–156
truck tires 10
  stopping distance and 10
tunnel vision 272, 300, 301
turning
  skidding and 58
  tiller ladders and 197
two-lane highways 253
  complex roadway geometry and 253
  lower light levels of 253
  nighttime driving on 253–254
  oncoming traffic on driver's line of sight and 253
  pedestrians and 254
  restricted roadway access and 254
  roadway markings and 254

U

underinflation. *See* inflation; tires
understeering 54, 68–70
  main factors of 68–69
  push and 68
  signs of 72

United States Air Force, P-18 rollover testing 40
United States Department of Transportation 81
United States Fire Administration 102
University of Michigan Transportation Research Institute 153
untripped rollovers 47, 54. *See also* rollovers
  calculating the speed of 47–49
  lateral g-force and 47
U.S. DOT Identification Number for tires 157

V

valve stem caps 152. *See also* tires
vehicles
  active suspension systems for 54
  ambient noise level in 78
  articulated 191
  caravaning of 91
  center of gravity and rollover threshold for 30
  crashes between fire apparatus 94
  crash investigation and 333–335
  data recorders and 52, 287–288
  dual tires on 52
  dynamics and rollovers 33
  g-force and 28
  governors for 52
  inspections of 319
    crash investigation and 333–335
  loading of 252
  maintaining a cushion when cornering 33–34
  maintenance, accidents involving 344
  off-road driving and 226–230
  path of travel, correcting 53
  radius of path of travel 33
  rollover warning devices and systems for 52, 54
  sound insulating and sirens and 77
  speed notification device for 52
  stability control system for rollover prevention in 31
  struts and stability 40

vehicles (*continued*)
  suspensions to prevent rollover 50
  traffic preemption devices and 96–99
  wet/dry switch 73
visibility
  adaption 255
  at night 250. *See also* sight distance; sight obstruction; stopping distance
  available forward sight distance 19
  backing and 246
  rain and 258
  spotters and 246
  of windshields 252–253. *See also* sight distance; sight obstruction; stopping distance
Volunteer Firemen's Insurance Services (VFIS) 184

W
Wail siren 79. *See also* sirens
water and driving 229
water tankers 243
  case studies on 43, 360, 382–385, 389–393
  rollovers involving 338–341, 359–360
wear rate
  tires and 155
  truck checks and 155
weather. *See* inclement weather
weight carrying hitch 208–209. *See also* hitches
weight certifications, annual 128
weight distributing hitch 208. *See also* hitches
weight distribution
  hitches 208–209
  trailer sway and 217
weight ratings 151
  gross combined vehicle 207
  gross trailer 207
  gross vehicle 207
weight shift 37–38
  center of gravity and 37
  lateral g-force and 38
  longitudinal g-force and 38
  rollover threshold and 37

suspension struts and 40
wet/dry switch 73
wet roadways
  coefficient of friction for 12
  friction circle and 62–63
wheel chocks 281–286
  proper use of 286–287
  selection of 283
  studies on 283
  types of 283
wheel diameter and tires 157
width of tires 156. *See also* tires
wildland firefighting operations 230
winches 233–235
  calculation of 240
  driver training and 322
  ground factor and 240
  maximum pulling capacity of 234
  selection of 234
  usage in recovery operations 237–238
wind
  driving and 265–266
  inclement weather and 265–266
  trailer sway and 218
window fog 260
  nighttime driving and 260
  rain and 260
windows
  noise and 82
  sirens and 82
windshields 252
  inclement weather and 253
  nighttime driving and 252
  visibility and 252–253
  wipers 260–261

Y
yaw 50, 68, 72
  case studies and 341–342
  marks 328–329
  oversteering and 68

Z
Zero-Tolerance for Alcohol & Drinking in the Fire & Emergency Service 165–166